KU-640-167

ADVANCED LIGHT MICROSCOPY

2. SPECIALIZED METHODS

QMW LIBRARY
(MILE END)

Advanced Light Microscopy

Advanced Light Microscopy

Volume 2

Specialized Methods

Maksymilian Pluta

Professor of Applied Optics
Head of the Physical Optics Department
Central Optical Laboratory, Warszawa

Elsevier
Amsterdam–Oxford–New York–Tokyo

PWN–Polish Scientific Publishers
Warszawa

1989

Cover design by *Zygmunt Ziemka*

Distribution of this book being handled by the following publishers:

for the USA and Canada
ELSEVIER SCIENCE PUBLISHING CO., INC.
52, Vanderbilt Avenue
New York, N.Y. 10017

for Albania, Bulgaria, Cuba, Czechoslovakia, German Democratic Republic, Hungary, Korean People's Democratic Republic, Mongolia, People's Republic of China, Poland, Romania, the USSR, Vietnam and Yugoslavia
ARS POLONA
Krakowskie Przedmieście 7, 00-068 Warszawa, Poland

for all remaining areas
ELSEVIER SCIENCE PUBLISHERS
25, Sara Burgerhartstraat
P.O. Box 211, 1000 AE Amsterdam, The Netherlands

Library of Congress Cataloging in Publication Data
Pluta, Maksymilian.

    Advanced light microscopy.

    Bibliography: p.
    Includes index.
    Contents: v.1. Principles and basic properties.
            v.2. Specialized methods.
    1.   Microscope and microscopy—Technique—
Collected works.   I.   Title.
QH207.P54   1988   502'.8'2   87-24605

ISBN 0-444-98939-0 (Vol. 1)
ISBN 0-444-98918-8 (Vol. 2)
ISBN 0-444-98940-4 (Series)

Copyright © by PWN—Polish Scientific Publishers—Warszawa 1989

All rights reserved
No part of this publication may be reproduced, stored in retrieval system, or transmitted in any form or by any means, electronic, mechanical, photocopying, recording, or otherwise, without the prior written permission of the copyright owner

Printed in Poland by D.R.P

Contents

Introduction

This volume, the second of the present three-volume monograph, covers specialized microscopical methods with particular reference to those techniques which are widely used; however, methods less known in practice are also discussed to an extent necessary to give full insight into the state of the art in modern light microscopy.

The emphasis is here on qualitative and descriptive methods, but a few selected aspects of quantitative microscopy are also discussed. However, the most significant quantitative methods and measuring techniques are omitted since their full treatment is to be found in Volume 3.

The volume begins with Chapter 5, devoted to phase contrast microscopy. After a concise theoretical introduction and a recapitulation of the imaging properties of this method, the present-day phase contrast instrumentation is discussed, priority being given to non-standard phase contrast systems. Consequently, highly sensitive phase contrast devices with soot phase rings, arrangements with alternating image contrast, and variable phase contrast systems with anisotropic optical elements are discussed in the first place. Next, instruments for interference phase contrast microscopy and a stereoscopic phase contrast microscope are described. The chapter ends with a brief presentation of phase contrast microrefractometry and other fields of application of the phase contrast method.

To obtain better image contrast has been one of the main purposes in the progress of microscopical research and technology. The discovery of the phase contrast by F. Zernike was a milestone on the road leading to the present-day high level of light microscopy—as regards both research and technology. Subsequently, other valuable image contrasting techniques were invented or improved. A survey of these, starting with amplitude contrast and ending with modulation contrast, is given in Chapter 6. Another basic tendency has been to work towards increasingly precise identification of substances, microobjects, and microstructures. For this purpose, biologists widely use histo- and cytochemical staining. Chemical colouring, however, cannot be used as effectively in differentiating inorganic materials as it is in distinguishing biological cells and tissues. Fortunately, micro-scopists who deal with microcrystals, asbestos fibres, mineral powders, and other inorganic colourless particles can use optical staining to aid the identification of solid microobjects. The most effective optical staining is based on both the spectral dispersion of the refractive index and the Christiansen effect. In particular, this technique has been improved and popularized by W. C. McCrone and his co-workers. Its use both conventional and in combination with the highly sensitive negative phase contrast occupy the central part of the same chapter, in which oblique illumination, the dark-field method, microstrioscopy, and other related techniques are also included.

Since the late 1960s, differential interference contrast (DIC) has become increasingly popular. Today, Nomarski's DIC system is part of the basic equip-ment of advanced microscopes manufactured by all the leading firms in Europe, Japan, and USA. This system and its modifications with a variable amount and direction of wavefront shear (VADIC), which have been developed by the author of this book, are widely discussed in Chapter 7.

Recently, essential improvements have been introduced in reflection contrast microscopy, which is dealt with in Chapter 8. This specialized method was mainly directed towards the study of living cells cultured in vitro. Recent work of several researchers on the enhancement of the image contrast of low-reflecting objects with the use of Antiflex objectives has resulted in the formation of properly contrasted images without resolution loss due to diffuse light. Reflection contrast microscopy seems therefore to be of particular value to cell biologists who study different problems of the growth and movement of cancer cells cultured in vitro.

More than any other microscopical techniques, fluorescence microscopy has undergone continuous improvements and has had an ever widening range of applications, especially in biology and medicine. In the last two decades, its development has been influenced mainly by the growing potentialities of immuno-fluorescence, which is now regarded as an extremely useful research method in biological sciences and is a routine technique in a large number of medical and veterinary diagnostic procedures. In its routine use, the fluorescence micro-scope serves as a tool for identifying various diseases caused by pathogenetic viruses, bacteria, and parasites. The state of the art in fluorescence microscopy is presented in Chapter 9, where the emphasis is on sources of light, excitation and barrier filters, condensers and objectives, and Ploemopak devices for immuno-fluorescence investigations. Moreover, some systems for combining fluorescence microscopy with other techniques (phase contrast, DIC) are described. The fluorescence rectomicroscope is also presented in that chapter.

Originally, fluorescence microscopy was closely connected with ultraviolet microscopy, but the latter has made little progress, and today it is seldom used. Nevertheless, UV microscopy is a useful technique, especially in connection with the microspectrophotometry of biological objects that are transparent in visible light, but show varied absorption in the different parts of the ultraviolet region of the spectrum. More or less the same can be said about infrared microscopy, although IR-microscopes seem to be useful for non-destructive testing of semiconductor materials and devices. The principles, the optical systems, and the applications of UV-, IR- and thermal microscopy are dealt with in Chapter 10.

Holography is a process in which both the amplitude information and the phase information about the object wavefront are recorded as an interference pattern resembling a diffraction grating. Photographic plates of high resolution are usually employed as a recording material from which, after exposure and development, the original object wavefront is reconstructed. This two-step imaging process (hologram recording and image reconstruction) was invented by D. Gabor in 1948 and then applied by several researchers in light microscopy. A highly coherent source of light, such as a He–Ne laser, is indispensable for this process, but laser light gives a granular (speckled) appearance to the image reconstructed from the hologram. This harmful effect hinders the application of holographic microscopy on a wide scale, although several methods of reducing speckles and coherent noise have been developed. Today, holographic microinterferometry seems to be more useful than holographic microscopy. These and other problems of holographic microscopy are discussed in Chapter 11.

The invention and commercial manufacture of the laser have made it possible to construct a number of quite new microscopical systems in the last two decades. The scanning optical microscope (especially its confocal version) and the laser projection microscope with the brightness amplifier are the most interesting. These and some other systems are described in Chapter 12, which ends this volume.

5. Phase Contrast Microscopy

Phase contrast was discovered by the Dutch physicist Frits Zernike (Fig. 5.1) probably in 1932. The principle on which the first phase contrast microscope was based was published by him in 1935 [5.1] and embodied in practice by C. Zeiss Jena six years later [5.2]. However, the first commercially available microscopes with phase contrast equipment did not appear until 1946. The hey-day of phase contrast microscopy was the decade 1950–1960, and for his discovery F. Zernike was awarded the Nobel Prize for Physics in 1953. In that decade, the phase contrast method was used to especially good effect by researchers in biology and experimental medicine, above all cell biologists. Living cells and many other biological specimens belong to the class of phase objects that are difficult to study by using an ordinary microscope. Before the advent of phase contrast microscopy, such objects were rendered visible by artificial staining. It is not, however, possible to effectively stain living cells and tissues without causing their death or changing their structure[1]. Alternative methods of observation included the use of dark-field or oblique illumination, defocusing procedures or schlieren microscopy (see Chapter 6). However, these methods were unreliable in providing information about cell structure.

5.1. General principles

Let us consider an ordinary microscope (Fig. 5.2), consisting of the objective, ocular, and condenser, by means of which a thin phase object is investigated in transmitted monochromatic light issuing from a small source located in the front focal point of the condenser. The source S may be simply a diaphragm with a small aperture, illuminated by a microscope lamp. In this set-up, a beam of parallel rays leaves the condenser C, passes through the object plane Π, is focused in the back focal point F'_{ob} of the objective Ob, and reaches the image plane Π' as an

---

[1] Histochemical staining is, of course, still widely used today but largely for other purposes.

Fig. 5.1. Frits Zernike (1888–1966).

extended light beam (B_u). This beam is called the *direct* (or *undiffracted*) *beam*. In Fig. 5.2 it is marked by semicircular and rectilinear hatching. A portion of the primary light wave is, however, perturbed by the object O, which diffracts and deflects an amount of light in a cone that widens out as the object decreases in size. In Fig. 5.2 the dotted area (B_d) shows only that portion of the perturbed light which enters the objective. This light creates a beam called the *diffracted beam*, which, in contrast to the direct beam, fully occupies the exit pupil of the objective and reaches the image plane II' as a focused light beam. The image O'

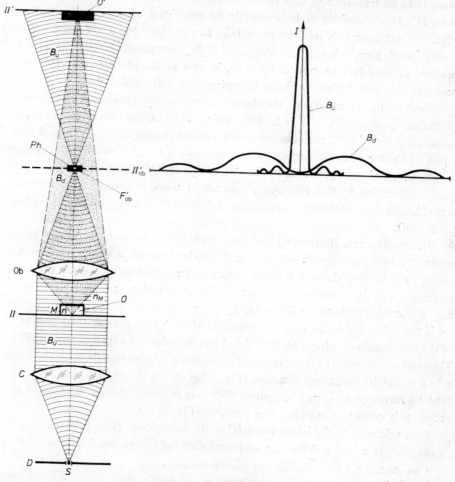

Fig. 5.2. Principle of phase contrast microscopy.

of the object O is a result of the interference of the direct and diffracted beams. Normally it is invisible because the intensity of the diffracted beam is small by comparison with the intensity of the direct beam, and also because phase relations between light vibrations of both beams are, as will be shown later, inadequate for producing effective constructive or destructive interference.[2]

Broadly speaking, the Zernike's phase contrast method depends on an amplitude and phase modification of the direct relative to the diffracted beam, which allows invisible optical path differences (or phase shifts) occurring in the object plane Π to be transformed into visible differences of light intensity in the image plane Π'. This transformation is possible by modifying of the exit pupil of the objective, and especially its focal plane Π'_{ob}. In this plane the direct beam covers a very small area, whereas the diffracted light overspreads amply. Thus it is possible to introduce an optical filter Ph, known as the *phase plate*, which acts only on the direct light without changing the diffracted light.

The exact theory underlying the phase contrast method requires some formulae of higher mathematics [1.1, 1.2, 3.4], which are not indispensable, however, for an analysis of basic phenomena. For a general insight into the physical principles and some practical aspects of this method, a rough vector representation is sufficient. This was proposed by Zernike [5.3] and developed in detail by Barer [5.4].[3] According to this representation, a light wave is described by a vector whose length denotes the wave amplitude and direction corresponds to the angular phase shift.

Vector diagrams illustrating the most typical situations occurring in phase contrast microscopy are shown in Fig. 5.3, where vector $\mathbf{p} = OQ$ corresponds to the light passing through a phase object under examination, vector $\mathbf{u} = OM$ represents the light passing only through the surrounding medium, and vector $\mathbf{d} = MQ$ corresponds to the diffracted light.

If the refractive index n of the object O (Fig. 5.2) is greater than that (n_M) of the surrounding medium M, then the object introduces a negative phase shift φ. This shift corresponds to the rotation of the vector $\mathbf{p}$ relative to $\mathbf{u}$ through the angle φ in a clockwise (negative) direction (Fig. 5.3a). When the object is less refractile than its surround ($n < n_M$), the phase shift φ is positive, and the relation of the vector $\mathbf{p}$ is counter-clockwise, i.e., positive (Fig. 5.3b).

With reference to the image plane Π' of the microscope (Fig. 5.2), the lengths of the vectors $\mathbf{u}$ and $\mathbf{p}$ define the amplitudes of light from the background and

[2] See Section 1.6 for definition and basic properties of light interference.

[3] Recently, Goldstein added a contribution to the Barer's vector theory, not restricted to objects of low phase retardation [5.5].

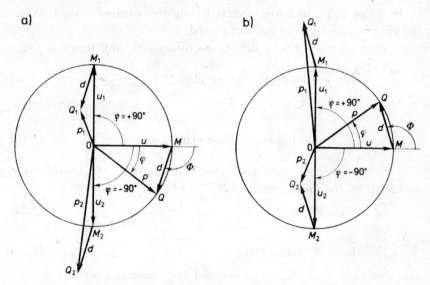

Fig. 5.3. Vector representation of the phase contrast method: a) phase-retarding object, b) phase-advancing object.

object image, respectively. If the light transmittance of the object is the same as that of the surrounding medium, the vectors **u** and **p** have the same length, thus the intensity within the object image is equal to the background intensity, and the object is invisible. In accordance with the diffraction theory of image formation in the microscope and Fourier optics, the vector **p** can be treated as the sum of vectors **u** and **d**. The latter is shifted in phase relative to **u** by an angle Φ equal to $90° + \varphi/2$ (Fig. 5.3b) or $-90° + \varphi/2$ (Fig. 5.3a). For phase objects producing very small phase shifts φ, the angle $\Phi \approx \pm 90°$.

As was mentioned earlier, the phase contrast method allows direct and diffracted light waves to be separated, so that one of these waves can be modified in phase and/or amplitude without affecting the other. This is done by means of a condenser diaphragm D and phase plate Ph (Fig. 5.2). In the typical phase contrast microscope the direct light beam is shifted in phase by an angle $\psi = +90°$ or $-90°$. Then, for $\psi = +90°$, the vector **u** is transformed into $\mathbf{u}_1 = OM_1$. This latter and **d** give a new resultant vector $\mathbf{p}_1 = OQ_1$, which is shorter than $\mathbf{u}_1$ for phase-retarding objects (Fig. 5.3a) and longer for phase-advancing objects (Fig. 5.3b). Consequently, these latter objects appear brighter and the former darker than the background. If a $-90°$ phase plate is used, the vector **u** is transformed into $\mathbf{u}_2 = OM_2$, which together with **d** gives a resultant vector $\mathbf{p}_2 = OQ_2$.

Now the image of phase-retarding objects is brighter, whereas that of phase-advancing objects is darker than the background.

Let the image contrast (C') be defined by formula (3.64b), then

$$C' = \frac{|\mathbf{u}_1|^2 - |\mathbf{p}_1|^2}{|u_1|^2} \quad \text{for a positive phase plate,} \tag{5.1a}$$

and

$$C' = \frac{|\mathbf{u}_2|^2 - |\mathbf{p}_2|^2}{|\mathbf{u}_2|^2} \quad \text{for a negative phase plate.} \tag{5.1b}$$

Calculating $|\mathbf{p}_1|^2$ and $|\mathbf{p}_2|^2$ by trigonometry, and assuming $|\varphi| \ll 90°$, the following approximate values for image contrast can be obtained:

$$C' = -2\varphi \quad \text{for } +90° \text{ phase plate,} \tag{5.2a}$$

and

$$C' = +2\varphi \quad \text{for } -90° \text{ phase plate.} \tag{5.2b}$$

These relations show that the image contrast (or image intensity) is directly proportional to the phase change φ due to the object. This relationship, however, only holds when φ is small.

The image contrast C' of slightly dephasing objects can be additionally increased by weakening the intensity of the direct light beam. In the vectorial diagrams (Fig. 5.3), this corresponds to decreases in the length of vectors $\mathbf{u}_1$ and $\mathbf{u}_2$. Let N be a number by which the intensity of the direct light is divided. Then Eqs. (5.2) become

$$C' = -2\varphi\sqrt{N} \quad \text{for } +90° \text{ phase plate,} \tag{5.3a}$$

$$C' = +2\varphi\sqrt{N} \quad \text{for } -90° \text{ phase plate.} \tag{5.3b}$$

If, for example, $N = 4$, the phase plate transmits only one-fourth of the direct light, and in relation to a transparent phase plate ($N = 1$) the image contrast is multiplied by $\sqrt{4} = 2$.

In Chapter 3 it was mentioned that phase contrast microscopy is a technique which employs the principles of the Fourier spectrum filtering. According to this approach, a phase specimen can be generally characterized by a transmittance function

$$f(x, y) = \exp[i\varphi(x, y)], \tag{5.4}$$

where x, y are the Cartesian coordinates in the object plane Π of the microscope objective (Fig. 5.2), and $\varphi(x, y)$ represents a phase variation of the plane wavefront of the monochromatic light passing through the specimen. Here small phase variations are taken into consideration and in such cases the function (5.4) may be written

$$f(x, y) = 1 + i\varphi(x, y).$$ (5.5)

The amplitude distribution of light in the back focal plane Π'_{ob} of the microscope objective (Fig. 5.2) is the Fourier transform $F(u, v)$ of $f(x, y)$:

$$
\begin{aligned}
F(u, v) &= \int\!\!\!\int_{-\infty}^{+\infty} [1 + i\varphi(x, y)] \exp[-2\pi i(ux + vy)] dx \, dy \\
&= \int\!\!\!\int_{-\infty}^{+\infty} \exp[-2\pi i(ux + vy)] dx \, dy \\
&\quad + \int\!\!\!\int_{-\infty}^{+\infty} i\varphi(x, y) \exp[-2\pi i(ux + vy)] dx \, dy,
\end{aligned}
$$ (5.6)

where u, v are spatial frequencies related to coordinates x, y. In this relation, the limits of the double integrals are formally from $-\infty$ to $+\infty$, but in reality the integration is carried out over an area which is equal to the objective aperture.

The second double integral in Eq. (5.6) expresses the diffracted light due to the phase variation $\varphi(x, y)$, whereas the first double integral represents only the diffraction pattern of the plane wave limited by the edge of the objective aperture. This pattern is similar to the Airy image of a point light source. The right side of Fig. 5.2 illustrates this situation schematically.

Suppose a phase plate introduces a phase shift of $+\pi/2$ on the direct light beam, i.e., in this case on the zero frequency component of the Fourier spectrum $F(u, v)$. This operation is equivalent to the multiplication of the first double integral in Eq. (5.6) by $e^{i\pi/2} = i$, and now $F(u, v)$ takes the form

$$
\begin{aligned}
F'(u, v) &= \int\!\!\!\int i \exp[-2\pi i(ux + vy)] dx \, dy \\
&\quad + \int\!\!\!\int i\varphi(x, y) \exp[-2\pi i(ux + vy)] dx \, dy.
\end{aligned}
$$

The intensity distribution $I(x, y)$ in the image plane Π' of the microscope[4] is given by $h(x, y)h^*(x, y)$, where $h(x, y)$ is the inverse Fourier transform of $F'(u, v)$, and $h^*(x, y)$ is the complex conjugate quantity of $h(x, y)$. Thus

$$h(x, y) = i + i\varphi(x, y),$$

[4] For the sake of simplicity, it is assumed that the coordinates in the image plane are the same as in the object plane, and the microscope magnification is equal to 1.

and
$$I(x, y) = 1 + 2\varphi(x, y) + \varphi^2(x, y).$$

The quadratic term is very small relative to the remaining terms and can be ignored. Thus one can write

$$I(x, y) = 1 + 2\varphi(x, y). \tag{5.7a}$$

It will be seen that the intensity distribution in the object image is directly proportional to the phase variation $\varphi(x, y)$. If $\varphi < 0$ (phase-retarding object), then $I(x, y) < 1$ and the object image is darker than the background (background intensity is in this case equal to 1), and conversely, the object image is brighter than the background if $\varphi > 0$ (phase-advancing object).

Now suppose the phase plate introduces a negative phase shift $\psi = -\pi/2$. This operation is equivalent to the multiplication of the first double integral in Eq. (5.6) by $e^{i(-\pi/2)} = -i$, and in this case

$$I(x, y) = 1 - 2\varphi(x, y). \tag{5.7b}$$

If $\varphi < 0$, the object image is brighter than the background, whereas it is darker than the background when $\varphi > 0$.

The above relations are derived for a transparent phase plate. If, however, the phase plate absorbs the direct light in addition to having the phase shift of $\pi/2$ or $-\pi/2$, then the zero frequency component of the Fourier spectrum is attenuated. This attenuation is equivalent to the multiplication of the first double integral in Eq. (5.6) by $\sqrt{\tau} = 1/\sqrt{N}$, where τ is the transmittance of the phase plate. In this situation, Eqs. (5.7) become

$$I(x, y) = \tau + 2\sqrt{\tau}\varphi(x, y), \quad \text{for } \pi/2 \text{ phase plate}, \tag{5.8a}$$

and

$$I(x, y) = \tau - 2\sqrt{\tau}\varphi(x, y) \quad \text{for } -\pi/2 \text{ phase plate}. \tag{5.8b}$$

The background intensity is now equal to τ, and according to the formula (3.64b) the contrast C' is given by

$$C' = \frac{\tau - \tau - 2\sqrt{\tau}\varphi(x, y)}{\tau} = -\frac{2\varphi(x, y)}{\sqrt{\tau}} \quad \text{for } +\pi/2 \text{ phase plate}, \tag{5.9a}$$

and

$$C' = \frac{\tau - \tau + 2\sqrt{\tau}\varphi(x, y)}{\tau} = +\frac{2\varphi(x, y)}{\sqrt{\tau}} \quad \text{for } -\pi/2 \text{ phase plate}. \tag{5.9b}$$

These formulae are the same as Eqs. (5.3), so that in conclusion one may say that by introducing a phase plate (in general, a phase filter) in the Fourier transform plane the object phase variations can be transformed into intensity variations in the object image.

Up to now it has been assumed $|\varphi| \ll 90°$. If we take into account phase objects giving larger phase shifts φ, the relationship between the contrast C' and φ is not linear but resembles a sine function. For non-absorbing quarter wavelength phase plates, the relation $I_o(\varphi)$ extended over the whole periods of phase angle $\varphi = 0$ to $\pm 360°$ is shown in Fig. 5.4. It will be seen that both

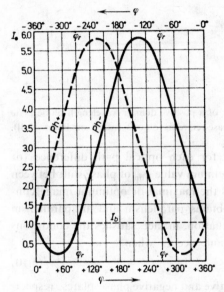

Fig. 5.4. Relationship between the object phase shift φ and intensity of object image (I_o) for non-absorbing quarter wavelength phase plates: positive (curve $Ph+$) and negative (curve $Ph-$).

positive and negative phase plates can give positive and negative image contrast depending on the value of φ. Following φ from zero to $\pm 360°$, the difference $I_b - I_o$ between the background intensity (I_b) and the object image intensity (I_o) reaches a maximum, then diminishes, and for some value $\varphi = \varphi_r$ it becomes zero, and then its sign changes. The phase φ_r defines the *range of unreversed contrast*. For phase-retarding objects, $\varphi_r = -90°$ and $-270°$ for positive and negative quarter wavelength phase plates, respectively. For phase-advancing objects, the opposite is true: values of φ_r are $+270°$ ($+\pi/2$ phase plate) and $+90°$ ($-\pi/2$ phase plate). If the phase plate absorbs some direct light, the range of unreversed contrast diminishes as N increases. This is illustrated by Fig. 5.5 for phase-retarding objects and positive $\pi/2$ phase plate.

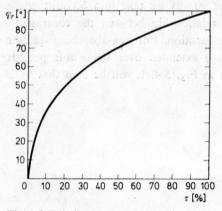

Fig. 5.5. Relationship between the transmittance τ of a positive quarter wavelength phase plate and the range of unreversed contrast (φ_r) for phase-retarding objects (based on Barer [5.4]).

Vector diagrams (Fig. 5.3) show that for each optical path difference (or phase shift φ) in the specimen there is an optimum value ψ_{op} of phase shift between the direct and diffracted light, which gives the maximal or optimum image contrast. This value is that for which one obtains parallelism and antiparallelism between vectors $\mathbf{d}$ and $\mathbf{u}_1$ or $\mathbf{u}_2$. Such a situation, in fact, satisfies the conditions for the most effective light interference and occurs when

$$\psi = \psi_{op} = \pm 90° + \tfrac{1}{2}\varphi, \qquad\qquad (5.10)$$

where signs "$+$" and "$-$" relate to the positive and negative phase plates, respectively.

Analysis of the vectorial diagrams also shows that for a given phase shift φ in the specimen there is an optimum value of transmittance τ of the phase plate. This is equal to

$$\tau = \frac{1}{N} = 4\sin^2\left(\tfrac{1}{2}\varphi\right), \qquad\qquad (5.11)$$

or

$$\tau = \frac{1}{N} = 4\cos^2\psi_{op}, \qquad\qquad (5.12)$$

where ψ_{op} is the optimum phase shift of the phase plate (Eq. 5.10). When $N > 1$, the direct light is reduced, and if $N < 1$, the diffracted light should be weakened. Relations (5.10) to (5.12) are usually referred to as *Richter's condition* [5.6].

5.2. Typical phase contrast systems

A typical phase contrast system for transparent specimens resembles a conventional bright-field microscope except that an annular diaphragm D (Fig. 5.6) and ring-shaped phase plate Ph are arranged in or near the focal planes of the condenser C and objective Ob, respectively.[5] The light issuing from the transparent annulus A of the condenser diaphragm is diffracted by a phase object O under examination, placed in the object plane II of the objective. The diffracted light B_d does not pass through the phase ring R but through its surrounding areas in the exit pupil of the objective. The portion of this pupil occupied by the image of the condenser annulus is called the *conjugate area*, whereas the remaining portion is called the *complementary area*. The former is completely covered by the ring R which changes the phase of the direct (undiffracted) light B_u with respect

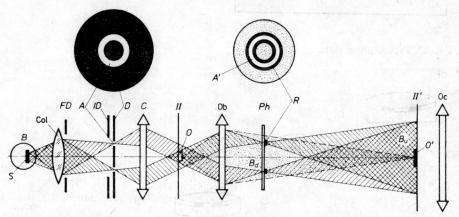

Fig. 5.6. Schematic optical system for typical phase contrast microscopy.

to the diffracted beam B_d by $+90°$ or $-90°$. The phase contrast image O' of the object O appears in the image plane II' as a result of interference between the direct and diffracted light beams and is observed through an eyepiece Oc. For most phase specimens, the intensity of diffracted light is low, and there is a great difference in the amplitudes of these two interfering beams. In order to make their

[5] The annular shape of the condenser diaphragm and phase plate is not obligatory, but is generally preferred from among other possible shapes. They are discussed by Françon in his book [5.9], and here it is only worth mentioning that the reduced circular diaphragm and small circular phase plate, as shown in Fig. 5.2, are not used in practice because of the much reduced resolving power of the microscope.

interference more effective, the intensity of the direct light is reduced by adding an absorbing thin film to the conjugate area. In typical phase contrast systems, this reduction is 75–90%.

The phase plates are nowadays made by vacuum deposition of thin dielectric (*DF*) and metallic (*MF*) films onto a glass plate (Fig. 5.7) or directly onto one of the lenses of the microscope objective (Fig. 5.8). The dielectric film shifts the phase of light, whereas the metallic film attenuates the light intensity. The phase ring *R* is usually glued between two glass plates (or two lenses) by means of a suitable optical cement *M*. The thickness *t* and refractive index of the dielectric

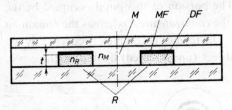

Fig. 5.7. Typical phase plate.

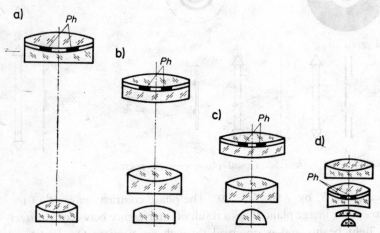

Fig. 5.8. Location of phase rings (*Ph*) in cementing layer of PZO achromatic objectives of magnification/aperture: a) $10 \times /0.24$, b) $20 \times /0.40$, c) $40 \times /0.65$, d) $100 \times /1.25$ (oil immersion).

film, as well, as the refractive index of the cement, are chosen so as to obtain a required phase shift ψ between the complementary and conjugate areas. This phase shift is defined as

$$\psi = 2\pi \frac{\delta}{\lambda} = \frac{2\pi}{\lambda}(n_M - n_R)t, \tag{5.13}$$

where n_M and n_R are the refractive indices of the complementary and conjugate areas, respectively. The metallic film usually introduces a small additional phase shift, which can be ignored or reduced to a zero value by using a suitably thicker or thinner dielectric film than calculated from Eq. (5.13). The phase plate is *positive* ($\psi > 0$) if $n_M > n_R$; the direct light is then advanced in phase compared to the diffracted light. Conversely, the phase plate is *negative* ($\psi < 0$) if $n_M < n_R$, and the direct light is then retarded in phase relative to the diffracted light. Phase plates which change the phase of the direct light relative to the diffracted light by $+90°$ or $-90°$ will be called *quarter wavelength phase plates* because they produce an optical path difference δ equal to $\lambda/4$.

Typical phase contrast systems are not manufactured as specialized microscopes but usually as phase contrast devices to be attached to ordinary microscopes. In general, a standard phase contrast device for transmitted light microscopes consists of a set of phase objectives and a special condenser incorporating a number of annular diaphragms (Fig. 5.9). In principle, phase objectives with different magnifications require annular diaphragms of different sizes. Usually, these diaphragms are located in a revolving disc. An essential feature of any phase condenser is an adjustment for centring the image A' of the condenser diaphragms A on the objective phase rings R (Fig. 5.6). This adjustment is controlled by replacing the eyepiece Oc with an auxiliary lunette and viewing the exit pupil of the phase objective Ob. For phase contrast observation, a Köhler illuminator consisting of a low-voltage bulb B, collector Col, and field diaphragm FD is commonly used.

Any phase contrast device can normally be used for bright-field microscopy as well. The most popular phase condensers are provided with a rotatable disc containing three to five annuli for phase contrast and one or two clear openings for bright-field observation. The phase ring in the objectives does, of course, cause some disturbance of bright-field images but frequently this defect is insignificant and the same phase objective can be used for both purposes.

A turret phase condenser can also be used for dark-field or rather quasi dark-field microscopy with low- and medium-power objectives. In this case, the condenser annulus is selected so that its image in the back focal plane of the objective lies completely beyond the edge of the objective exit pupil.

For simultaneous phase contrast, bright-field, and dark-field microscopy, *universal phase condensers* with a continuously variable diameter of the annulus are more suitable. Among such systems will be found the Heine condenser manufactured by E. Leitz Wetzlar, the Polyphos condenser produced by C. Reichert Wien, and the pancratic condenser available from C. Zeiss Jena.

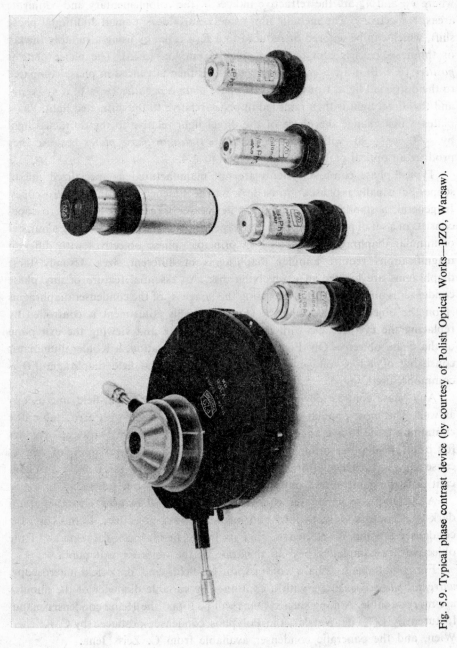

Fig. 5.9. Typical phase contrast device (by courtesy of Polish Optical Works—PZO, Warsaw).

It should be added that recently some new facilities have been introduced for inverted biological microscopes. Nikon, for instance, offers the Diaphot microscope with a long working-distance $40 \times /0.55$ phase contrast objective with a coverglass correction collar to allow a thickness correction from 0 to 2 mm. This correction also allows for a variation in the working distance from 0.18 mm to 2.39 mm. The microscope is also fitted with a low-power $4 \times$ phase contrast objective of a kind never produced before, which greatly facilitates rapid low-magnification screening under phase contrast.

Typical phase contrast devices are described in many books. They are well known in practice, and their detailed presentation here is superfluous. Only a survey of some non-standard phase contrast devices will be found in the following sections of this chapter.

5.3. Imaging properties

Phase contrast microscopy has some properties which cannot be explained on the basis of vector representation and other simplified theories. An exact explanation can be given only in terms of Fourier optics and spatial filtering [3.4, 5.7, 5.8]. The mathematical formalism of this approach is, however, complicated and beyond the scope of this book. Therefore, a verbal description supported by some experiments relating to the quality and basic defects of phase contrast imagery will be given in this section.

5.3.1. Halo, shading-off, and image fidelity

So far it has been assumed that direct and diffracted light are completely separate in the objective exit pupil. In principle, this condition could be approached if both the clear area of the condenser diaphragm and its conjugate area on the phase plate were made extremely small. In practice, these elements must have a finite size and suitable shape which are determined by various factors such as adequacy of illumination, light coherence, image quality, and resolution. The annular diaphragm and ring-shaped phase plate have evolved as a practical compromise (Fig. 5.6). However, with such elements one cannot completely separate direct from diffracted light because microscopic specimens include structures corresponding to various spatial frequencies, which diffract the light over the whole aperture of the objective. Some of the diffracted light inevitably passes through the phase ring and is modulated in the same way as the direct light. In any case, this defect is inevitable because for certain practical reasons the phase ring must be wider (usually about 25%) than the conjugate area defined by the image of the condenser annulus. On the other hand, some of the direct light tends to spread into the region of the complementary area. All these circumstances create certain optical effects known as *halo* and *shading-off*.

In phase contrast microscopy, the halo is seen in the background as bright or dark zones surrounding the object images. The bright haloes are attached to dark images against the lighter background and vice versa, dark haloes accompany bright images against the darker background. The dark haloes are in general less disturbing than the bright ones.

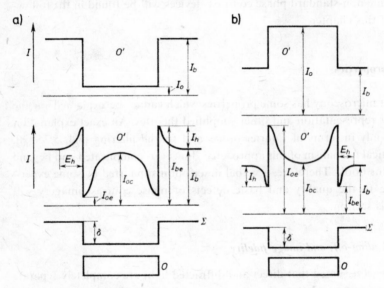

Fig. 5.10. Halo and shading-off effects: a) positive (dark) phase contrast image, b) negative (bright) phase contrast image.

The shading-off effect occurs in the object image and exhibits itself as brighter and darker central areas in dark and bright images respectively. As a result of this effect, the distribution of light intensity in the image does not correspond to the distribution of optical path difference in the object. This can be illustrated by means of a thin phase object of uniform optical path difference δ as in Fig. 5.10. If there were no shading-off and halo effects, the image of the object O would appear uniformly dark or bright as shown in the top of Fig. 5.10a and 5.10b respectively.

Taking into account the halo and shading-off effects, one must distinguish four different light intensities, I_b, I_{be}, I_{oe}, and I_{oc}, related respectively to the distant background, the halo region near the edge of the object image, the object image near its edge and the object image in its centre. A difference (I_h) in intensities I_b and I_{be} characterizes the *halo intensity* ($I_h = I_b - I_{be}$). If $I_b < I_{be}$, the halo is bright, and it is dark when $I_b > I_{be}$. Apart from I_h, it is also useful to speak about the *halo extension* (E_h), i.e., the width of a background zone of increasing or decreasing intensity which surrounds the object image. A difference (I_s) in intensities I_{oc} and I_{oe} characterizes the shading-off effect. The visibility of the phase object is mainly defined by the edge contrast (C_e), which can be expressed, for instance, as $C_e = (I_{be} - I_{oe})/I_{be}$, whereas the difference in intensities I_{oe} and I_{oc}, as well as, to some degree, that between intensities I_b and I_{be}, characterize the *image fidelity* (or *image faithfulness*). This feature can be quantitatively evaluated by means of a suitably constructed specimen containing thin strips of a transparent material, e.g., vacuum evaporated onto a glass slide.

The halo and shading-off effects are in general unwanted and together with the non-linearity between the image intensity and optical path difference (δ) in the specimen (linearity occurs only for small δ) make a correct interpretation of images of more complicated objects difficult. On the other hand, the halo often emphasizes the contrast between the object image and its surrounding background and accentuates thin edges and border details of some objects, e.g., biological cells. This is true, in particular, of negative phase contrast when a dark halo appears around the image detail (see, e.g., Figs. X and XIa).

The halo and shading-off effects depend on both the geometrical and optical properties of the phase plate and objects under examination. For phase objects of given optical path difference and dimensions, the effects greatly depend on the width and transmittance of the phase ring. In general, the wider the phase ring and the smaller its transmittance, the more obtrusive the halo and shading-off effects become, whereas moderate alterations in phase shift (ψ) and diameter of this ring are less important. For a given phase ring, e.g., for a positive or negative quarter wavelength ring, the halo and shading-off effects depend, first of all, on the optical path difference and size of phase objects, as well as their shape and structure. Moreover, these effects, and especially the shading-off, greatly depend on the magnifying power of the objective. This and other factors are illustrated by pictures and microdensitometric graphs in Figs. 5.11–5.14, and summarized in Table 5.1. This summary has been prepared on the basis of several research papers, primarily papers [5.10]–[5.13]. As can be seen, the fidelity of phase contrast imaging depends on (1) the width and light absorption of the

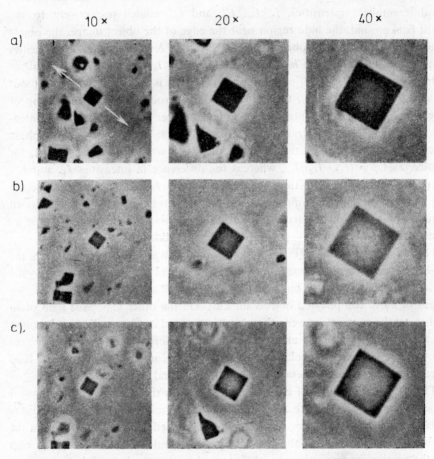

Fig. 5.11. A sequence of photomicrographs showing the dependence of the halo and shading-off effects on the objective magnification and on geometrical and optical properties of the phase plate. The photomicrographs are taken with three different sets: a, b, and c, of positive phase objectives of magnification $10\times$, $20\times$, and $40\times$. The phase rings in these objectives had the following characteristics: a) $\psi = +90°$, $\tau = 15\%$, $d/d_{EP} = 0.28$, $w/d_{EP} = 0.045$, $a/a_{EP} = 0.055$; b) $\psi = +90°$, $\tau = 25\%$, $d/d_{EP} = 0.55$, $w/d_{EP} = 0.07$, $a/a_{EP} = 0.14$; c) $\psi = +90°$, $\tau = 15\%$, $d/d_{EP} = 0.5$, $w/d_{EP} = 0.1$, $a/a_{EP} = 0.17$; where ψ, τ, d, w, and a are the phase shift, transmittance, diameter, width, and area of the phase ring, respectively; d_{EP} and a_{EP} are the diameter and area of the exit pupil of objective, respectively. The square flat object in the centre of the photomicrographs has the dimensions 0.01×0.01 mm, and phase shift between light passing through this object and its surrounding medium is about $-8°$ [5.11].

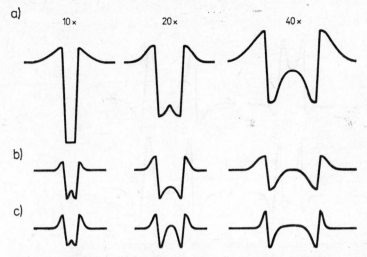

Fig. 5.12. Microdensitometric graphs across the images of the square object in Fig. 5.11. The curves a, b, and c relate to the photomicrographs a, b, and c in Fig. 5.11, respectively [5.11].

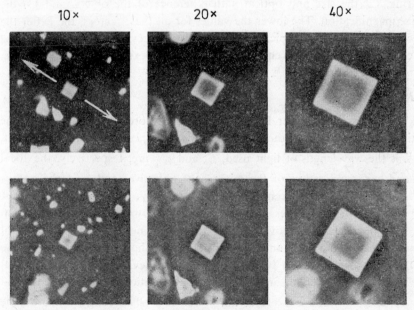

Fig. 5.13. As in Fig. 5.11, but photomicrographs are taken with objectives having negative phase rings made of soot, the parameters of which are as follows: a) $\psi = -90°$, $\tau = 3$ to 4%, $d/d_{EP} = 0.55$, $w/w_{EP} = 0.065$, $a/a_{EP} = 0.13$; b) $\psi = -90°$, $\tau = 3$ to 4%, $d/d_{EP} = 0.5$, $w/w_{EP} = 0.1$, $a/a_{EP} = 0.17$.

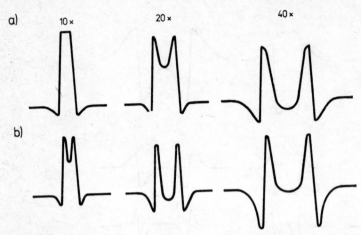

Fig. 5.14. Microdensitometric graphs across the images of the square object in Fig. 5.13. The curves a and b relate to the photomicrographs a and b in Fig. 5.13 [5.11].

phase ring, (2) the size and optical path difference of the object, and (3) the objective magnification. The lower the values for all these factors, the better the phase contrast imaging. In typical phase contrast devices, the width of the phase ring is chosen so that its area is equal to one-tenth of the exit pupil area of objective.

5.3.2. Resolution

Let us for a moment consider Figs. 5.2 and 5.6. According to Eqs. (3.34), the resolution is $\varrho_1 = f'_{ob} \lambda / r_{E'_{ob}}$ in the first case, and $\varrho_2 = f'_{ob} \lambda / (r_{E'_{ob}} + r)$ in the second, where λ is the wavelength of light used, f'_{ob} and $r_{E'_{ob}}$ are, respectively, the focal length and exit pupil radius of the microscope objective, and r is the radius of a phase ring. As can be seen, ϱ_2 is smaller than ϱ_1, hence a phase contrast system with a phase ring possesses a better resolving power than the same system with a small phase disc. Moreover, the greater r, the smaller ϱ_2. Due to optical aberrations, the phase ring cannot, however, occupy the marginal zones of the exit pupil of the objective. In typical phase contrast devices, r is usually equal to $r_{E'_{ob}}/2$. This value is a compromise between the resolution, contrast, and general quality of imaging because the resolving power of a phase contrast microscope is perturbed by the halo which in general makes resolution worse. This defect is more noticeable in positive phase contrast (bright halo) than in negative phase contrast (dark halo). But a properly constructed phase contrast microscope

TABLE 5.1

Phase contrast imaging properties depending on phase ring parameters, object parameters, and objective magnification*

Specification of parameters	Image fidelity	Edge contrast	Halo intensity	Halo extension	Shading-off effect	Resolution
Phase ring parameters—variable; object parameters—constant; objective magnification—constant						
$r_{A'} = r$ increases $w_{A'} = w$ constant ψ constant, τ constant	increases	decreases	slowly decreases	slowly increases	decreases	ameliorates
$w_{A'} = w$ increases $r_{A'} = r$ constant ψ constant, τ constant	decreases	decreases	increases	decreases	increases	ameliorates
w increases $w_{A'}$ constant $< w$ $r_{A'} = r$ constant ψ constant, τ constant	decreases	decreases	increases	decreases	increases	deteriorates
τ decreases remaining parameters constant	decreases	increases	increases	increases	increases	deteriorates
Object parameters—variable; phase ring parameters—constant; objective magnification—constant						
d_o increases φ constant	decreases	decreases	increases	decreases	increases	—
$|\varphi|$ increases, but $|\varphi| < |\varphi_r|$ d_o constant	decreases	increases, next decreases	increases	increases	increases	deteriorates
Objective magnification β_{ob}—variable; phase ring parameters—constant; object parameters—constant						
β_{ob} increases	decreases	decreases	increases	decreases	increases	ameliorates

* r, w, ψ, and τ—radius, width, phase shift and transmittance of the phase ring, respectively; $r_{A'}$ and $w_{A'}$—radius and width of the annular diaphragm image, respectively; d_o and φ—diameter and phase shift of a circular phase object.

may be said to possess a resolving power comparable to that of the ordinary microscope.

Resolution in phase contrast microscopy is discussed in greater detail by Bennett, Jupnik, Osterberg, and Richards in their book *Phase Microscopy* [5.14].

5.3.3. Sensitivity

One of the most important aspects of the phase contrast method is its *sensitivity*, defined as the smallest optical path difference δ_m (or phase shift φ_m) that can be detected by the observer. Here we are only concerned with very small phase shifts φ so that Eqs. 5.3 apply, and the minimum image contrast (or contrast threshold) C_m perceived by the human eye can be expressed as

$$C_m = 2\varphi_m \sqrt{N} = 2\frac{2\pi}{\lambda}\delta_m \sqrt{N},$$

and thus

$$\delta_m = \frac{C_m \lambda}{4\pi \sqrt{N}}. \tag{5.14}$$

The contrast threshold C_m depends on the shape and size of the object, brightness of the background, state of accommodation of the eye, and other factors. In the case of two areas separated by a straight-line boundary, $C_m = 0.02$ is usually accepted. Then for a non-absorbing phase plate ($N = 1$), the sensitivity $\delta_m = 1$ nm, and for a phase plate which absorbs 96% ($N = 25$) of the direct light, $\delta_m = 0.2$ nm. In favourable circumstances, these theoretical limits may be obtained in practice.

Experimentally, the sensitivity of a phase contrast microscope can be precisely evaluated by using a test object consisting of narrow strips of thin dielectric films evaporated in vacuum on a glass slide and next covered with a high dispersion

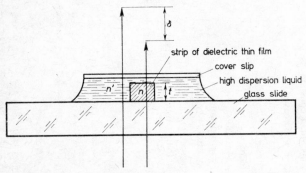

Fig. 5.15. Test specimen for measurement of the sensitivity of a phase contrast microscope.

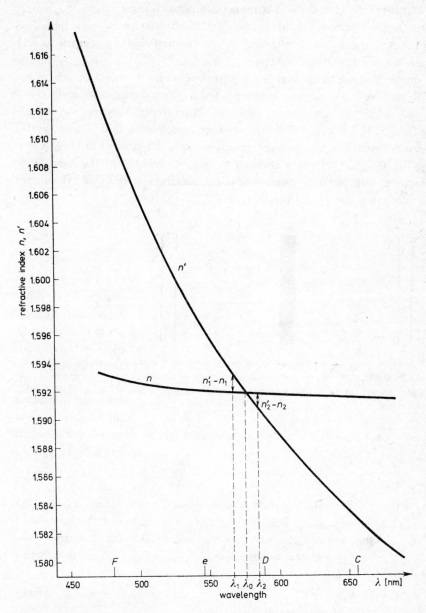

Fig. 5.16. Principle of measurement of phase contrast microscope sensitivity by using the test specimen as shown in Fig. 5.15.

liquid[6] and cover slip (Fig. 5.15). The immersion liquid is chosen in such a manner that for a light wavelength λ_0 the optical path difference δ between light rays passing through the dielectric strip and its surrounding medium is equal to zero in the middle of the visible light spectrum.

Let us suppose the phase contrast microscope under examination has objectives with positive phase rings. It is sharply focussed on the dielectric strip transilluminated with monochromatic light of continuously varying wavelength Starting from λ_0 (Figs. 5.16 and 5.17a), two such wavelengths $\lambda_1 < \lambda_0$ and $\lambda_2 > \lambda_0$ can be fixed, for which the observer perceives the smallest positive (Fig. 5.17b) and negative (Fig. 5.17c) phase contrast images of the strip. If the microscope has objectives with negative phase rings, the situation is reversed: the image contrast is negative for λ_1 and positive for λ_2.

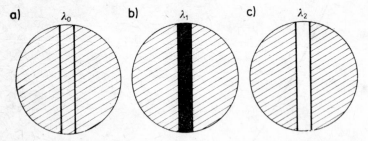

Fig. 5.17. Thin film strip images related to the phase contrast miscroscope sensitivity measurement as shown in Fig. 5.16: a) zero contrast image in light of wavelength λ_0, b) positive contrast image hardly visible in light of wavelength $\lambda_1 < \lambda_0$, c) negative contrast image hardly visible in light of wavelength $\lambda_2 > \lambda_0$.

The smallest optical path differences detected by the observer are

$$\delta_1 = (n'_1 - n_1)t, \tag{5.15a}$$

$$\delta_2 = (n'_2 - n_2)t, \tag{5.15b}$$

where n_1 and n_2 are the refractive indices of the dielectric strip for the wavelengths λ_1 and λ_2, respectively, n'_1 and n'_2 are the refractive indices of the immersion liquid for the same wavelengths, and t is the thickness of the dielectric strip. Practically, $|\delta_1| = |\delta_2|$, and the sensitivity $\delta_m = |\delta_1|$ or $\delta_m = |\delta_2|$, but a better value is

$$\delta_m = \tfrac{1}{2}(|\delta_1| + |\delta_2|). \tag{5.16}$$

[6] Available, e.g., from Cargille Laboratories, Inc., 55 Commerce Road, Cedar Grove, N.J. 07009, USA.

Knowing the thickness t of the dielectric thin film strip (Fig. 5.15) and dispersion curves $n(\lambda)$ and $n'(\lambda)$ of the strip and its surrounding liquid (Fig. 5.16), one can determine the refractive index differences $n'_1 - n_1$ and $n'_2 - n_2$, and then calculate the optical path differences δ_1 and δ_2 from Eqs. (5.15). As can be seen, the procedure is very simple.

To obtain δ_m for the middle of the visible light spectrum, the differences $\lambda_0 - \lambda_1$ and $\lambda_2 - \lambda_0$ should be small (not greater than 25 nm). They can be properly adjusted by preparing thin film strips of various thickness. The most suitable width of the strips is 0.1 mm/M_{ob}, where M_{ob} is the magnifying power of the objective.

In experiments performed by the author of this book, strips of thin film were deposited onto a glass slide by reactive vacuum evaporation of ceric fluoride. The slide was masked by using a photoresist layer previously spread over the slide and adequately photochemically processed before and after the vacuum deposition of the thin film [5.15]. Depending on the conditions of the vacuum process, the refractive index n_D of the ceric fluoride strips equals 1.59 to 1.60, and its spectral dispersion is very small ($n_F - n_C = 0.001$). The desired effect is obtained by using the Cargille high dispersion liquid with $n_D = 1.590$ or $n_D = 1.600$, and $n_F - n_C = 0.026$ (Fig. 5.16).

In these experiments, the phase contrast microscope to be examined was fitted with a monocular head and eyepieces of magnifying power $5\times$ or $10\times$. Monochromatic light of continuously variable wavelength was obtained from a halogen illuminator by inserting an interference monochromator (wedge interference filter) under the condenser. Instead of this monochromator, a set of typical interference filters with a half power width smaller than 10 nm can also be used. However, the spectral intervals between peak wavelengths of succeeding filters should be not greater than 5 nm.

It was stated earlier that experimental values obtained for the sensitivity of many phase contrast devices agree with those resulting from Eq. (5.14) but only at the centre of the field of view of the microscope. Due to stray light, sensitivity and image contrast at the peripheries deteriorate as the distance from the centre of the field of view increases. Some experimental data concerning the sensitivity of different phase contrast device are given in Table 5.4 (see Subsection 5.5.2).

5.3.4. Influence of stray light

As was mentioned earlier, typical phase contrast rings consist of a dielectric thin film for shifting the phase of the direct light beam by $+90°$ or $-90°$ and a metallic film for decreasing the intensity of the beam. Both films are usually deposited in vacuum onto a glass plate or lens surface located in the back focal plane of the objective. Such phase rings have a serious defect: the metallic film reduces the direct light intensity not only by absorption but also by reflection. The reflected light returns to the front lens of objective as well as to the cover slip, is reflected again, enters the objective, and reaches the image plane as *stray light* or *glare*. This results in some reduction of image contrast and of phase contrast sensitivity. The smaller the transmittance (τ) of the phase ring, the greater the stray light intensity. Moreover, if the phase ring is deposited on a flat plate (Fig. 5.7), it produces more glare than when deposited on a spherical surface (Fig. 5.8), as in the latter case a greater portion of the direct light reflected by the metallic film spreads beyond the field of view of the phase contrast microscope. All these factors mean that metallic substances, such as aluminium or chromium, cannot be effectively used for manufacturing highly sensitive phase contrast devices, which ought to have greatly absorptive phase rings ($\tau < 10\%$), unless steps are taken to eliminate reflected light. One preventive measure is to use phase rings with antireflecting coatings. However, the problem is not a simple one because an efficient antireflection effect is achieved only when a reflecting substrate is coated with a transparent thin film in such a manner that rays reflected from its outer surface and those reflected from the substrate interfere destructively. The destructive interference is complete if the interfering rays have a phase difference of half a wavelength and equal intensities. However, the reflectance (ϱ) of metallic surfaces is extremely strong ($\varrho > 30\%$) compared to that of common dielectric surfaces ($\varrho = 2-14\%$) and therefore difficult to eliminate by conventional antireflecting coatings as described in Subsection 1.6.3.

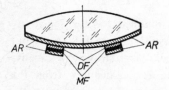

Fig. 5.18. Reichert's anoptral phase ring: *AR*—antireflecting films, *DF*—dielectric film, *MF*—metallic film [5.13].

An efficient solution of the problem has recently been put forward by Yamamoto and Taira [5.12], although the first phase rings with antireflecting coatings were applied commercially by C. Reichert (Vienna, Austria) in their so-called *anoptral contrast device.*[7] An anoptral ring is the negative phase ring ($\psi = -90°$), which transmits approximately 8% of direct light. The exact construction of anoptral phase rings has not been published, but Françon in his book [1.31] suggests that the structure of these rings is as shown in Fig. 5.18. The anoptral contrast device produces a good image contrast without fogginess (stray light) in the field of view of microscope. Both monochromatic (green) and white light can be used for observation of many phase specimens, especially living cells and their structure [5.16]. In white light the observer sees an agreeable golden-brownish background.

The Yamamoto and Taira *antireflecting phase ring* is more complicated than the anoptral ring. An example consisting of six layers is shown in Fig. 5.19. Two layers are thin metallic films of desired absorbance, and four layers are dielectric films of different refractive indices and appropriate thickness. In their paper [3.12], the authors give comprehensive theoretical grounds for such a multilayer structure. Its antireflection effect is shown in Fig. 5.19 by curve a. For

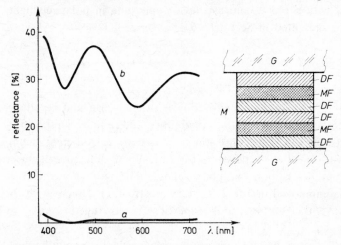

Fig. 5.19. Structure and antireflecting effect of a multilayer phase ring after Yamamoto and Taira [3.12]: *DF*—dielectric thin film, *MF*—metallic thin films, *G*—glass, *M*—cement.

comparison, curve b shows the reflectance of a typical two-layer phase ring consisting of single dielectric and metallic films. As can be seen, the antireflection effect of the new phase ring is extremely high over the entire visible spectrum.

Antireflecting phase rings based on the design by Yamamoto and Taira are manufactured by the Olympus Optical Co., Ltd. (Tokyo, Japan).

[7] "Anoptral" is a trade mark and means non-reflecting.

Another, simpler approach to the elimination of stray light and glare from phase contrast imaging was noticed by Wilska [5.16, 5.17] and developed by the author of this book [5.18]. It depends on making phase rings of a highly absorbent substance which reduces the intensity of light only by absorption. Such a substance is soot. As was stated earlier [5.18], soot layers obtained by exposing a glass plate or lens to stearin, kerosene, or illuminating gas flames have very interesting optical properties relating to light absorbance and phase shift. The relationship between these quantities is very advantageous for making negative phase contrast rings. In particular, it makes possible the production of highly sensitive negative phase rings consisting of a single soot layer.

The optical properties of different soot layers, as well as negative and positive phase contrast devices with soot phase rings, are covered by Section 5.5 and Subsection 5.6.2.

The light reflected by a phase ring and glare can be also more or less reduced in microscopes fitted with a phase contrast device operating in polarized light. Such a device will be presented in Section 5.6.

5.4. Nomenclature

The nomenclature of phase contrast microscopy is not uniform and results in some confusion about the optical properties, capabilities, and commercial specifications of phase contrast devices. This confusion is caused by (1) different definitions of image contrast, (2) various expressions for optical path difference or phase shift, and (3) different descriptions of phase plates. The basic nomenclature used in this book is summarized in Table 5.2. It is mostly in agreement with the nomenclature of many other authors, and in particular with that proposed by Zernike [5.19].

All phase contrast devices having objectives with quarter wavelength phase plate decreasing the intensity of the direct light to the range of about 25% to 10% will be designated as *standard phase contrast devices*. They are commercially manufactured by many optical firms in Europe, USA, and Japan as equipment for typical biological microscopes. Each comprises a set of phase contrast objectives with positive or negative phase rings, a phase condenser with annular diaphragms arranged in a turret (this condenser usually also serves for bright-field illumination), and an auxiliary telescope for observing the objective exit pupil while centring the image of the condenser annulus on the phase ring. An example of such a device is shown in Fig. 5.9. It is intended for common biological micro-

TABLE 5.2

Basic nomenclature and relations between optical properties of the object, type of phase plate, and image contrast*

Object		Phase plate		Image contrast	
Relation between n and n_M	Sign of phase shift	Relation between n_C and n_R	Sign of phase plate	Relation between I_{oe} and I_b	Sign of contrast
$n > n_M$	Negative (−)	$n_C > n_R$	Positive (+)	$I_{oe} < I_b$	Positive (+)
$n > n_M$	Negative (−)	$n_C < n_R$	Negative (−)	$I_{oe} > I_b$	Negative (−)
$n < n_M$	Positive (+)	$n_C > n_R$	Positive (+)	$I_{oe} > I_b$	Negative (−)
$n < n_M$	Positive (+)	$n_C < n_R$	Negative (−)	$I_{oe} < I_b$	Positive (+)

* n and n_M—refractive indices of the object and its surrounding medium; n_C and n_R—refractive indices of the complementary and conjugate areas of the phase plate; I_{oe} and I_b—light intensity (or brightness) of the object image and its background.

scopes, whereas inverted biological microscopes need long working-distance phase condensers. Somewhat differently constructed are standard phase contrast devices for metallurgical microscopes, which have quarter wavelength phase rings arranged in a slide or turret behind the objective and annular diaphragms located (also in a slide or turret) in an epi-illuminator.

Consequently, *non-standard phase contrast devices* comprise all other special systems and some which resemble standard ones but incorporate non-typical phase rings (e.g., highly or weakly absorbing, with small or zero phase shift), and phase plates termed by Bennett *et al.* [5.14] B-type (intensity of diffracted light is reduced relative to that of direct light), as well as some devices incorporating non-ring-shaped phase plates.

Many different non-standard phase contrast systems are described in scientific or technical journals, as well as in patents, and it is impossible to present all of them in a single chapter of limited scope. Thus, in the first place a survey will be given of some non-standard devices which are or have been commercially available.

5.5. Highly sensitive phase contrast devices

In 1956, the author of this book undertook experiments to find a suitable industrial process for making soot phase rings. First it was necessary to find a method of hardening and fixing soot layers on glass surfaces. It was discovered that this is attained by moistening with a wetting liquid such as alcohol or ether. When the liquid evaporates, the soot layer becomes harder, adheres to the glass surface better, and can be recessed with great accuracy by means of a suitable cutting tool. Most important of all, a ring made of such a soot layer can be cemented without damage. Finally, as a result of this procedure and studies of the optical properties of different soot layers, there were designed two *highly sensitive phase contrast devices* with negative and positive phase rings [5.20—5.23], and an *amplitude contrast device* [5.18], which will be described in Chapter 6. The optical properties of some soot layers and two phase contrast devices with soot phase rings are presented below.

5.5.1. *Optical properties of soot layers*

A thin layer of soot obtained, e.g., by exposing a glass slide to the flame of a stearin candle, changes both the amplitude and phase of the transmitted light. Light passing through a strip of a partially transparent soot layer SF is considerably retarded in phase relative to light passing outside the strip through the surrounding air medium M (Fig. 5.20). On the analogy of a transparent object, one may write $\delta = (n_M - n_S) t$, where n_M is the refractive index of the medium, n_S is the effective refractive index of the soot strip, and t its thickness. The optical path difference δ depends, in general, on the light transmittance τ of the soot layer. This dependence has been measured by using a double refracting interference microscope (MPI-5 model) described in Chapter 16. The results for soot layers obtained from illuminating gas, stearin, and kerosene flames are presented in Fig. 5.21. Curve 2 shows that for the stearin soot layer of about 10% transmittance, the optical path difference $\delta = -0.25\lambda$. Hence, a ring made of a stearin soot layer which absorbs 90% of the light constitutes a negative phase ring which shifts the phase of the direct light by $-90°$. This satisfies the condition of optimal negative phase contrast for slightly refractile phase objects.

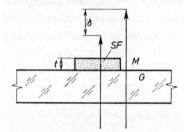

Fig. 5.20. Strip of a soot layer SF on a glass slide G, and its phase retarding effect.

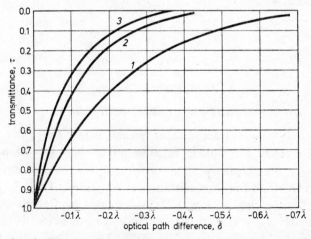

Fig. 5.21. Optical path difference (measured in relation to air medium) as a function of light transmittance for soot layers obtained by exposing a glass plate to flames of illuminating gas (curve 1), stearin (curve 2), and kerosene (curve 3) [5.18]

The interferometric technique of a double immersion medium (see Subsection 16.5.3) was used to determine the thickness t and effective refractive indices n_S of various soot layers differing in transmittance of light, then the dependence between the optical density D (or $\log \tau^{-1}$) and thickness t was examined, and it was stated that the functions $t(D)$ and $\delta(D)$ are linear (at least over the density interval $D = 0.1$ to 1.1) for different soot layers, thus showing that Lambert's law $\tau = e^{-Kt}$ is obeyed for them. In view of this, the following relationship has been established:

$$\delta = -2.3(n_M - n_S) \frac{\log \tau}{K}. \tag{5.17}$$

Knowing the absorption coefficient K and refractive indices n_S and n_M, one can calculate δ for a given τ. The determined refractive index n_S and absorption coefficient K for a few soot layers are given in Table 5.3. The optical and other properties of soot layers are presented in more detail in papers [5.11], [5.18]. The data given here are sufficient to understand the following text.

TABLE 5.3

Refractive index n_S and absorption coefficient K of soot obtained from the flames of stearin, kerosene, and illuminating gas (for wavelength $\lambda = 546$ nm)

Type of soot	n_S	K [10^5 cm^{-1}]
Stearin	2.32	2.01
Kerosene	2.26	2.37
Illuminating gas	2.12	0.95

5.5.2. Highly sensitive negative phase contrast device (KFA)

Soot layers have a high refractive index, greater than 2 (see Table 5.3), and are thus very convenient for making negative phase rings. For thin and slightly refracting phase objects, such a ring must retard the phase of direct relative to diffracted light by $-90°$. From the values n_S and K given in Table 5.3 and assuming that the soot phase ring R (Fig. 5.22) is cemented by Canada balsam having a refractive index $n_M = 1.53$, a transmittance corresponding to the phase shift $\psi = -90°$ can be calculated from Eq. (5.17). This transmittance will be 3.2%, 1.2%, and 11% for soot rings of stearin, kerosene, and illuminating gas flames, respectively.

For highly sensitive phase contrast microscopy, therefore, which needs as great a reduction of direct light as possible, stearin or kerosene soot deposits are most suitable. After several trials, stearin soot was decided on. It was found experimentally that the tolerance for light transmittance τ can be up to $\pm 1\%$ and ultimately $\tau = 3$ to 4% was selected. For this transmittance, the phase retardation ψ, according to Eq. (5.17), should remain within the range of $-83°$ to $-91°$.

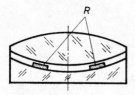

Fig. 5.22. Soot phase ring (R) cemented between two lenses of an objective doublet.

Using these data, a highly sensitive negative phase contrast device was constructed for biological microscopes [5.20–5.22]. This is known as KFA and is commercially available from the Polish Optical Works (PZO), Warsaw.

The KFA equipment comprises four objectives (see Table 5.4), a phase condenser with four annular diaphragms arranged in a turret, and an auxiliary centring telescope. Each objective incorporates a soot phase ring. This ring is located between two lenses (Fig. 5.22) nearest the back focal plane of the objective (Fig. 5.8). The geometrical characteristics of the phase rings are as follows:

$$d/d_{EP} = 0.5, \quad w/d_{EP} = 0.1, \quad a/a_{EP} = 0.17,$$

where d, w, and a are the diameter, width, and area of the phase ring, d_{EP} and a_{EP} are the diameter and area of the exit pupil of the objectives, respectively.

The variation in optical properties of the soot rings, i.e., phase shift and light transmittance, has been measured for a number of different objectives selected from among manufactured equipment. Measurements were made by means of a special shearing polarizing interferometer (see Section 4.3 and Fig. 4.11). The results were within the range of manufacturing tolerance to the theoretical values.

The sensitivity and range of unreversed contrast were determined using test objects consisting of narrow strips of dielectric thin films evaporated in vacuum on a glass slide (Fig. 5.15). Applying different immersion liquids between a cover slip and the slide, various optical path differences δ were obtained. The range of unreversed image contrast was established to be 0 to $-150°$ for phase-retarding objects and 0 to $+20°$ for phase-advancing objects. Using the method described earlier (see Subsection 5.3.3), it was established that this device is able to reveal

TABLE 5.4

Principal characteristics of the phase contrast devices available from PZO, Warsaw

Phase contrast device type	Magnification and numerical aperture of objectives	Phase ring			Optical path difference δ [nm] best suited for observation of objects with refractive index		Sensitivity δ_m [nm]
		Type	Phase shift ψ [°]	Transmittance τ [%]	higher than that of the surrounding medium	lower than that of the surrounding medium	
KF	10×/0.24 20×/0.40 40×/0.65 100×/1.30 OI*	Positive, dielectric-metallic	+90	15	15–140	15–150	1.0–1.2
KFA	10×/0.24 20×/0.40 40×/0.65 100×/1.30 OI	Negative, soot	−90	3–4	4–100	4–30	0.3–0.5
KFS	10×/0.24 20×/0.40 40×/0.65 100×/1.30 OI	Positive, dielectric-soot	+90	10	10–90	10–130	0.5–0.7
KFZ	10×/0.24 20×/0.40 40×/0.65 100×/1.30 OI	Positive, dielectric-metallic, and negative, soot	+90 −90	15 3–4	8–160 2–110	8–180 2–70	0.6–0.8 0.2–0.4

* OI—oil immersion.

optical path differences of about 0.2 nm. However, such high sensitivity is not good for observing most common phase objects as their image is overcontrasted. For such objects, a sensitivity equal 0.3 to 0.5 nm has been accepted as most effective and this is the obligatory tolerance in the manufacturing process of the KFA devices.

The soot ring is practically achromatic and its phase shift is almost constant over the whole visible spectrum. Thus, the KFA device gives very good image contrast in both white light and monochromatic light of different wavelengths. When white light is used, the field of view of the microscope presents an agreeable brownish colour. This colour is the result of the weak filtering effect of the soot layer, which transmits the longer wavelengths somewhat better than short wavelengths of the visible spectrum.

Due to its high sensitivity, the KFA device is suitable for observation of fine details, small particles, close-packed structures (Figs. X, XI), and very thin completely transparent objects whose refractive indices differ only slightly from that of the surrounding medium. In particular, it reveals very subtle structures, small granules and inclusions in living cells and tissues. Because of the large range of unreversed contrast for phase-retarding objects, the KFA device is also convenient for the examination of any other phase objects whose refractive index differs more markedly from that of the surround. In the observation of phase-advancing objects, this device is of only limited use as the range of unreversed contrast for these objects is small (see Table 5.4). The majority of microscopic specimens, however, and especially living tissues and cells (as well as their inclusions) behave as phase-retarding objects.

In illustration, photomicrographs of epithelial cells from oral mucosa and human blood corpuscles are shown in Figs. XII–XVI. These were taken with objectives of high magnification. Figure XII shows an abundance of protoplasmic granules characteristic of most oral epithelial cells; these are some highly refractile particles. Another characteristic feature is a peculiar ridged and microfolded structure of the membrane (Fig. XIII). This structure, which is at the limit of the resolving power of the light microscope, can be seen particularly well with this device. Figure XIV shows a fragment of another epithelial cell with somewhat finer membrane structure covered by a cluster of bacteria. In this case, the cell membrane appears as a delicate spongy network. This network and the highly refractile bacteria are both well seen. The ability to produce clear images of fine structures and of highly refractile phase-retarding particles at the same time is a basic advantage of this device.

This feature is also seen in photomicrographs of living leucocytes taken with the KFA device. Large structural elements such as nuclei and the small cytoplasmic granules are both well displayed. Figure XV shows neutrophil and eosinphil granulocytes among some red cell "ghosts". Some of the granules undergo Brownian movement and do not always appear sharp in photomicrographs taken with an exposure of a few seconds. Figure XVI shows two neutrophils, one of which burst after a few minutes (lower picture) when the Brownian movement ceased and the granules appeared sharp.

The KFA optics is especially suitable for studies of cells grown in tissue culture and has been successfully used for this purpose by Veselý [5.24, 5.25]. He studied the surface motility of cultured cells by means of an apparatus for time-lapse phase contrast ciné light micrography with improved spatial and temporal resolution. The KFA objectives appeared to be the best optics for the imaging of tiny cell surfaces structures. Among other things, it was possible to observe the ruffling movement of the cell membrane, the behaviour of microvilli, and the opening of pinocytotic channels onto the surface.

In addition, the KFA device is very suitable for highly accurate immersion refractometry (see Subsection 5.11.3). Experiments showed that refractive index differences of about 0.0002 can be detected when small objects such as thin strips of transparent material a few μm thick are immersed in a HD liquid according to conditions shown in Fig. 5.16.

5.5.3. Highly sensitive positive phase contrast device (KFS)

An essential difference between this device and the KFA equipment is to be found in the phase rings. Instead of negative phase rings consisting of a single soot layer, positive quarter wavelength phase rings consisting of soot and dielectric films are used (Fig. 5.23).

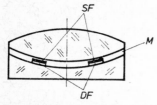

Fig. 5.23. Positive phase ring consisting of soot film (*SF*) and dielectric film (*DF*) cemented between two lenses of an objective doublet.

In order to make such a phase ring, it is necessary to compensate the negative phase shift produced by the soot layer. This compensation cannot be achieved by embedding the soot ring in a suitably matched medium since the refractive index of soot layers is greater than 2 (see Table 5.3), and there is no optical cement with so high a refractive index. The only way therefore of compensating the negative phase shift of the light passing through the soot ring SF (Fig. 5.23) is to underlay this ring with a transparent dielectric film DF, whose refractive index n is considerably lower than the refractive index n_M of the medium M (optical cement), which serves to glue the ring between two glass plates or lenses. In order to obtain the desired phase shift equal to $+90°$ between the light passing through the ring SF and light passing outside it, the thickness t of the dielectric film DF should be selected so as to fulfil the following condition:

$$t = \frac{|\delta|}{n_M - n} + \frac{\lambda}{4(n_M - n)}, \tag{5.18}$$

where λ is the wavelength and δ is the optical path difference produced by the soot ring (see Eq. (5.17)).

In preliminary experiments it was established that the transmittace τ of this phase ring should be 10–12%. This is much higher than in the case of the KFA device, and it may be questioned whether a phase plate which transmits 10–12% of the direct light can be termed "highly sensitive". The answer is in the affirmative if the properties of positive phase contrast are taken into account. An essential objection to $\tau < 10\%$ is in this case a small range of unreversed contrast for phase objects, which exceed their surround in refractive index. No such objections exist for objects whose refractive index is smaller than that of the surrounding medium, but such objects are in a minority.

The sensitivity of the KFS device is somewhat less than that of the KFA device, and as measured experimentally using the procedure described in Subsection 5.3.3 is equal to 0.4–0.5 nm, whereas the manufacturing tolerance is 0.5–0.7 nm. When white light is used, the field of view has an agreeable brownish tint, resembling that obtained with the KFA device. However, the achromatism of the phase ring of soot and dielectric film in KFS objectives is not as good as that of the soot ring in KFA objectives, and image contrast may be somewhat increased by using green light.

To illustrate the performance of this device, two photomicrographs are shown in Figs. XVII and XVIII. The image contrast displayed by the KFS objective is significantly greater in comparison with that produced by an equivalent objective with a typical quarter wavelength positive phase ring made of dielectric (cryolite) and metallic (chromium) substances, provided that geometrical parameters, light transmittance of the phase rings, and the optical systems are identical in the two objectives. This results exclusively from the non-reflecting properties of the soot which decreases the intensity of the direct light by absorption only. On the other hand, the metallic film reduces the direct light intensity not by absorption alone but also by reflection. The reflected light causes an increase of stray light in the field of view, decreasing the image contrast. This is a problem which was discussed earlier (see Subsection 5.3.4).

Figure XVII shows an oral epithelial cell mounted in saliva. The picture was taken with a high power, $100 \times /1.2$ (oil immersion), objective focused at the top cell surface. The ridges and microfolds of the cell membrane are clearly visible.

Figure XVII refers to a specimen classed as a phase-retarding object, while phase-advancing objects are shown in Fig. XVIII. In relation to these, the KFS device presents almost the same advantages as KFA equipment does for phase-retarding objects.

In comparison with most standard commercial positive phase contrast systems, the KFS device is more sensitive, but has a somewhat smaller range of unreversed contrast; it yields more contrast images of minute and thin phase objects, and produces a brownish coloured background (when white light is used). It is described more in detail in the paper [5.23].

The KFS device is manufactured by PZO, Warsaw, as equipment for biological microscopes. Its basic technical parameters are given in Table 5.4.

5.6. Alternating phase contrast systems

Phase contrast microscopy suffers from certain inherent contradictions. For instance, as we increase image contrast and sensitivity we decrease the range of unreversed contrast (see Fig. 5.5); simultaneously, the halo and shading-off effects become more disturbing. In order to eliminate these defects and obtain a truer representation of the specimen, i.e., better imaging fidelity, some microscopists prefer objectives with low absorption phase rings. The advantages of such objectives for the study of biological cells have been discussed by Ross [5.26] and by other cytologists. Positive phase contrast objectives with 25% absorption of direct light were at one time manufactured by Watson and Sons, Ltd., England, whereas the American Optical Company, Buffalo, USA, markets phase objectives with zero absorption phase plates. They are recommended for highly refractile objects and specimens including both phase and amplitude structures.

The halo and shading-off effects can be reduced and the fidelity of imaging improved without diminution of contrast by using as narrow a phase ring as possible. In this case, however, a powerful source of illumination is needed and troublesome diffraction effects arise as a consequence of an over-narrow annular condenser diaphragm conjugate with the phase ring.

Another contradiction is that positive phase contrast is more acceptable to biologists than negative phase contrast, since the former produces, in general, dark images which resemble those of stained specimens in ordinary bright-field microscopy. From the physicist's point of view, on the other hand, negative phase contrast appears to be better, because its range of unreversed contrast for phase-retarding objects (which are in the majority) is much larger, dark haloes around bright object images are less disturbing than bright haloes around dark images (see Figs. X and XI), and thus negative phase rings can be made more sensitive than positive rings.

In the author's view, both types of phase contrast are generally acceptable and may be considered complementary. The use of positive or negative phase contrast depends largerly on the specimen to be examined. In practice, however, especially in biological and medical research, there are highly complicated specimens containing large and small microobjects, slightly or heavily refractile particles, structures of different spatial frequencies, and objects of diverse light absorption. Some are seen better in negative, others in positive phase contrast or even in bright field. It is thus unquestionably advantageous to be able to analyse an unfamiliar image using both types of phase contrast alternatively or simultaneously as well as systems with changeable or continuously variable image contrast. Some which have found practical application will be described below and in the following section.

5.6.1. Beyer's phase contrast device

As was mentioned earlier (see Table 5.1), the halo and shading-off effects depend on the size of the object under examination and the width of the phase ring. Let be r_o and w the radius of a circular thin phase object (measured in the object plane of the microscope) and width of a phase ring, respectively. When the product of these parameters, $r_o w$, diminishes, the halo spreads and becomes less intense in the immediate vicinity of the image. Simultaneously the shading-off abates and the image contrast becomes more uniform. Thus, the light distribution in the image corresponds more closely to variations of optical path difference in the object, and the general fidelity of phase contrast imaging improves. This fidelity was carefully examined both theoretically and experimentally by Beyer [5.10, 5.27]. He characterizes the fidelity of phase contrast imaging by a parameter Γ defined as follows

$$\Gamma = \frac{k}{f} r_o w, \tag{5.19}$$

where f is the focal length of the objective and $k = 2\pi/\lambda$ (λ is the wavelength of light used).

If the phase shift φ produced by the object is small ($|\varphi| \ll 90°$) and $r_o w \to 0$, the parameter Γ approaches zero, and the fidelity of imaging is ideal. This situation is only possible theoretically and not attainable in practice. If $0 < \Gamma \leqslant 1$, the image represents the object fairly well, but for $1 < \Gamma \leqslant 2.5$ the halo and shading-off effects are quite noticeable and become more evident the greater the product $r_o w$. When $\Gamma > 2.5$, the halo and shading-off effects are obtrusive and cause unacceptable disturbance to the object image. The fidelity of imaging of broader objects is then very poor.

To sum up, for a given diameter $2r$ of the phase object, the object image represents optical path difference variations the more accurately, the smaller is the width w of the phase ring. But a narrow phase ring has certain defects; in particular, it reduces resolution and image definition. This latter defect applies especially to fine details the dimensions of which are somewhat greater than the limit of resolution. For such objects, a wider phase ring is preferable.

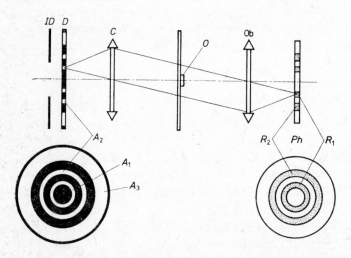

Fig. 5.24. Schematic diagram of Beyer's variable phase contrast device.

Taking these facts into account, Beyer has developed a phase contrast device which permits both small and large objects to be observed with variable contrast and improved image fidelity. The basic feature of this device (Fig. 5.24) is a combination of two concentric quarter wavelength phase rings R_1 and R_2, both either positive or negative, situated in the back focal plane of the objective Ob, and an aperture diaphragm D consisting of two annuli A_1 and A_2, located in the front focal plane of the condenser C. The outer phase ring R_2 is of such a size as to cover the image of the outer condenser annulus A_2. The diameter and width of these elements more or less correspond to those in a standard phase contrast device. The inner phase ring R_1 is arranged to cover the image of the inner condenser annulus A_1. These two elements are much smaller than R_2 and A_2. The gap between the phase rings R_1 and R_2 is fixed so that the diffracted light beams, the primary sources of which are the condenser annuli A_1 and A_2, are not greatly affected by the phase rings R_2 and R_1, respectively. This is achieved when the following relation is fulfilled:

$$G = \frac{k}{f} g r_o > 5, \tag{5.20}$$

where g is the desired gap between the two phase rings and the remaining symbols are as in Eq. (5.19).

Immediately in front of the annular diaphragm D there is an iris diaphragm ID. When the latter is entirely open, the light passes through both annuli A_1 and A_2. Consequently, both phase rings are simultaneously in action, but the inner R_1 contributes much less than the outer one to the production of the phase contrast image. In this case, the image is similar to that obtained with any normal positive phase contrast device. If, however, the iris diaphragm ID is set so that the outer annulus is obscured, then the inner phase ring R_1 acts alone and an image with spread halo and reduced shading-off effect is obtained.

In reality, the annular diaphragm D has three annuli, two as described, and the third one (A_3) lying outside A_2. The annulus A_3 is much larger than A_2 and serves for bright-field observation. By progressively closing the iris diaphragm ID, one can pass continuously from bright-field, through normal to modified phase contrast observation. Thus a specimen (O) which includes both phase and amplitude objects of different sizes and various characteristics can be studied more comprehensively than with a standard phase contrast microscope.

This device is manufactured commercially by C. Zeiss Jena, and supplied as an attachment to conventional microscopes. It includes four Phv objectives of magnification 10, 20, 40, and 90× (oil immersion) and an aplanatic condenser provided with a turret incorporating annular diaphragms. Another version of this device, without Phv objectives but with phase plates arranged in a slide, is supplied by C. Zeiss Jena as equipment for some specialized microscopes (e.g., Peraval–Interphako).

5.6.2. Device with both positive and negative phase rings (KFZ)

An ordinary microscope equipped with standard positive and negative phase contrast devices of the Zernike-type with quarter wavelength phase rings is adequate for investigating most phase objects. However, the change-over from positive to negative phase contrast and vice versa is inconvenient and slow. A device which enables the user to alternate rapidly between both types of phase contrast but using one piece of equipment has been developed by the author of this book [5.28]. It is designated "KFZ", and is commercially available from PZO, Warsaw.

The basic elements of this device (Fig. 5.25) are two concentric phase rings (positive R_1 and negative R_2) sandwiched between two cemented lenses of a doublet situated in (or near) the back focal plane of the microscope objectives Ob. An aperture diaphragm D with two transparent annuli A_1 and A_2 covered by polarized films P_1 and P_2 is inserted in (or near) the front focal plane of the substage condenser C. The transparent annuli A_1 and A_2 and the phase rings R_1 and R_2 are in conjugate planes, so that the image of A_1 is formed on R_1 and the image of A_2 on R_2. The disc-shaped polarizing film P_1 covers the annulus A_1 and the ring-shaped polarizing film P_2 covers the annulus A_2. The polarization planes of these films are perpendicular to each other. The third polarizer P_3 is placed below the aperture diaphragm D and covers both transparent annuli A_1 and A_2 (this polarizer can also be placed above the objective Ob or eyepiece). By rotating the polarizer P_3 around the axis of the microscope, it is possible to control the ratio of the intensity of the light passing through the transparent annuli A_1 and A_2. When the polarizer P_3 is crossed with the polarizer P_1, the light passes through the annulus A_2 alone and not through A_1 and the phase contrast obtained is that determined by the phase ring R_2 (negative). If, however, the polarizer P_3 is crossed with polarizer P_2, the light does not pass through the annulus A_2 but through A_1 only, and the phase contrast obtained is that determined by the phase ring R_1 (positive). In intermediate positions of the polarizer P_3, when the direct light passes through both annuli, a "mixed" phase contrast appears in which either bright or dark images prevail. Thus, by rotating the polarizer P_3, it is possible to alternate the image contrast rapidly and continuously

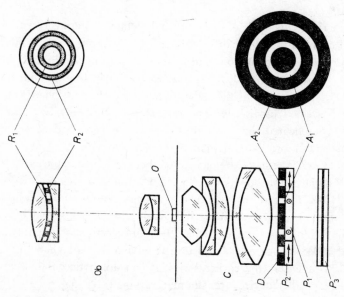

Fig. 5.25. Phase contrast device (KFZ) with both negative and positive phase rings [5.28]: left—its optical system, right—its connection with the Biolar microscope (photo by courtesy of PZO, Warsaw).

and to pass from positive phase contrast to negative (and vice versa) without interrupting the observation of the specimen O.

The dimensions of the phase rings as well as diameters and widths of the condenser annuli are so designed that there is no change in image sharpness during the transition from negative to positive phase contrast, and the diffracted light, the primary source of which is the annulus A_1, is not greatly disturbed by the phase ring R_2, while the diffracted light derived from the annulus A_2 is not greatly affected by the phase ring R_1. These requirements can be fulfilled when the diameters of the phase rings R_1 and R_2 are equal to 1/4 and 1/2 of the exit pupil diameter of objective and the widths of these rings are equal to 1/9 and 1/6 of their diameters, respectively. The principal geometrical characteristics of the phase rings are given in Table 5.5.

TABLE 5.5

Diameters (d), widths (w), and areas (a) of phase rings R_1 and R_2 (Fig. 5.25) relative to the diameter (d_{EP}) and area (a_{EP}) of the exit pupil of objective

Phase ring	d/d_{EP}	w/d_{EP}	a/a_{EP}
Positive R_1	0.28	0.045	0.055
Negative R_2	0.55	0.065	0.13

The positive phase ring R_1 is made of dielectric and metallic materials (cryolite and chromium) and changes the phase of the direct light by about $+90°$, whilst the negative phase ring R_2 is made of soot and alters the phase of the direct light by about $-90°$. The phase ring R_1 reduces the intensity of light to 15% and the phase ring R_2 to 3–4%. This latter has the same properties as the soot phase ring in the KFA device described previously. The procedure for making two such phase rings is more complicated than that of a single soot ring, and thus the KFZ device is somewhat more expensive than the KFA device.

The polarizing films P_1 and P_2 are cemented between the diaphragm D and an additional glass plate (not shown in Fig. 5.25). The quality of the polarizing films P_1, P_2 and polarizer P_3 need not be excellent. A degree of light polarization of 99.8% is quite sufficient. There are four ring diaphragms D with polarizing films P_1 and P_2 in the KFZ device (one for each of the four objectives of magnification 10, 20, 40, and 100×). They are placed in a revolving disc attached to an aplanatic-achromatic condenser (the earlier series has an aplanatic condenser). The basic parameters of this device are summarized in Table 5.4.

Apart from allowing a rapid change-over from one type of contrast to another, this device gives relatively good image fidelity since narrow phase rings are used. The sensitivity is high, as much as 0.2 nm for the negative and about 0.6 nm for the positive phase ring. The photomicrographs in Figs. XIX–XXI illustrate the performance of this system.[8]

Some experiments have also been carried out with objectives in which the negative phase ring R_2 (Fig. 5.25) was made of dielectric and metallic materials instead of soot [5.29]. However, their performance was not as good because of light reflection of the metallic film. This reflection is particularly troublesome in negative phase contrast as the background is dark and should not be brightened by stray light. On the other hand, the reflected light is less detrimental in positive phase contrast, especially as the ring R_1 occupies a small area (one third of the area of the negative phase ring R_2), so that it does not reflect as much light. Objectives in which the inner phase ring R_1 is negative (phase shift $\psi = -90°$) and the outer R_2 positive ($\psi = +90°$) have also been constructed and examined, but less satisfactory results were obtained; both negative and positive contrast in fact were of rather poor quality. The best alternating phase contrast system was achieved with an arrangement of phase rings as shown in Fig. 5.25.

5.7. Phase contrast system with continuously variable image contrast

The alternating phase contrast devices presented above use phase rings with constant phase shift (ψ) and transmittance (τ). Thus, they do not comply universally with Richter's conditions (Eqs. (5.10)–(5.12)), are not therefore suitable for the observation of all phase objects. Beyer's system takes into account the size of objects and not their phase shifts. Image contrast of the same sign, either positive or negative, is altered via the size (width and radius) of two concentric phase rings. On the other hand, the design of the KFZ device took into account the fact that some phase objects are better displayed in negative and some in positive phase contrast; this device alternates between two levels of image contrast of different signs (positive and negative).

Another group of alternating phase contrast devices complying with Richter's conditions will be presented below. Both groups include what is known as *variable phase contrast microscopy*.

---

[8] Additional documentation illustrating the versality of the KFZ device may be found in an article by Göke (see Ref. [5.30]).

5.7.1. *The Polanret system*

Polarizing, birefringent, and optically active elements are particularly suitable for variable phase contrast microscopy and allow one to construct systems with continuous variation of the phase and/or transmittance ratio τ_r between direct and diffracted light. This makes it possible to adjust the microscope for optimum image contrast according to Richter's conditions.

A great number of phase contrast systems with anisotropic phase plates have been described in the literature (a more detailed review is given, e.g., in Refs. [5.9] and [5.11]). One of the best known is the *Polanret system*[9] developed by Osterberg [5.31, 5.32]. The most general version is shown in Fig. 5.26. Its basic

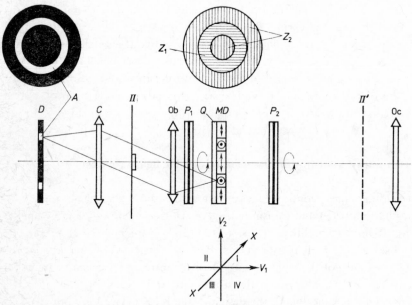

Fig. 5.26. Osterberg's Polanret phase contrast system.

element is a "micoid" disc *MD* placed in the back focal plane of the objective Ob. This disc is composed of a quarter-wave retarding plate Q made, e.g., of mica, and of two zonal polarizers, Z_1 and Z_2, made of polarizing film. The polarizer Z_1 covers the conjugate area, defined by the image of a condenser annulus A, whereas Z_2 covers the complementary area. The polarization planes of these polarizers are perpendicular to each other and their directions of light vibrations V_1 and V_2

[9] The name *Polanret* is an acronym of the terms *pol*arizer, *an*alyser, and *ret*ardation.

make an angle of 45° with the slow (or fast) axis X of the quarter-wave plate Q. A linear polarizer P_1 placed before the disc MD and an analyser P_2 inserted behind MD complete the basic Polanret system. Both P_1 and P_2 are rotatable around the objective axis. Rotation of the polarizer P_1 changes the phase ψ between the conjugate and complementary areas, whereas rotation of the analyser P_2 varies the light transmittance ratio τ_r of these two areas. Let the initial (zero) orientation of P_1 and P_2 be defined as follows: the direction of light vibration transmitted by P_1 is parallel to the X axis of the quarter-wave plate Q and the direction of light vibration transmitted by P_2 is parallel to the light vector V_2 transmitted by the zonal polarizer Z_2. Then the rotation of P_1 through an angle θ introduces a phase difference $\psi = 2\theta$ between the direct and diffracted beams, whereas rotation of the analyser through an angle α varies the transmittance ratio τ_r of the conjugate area relative to the complementary area so that $\tau_r = \tan^2\alpha$. Thus, in general, ψ can be varied from $-180°$ to $+180°$ and τ_r from zero (conjugate area completely dark) to infinity (complementary area completely dark).

The polarizer P_1 and quarter-wave plate Q in fact constitute a Sénarmont compensator. The transmission direction of the polarizer P_1 together with vectors V_1 and V_2 determine four *quadrants* (I, II, III and IV), of which two (I and III) are *neutral* and two *antineutral* (II and IV). The neutral quadrants contain the transmission direction of the polarizer P_1. When the transmission direction of the analyser P_2 passes from the neutral to antineutral quadrants, the phase shift ψ between the diffracted and direct light beams jumps suddenly by 180°. For example, if $\psi = 90°$ in the neutral quadrant, then in the antineutral quadrants ψ is equal to 270° (or $-90°$). Thus, by rotating the analyser, one can simultaneously vary the transmittance ratio τ_r and pass from positive to negative phase contrast. This important feature of the Polanret system holds both with and without the quarter-wave plate Q. When this plate is omitted, the required phase difference ψ (e.g., equal to 90°) between diffracted and direct light can be introduced in other ways, e.g., by coating the conjugate area with a thin dielectric film.

With a quarter-wave plate, the Polanret system constitutes a highly versatile system capable of giving a continuously variable phase shift ψ and transmittance ratio τ_r, whereas without this plate it has continuously variable τ_r and two constant, positive and negative, values of phase shift, e.g., $\psi = +90°$ and $-90°$.

The Polanret system (Fig. 5.26) can sometimes be used for measuring the optical path difference in specimens. This measurement is only possible for small objects that are not disturbed by the halo and shading-off effects, and depends on setting the rotatable polarizer and analyser at two positions for which first the image and then its surrounding background are made as dark as possible. Another interesting measuring feature of this system is its capacity for determining the area or effective radius of small, unresolvable, opaque particles [5.32–5.34].

Despite its versatility, Osterberg's Polanret system has found rather limited practical application as its manufacture is difficult and expensive. Moreover, as Osterberg himself stated [5.35], it has certain optical defects. In particular, it incorporates anisotropic elements in the image space of the objective where converging light beams pass. These elements produce non-uniform phase shifts for individual rays traversing the objective at different angles and cause the focused image to become blurred. Moreover, due to the elliptical polarization introduced by the lenses, the system fails to achieve the maximum possible image contrast and also suffers from troublesome halo effects.

The most recent commercial version of the Polanret system is shown in Figs. 5.27 and 5.28. It contains four plates of zonal polarizers arranged in a rotatable

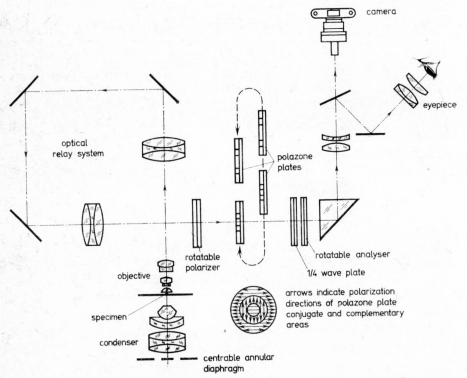

Fig. 5.27. Diagram of the most recent commercial version of the Polanret system (by courtesy of American Optical Corporation, Buffalo).

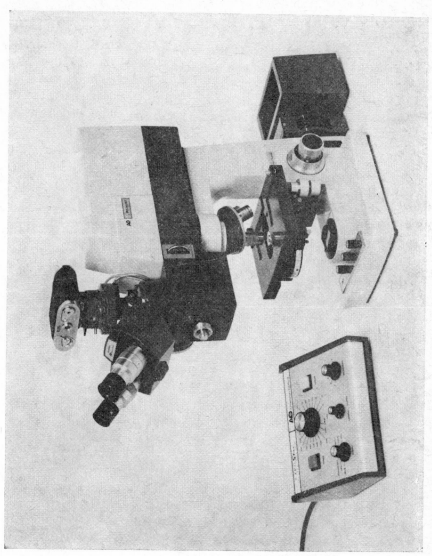

Fig. 5.28. Commercially available Polanret microscope (by courtesy of American Optical Corporation, Buffalo).

disc located in the secondary focal planes of planachromatic objectives of magnifying power $10\times$, $20\times$, $40\times$, and $100\times$ corrected at infinite image distance.
In Fig. 5.27 the quarter-wave plate is situated behind the zonal polarizers, while
in Fig. 5.26 in front of it. The only difference makes that the functions of the
rotatable polarizer and analyser are transposed. This version is presented in
greater detail by Richards [5.36]. A number of examples illustrating the use
of this instrument in studying a range of specimens is also given in the reference
cited above.

5.7.2. Nomarski's variable achromatic phase contrast system

In the Polanret method, as well as in some other variable phase contrast systems,
a basic phase difference ψ_b, usually equal to 90°, is introduced between the direct
and diffracted light, to which a variable component is added or subtracted by
means of a compensator. In the system shown in Fig. 5.26 the basic phase difference
$\psi_b = 90°$ is produced by the quarter-wave plate Q. This plate, as well as all the
other standard phase plates made of transparent refracting materials, are chromatic, i.e., ψ_b varies with wavelength.

Another defect of the Polanret method is that optimum contrast is achieved
by two separate manipulations. One step consists in matching the phase shift
by means of the polarizer P_1 (Fig. 5.26), the other consists in changing the amplitude by using the analyser P_2. This is an inconvenience which can lead to inaccu-
racies in phase matching.

These defects are eliminated in a variable phase contrast device developed by
G. Nomarski [5.37]. This device is similar to the Polanret system, but the arrangement of its individual polarizing and birefringent elements is basically
different. To begin with, the zonal polarizers Z_1 and Z_2 (Fig. 5.26) are set up so
that their directions of light vibration (vectors V_1 and V_2) make an angle of 45°
with each other. Secondly, the direction of light vibration transmitted by the
polarizer P_1 is parallel to the light vibrations from the conjugate area (zonal
polarizer Z_1). Thirdly, the quarter-wave plate Q is placed between the zonal
polarizers and analyser P_2 and is oriented so that its slow (or fast) axis is parallel
to the light vibrations of the conjugate area (zonal polarizer Z_1). The quarter-
wave plate and analyser now constitute a Sénarmont compensator. Other birefringent compensators can be used equally well.

The theory formulated by Nomarski [5.37] shows that in these conditions the
compensator alters both the phase ψ and intensity ratio τ_r of the direct light

simultaneously. The alteration, moreover, complies with Richter's condition (Eqs. (5.10)–(5.12)). Thus the image of a phase object becomes maximally dark (or bright) as a result of a single manipulation, which allows one to make more exact observations of some optical properties of the object.

Returning to the vector representation of phase contrast (Fig. 5.29), Nomarski's system takes advantage of the fact that a maximally dark image of a phase object can be obtained by adding to the vector of the direct light **u** a vector −**p** being opposed to the vector **p**, which represents the light deviated in phase by the object under examination. The sum of vectors **u**+(−**p**) gives a resultant vector **u**′, which now represents the modified background light. As can be seen, the vector **u**′ has the same length as vector **d**, which represents the diffracted light, but is antiparallel to **d**. The brightness of the image of the object now results from the sum of vectors **u**′ and **d**, but **u**′+**d** = **0**, so the object image is maximally dark. As can easily be proved, the angle between the vector **u**′ and **u** is the optimum phase shift, equal to $\psi_{op} = -90° + \varphi/2$, whereas the length of the vector **u**′ is equal to $2\sin(\varphi/2)$; thus the transmittance ratio $\tau_r = 4\sin^2(\varphi/2)$. These relations are in agreement with Eqs. (5.10) and (5.11).

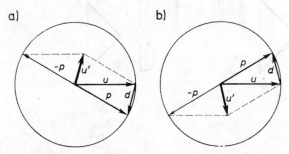

Fig. 5.29. Vector representation of Nomarski's variable achromatic phase contrast system.

The zonal polarizers in this system must not produce any additional phase shift between the diffracted and direct light and must therefore be made very carefully; their thickness should be absolutely identical. If this condition is fulfilled, the achromatic error $\Delta\psi$ of phase matching is very small and given by

$$\Delta\psi = \tfrac{1}{2}[\psi(\lambda) - \varphi(\lambda)],$$

where $\psi(\lambda)$ and $\varphi(\lambda)$ are the dispersions of the phase shift in the compensator and specimen, respectively.

5.7.3. *Nikon interference-phase contrast device*

An essential difference between this device (Fig. 5.30) and the Polanret system (Fig. 5.26) lies in the design of the polarization phase plate, which in the Nikon device consists of the zonal polarizer Z_2 only; the polarizer Z_1 is omitted. Thus, only the complementary area is covered by the polarizing filter and the conjugate area is clear. The remaining polarizing and birefringent elements are identical in both systems, but their arrangement is somewhat different. In the Nikon system, these elements are placed in a transferred-image unit attached to the microscope between the objective nosepiece and ocular head. The function of this

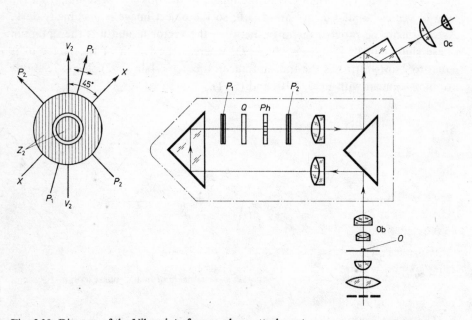

Fig. 5.30. Diagram of the Nikon interference-phase attachment.

unit is the same as in the **AOC** variable phase contrast microscope shown in Figs. 5.27 and 5.28. The polarization phase plate *Ph* is located in the reimaged back focal plane of the objective Ob (Fig. 5.30), and a rotatable polarizer P_1 and quarter-wave plate Q are placed immediately in front of this plate; behind the plate *Ph* there is a rotatable analyser P_2. The quarter-wave plate is oriented so that its slow (or fast) axis XX makes an angle of 45° with the direction of light vibrations (V_2V_2) transmitted by the phase plate Ph. Rotating the polarizer P_1 thus changes the phase shift ψ between the diffracted and direct light beams, while rotating the analyser P_2 varies the intensity ratio of these beams. This produces almost the same effect as in the Polanret system, but the physical basis of phase contrast imaging more closely resembles that of the "interphako" contrast (see Subsection 5.8.1).

When white light illumination is used, every small phase detail shows different interference colours produced by the mutual interference of the direct and diffracted light. The colours depend on phase differences φ in the specimen under examination, and the optical path difference δ between the object O and its surrounding medium can be evaluated if any difference in colour appears between the object image and its background. The most widely differentiated colours occur in the image when the background is red–purple. The technical description supplied by Nikon does not explain how this is achieved, but presumably a first-order red plate is added to the quarter-wave plate. A more accurate method of measuring the optical path difference δ is to use monochromatic (green) light and set the rotatable polarizer and analyser so that the object image appears maximally dark, and then rotate these polarizing elements until the background appears as dark as possible. The optical path difference δ is calculated from the formula

$$\delta = \frac{\theta_1 - \theta_2}{180°}\,\lambda, \tag{5.21}$$

where θ_1 and θ_2 are the readings (in degrees) of the settings of the polarizer at maximum darkness of the object image and background, respectively.

However, it must be stressed that this device does not have all the measuring capacity of a true interference microscope. It is, in fact, a variable phase contrast accessory and has defects typical of the phase contrast method. In particular, halo and shading-off effects appear, so that only very small objects (less than 0.01 mm) can be measured. In the case of 2 μm or smaller objects, it is possible to obtain an accuracy of optical path difference measurement of about $\lambda/50$.

The Nikon system is available commercially in the form of an attachment for Nikon transmitted-light microscopes. It consists of a special body tube inserted between the nosepiece and ocular head. The body tube carries a slider with four polarizing phase plates corresponding to 10, 20, 40, and 100× standard objectives. A conventional turret condenser with annular diaphragms, centring telescope, and monochromatic filters are also provided.

5.7.4. Variable phase contrast device with a single polarizing phase ring

This is the simplest device among other continuously variable phase contrast systems described to date. Its idea was formulated by the present writer in 1974 and described in detail some ten years later [5.38, 5.39]. For the sake of briefness, it will be designated "KFV". Its main optical elements are shown schematically in Fig. 5.31.

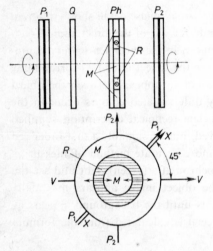

Fig. 5.31. Diagram of variable phase contrast device developed by the author of this book [5.38].

Instead of three (Figs. 5.26 and 5.27) or two (Fig. 5.30) zonal polarizers, the KFV device contains only one such polarizer in the shape of a ring, which covers the conjugate area (image of condenser annular diaphragms). This polarizer is made of a polarizing foil cemented between two glass plates by means of a suitably chosen optical cement whose refractive index (n_M) is equal to that (n_R) of the polarizing foil or, in general, is matched so that it satisfies the condition

$$\delta_{MR} = (n_M - n_R)t = m\lambda,$$

where δ_{MR} is the optical path difference between the light passing through the polarizing ring R (Fig. 5.31) and light passing outside it, t is the the thickness of this ring, λ is the wavelength of light, and $m = 0, \pm 1, \pm 2, \dots$ The best situation is if $m = 0$ (then $n_M = n_R$), but experiments have proved that one can also obtain quite satisfactory results when $|m| = 1$ or even 2. Due to the refractive index dispersion, $|m|$ greater than 2 are rather undesirable, in particular, for phase contrast observation in white light.

Such a polarizing ring R and its surrounding medium M (optical cement) sandwiched between two glass plates constitute a phase plate (*Ph*) with the zero phase shift ψ (or equal to 2π, 4π, ..., but these values are equivalent to $\psi = 0$). A completely different situation arises when this plate is preceded by a quarter-wave plate Q and both are placed between rotatable polarizing filters: polarizer P_1 and analyser P_2. The phase shift ψ and transmittance ratio τ_r of the direct relative to diffracted light can then be varied continuously. To produce this variation, the quarter-wave plate Q is oriented so that its slow (or fast) axis XX makes an angle of 45° with the direction of light vibrations (VV) transmitted by the polarizing ring R. Let the initial (zero) orientations P_1P_1 and P_2P_2 of the polarizer and analyser, respectively, be defined as follows: P_1P_1 is parallel to XX, and P_2P_2 is perpendicular to VV. In such a situation, the phase shift ψ between the direct and diffracted light beams is equal to zero, and the conjugate area is completely extinguished. Hence, central dark-ground microscopy is obtained. The rotation of the analyser P_2 through an angle α varies the transmittance ratio τ_r of the conjugate area relative to the complementary area, but for four particular values of α the ratio $\tau_r = 1$ and hence there appear conditions equivalent to bright-field microscopy. On the other hand, rotating the polarizer P_1 only varies the phase shift ψ between the conjugate and complementary areas. When this element is rotated through 360°, starting from its initial (zero) position, the sign of the phase contrast alters four times: twice from positive to negative, and twice from negative to positive. Within each of these quadrants the image contrast attains a maximum (or optimum) value depending on the phase shift φ in the specimen under examination. The image contrast also changes four times, but suddenly, during the rotation of the analyser through 360° (if the polarizer P_1 is not set at the zero value of the phase shift ψ).

As can be seen, the KFV device has the same versatility in qualitative observation of microscopical specimens as the Polanret system devised by Osterberg. The polarizer P_1 together with the quarter-wave plate Q acts as a Sénarmont compensator, thus allowing the measurement of the optical path difference of some small objects. The construction of the KFV device is, however, much simpler and lends itself more easily to commercial manufacture. In comparison with the Polanret and Nomarski systems and the Nikon device, the KFV attachment is also more economical in the use of light.

The latter statement needs a commentary. In the original Polanret system (Fig. 5.26), the diffracted light passes through three polarizing filters: P_1, Z_2, and P_2; the direct light also traverses similar set of filters: P_1, Z_1, and P_2. Let us suppose they are made of a polaroid foil whose transmittance is equal to about 30%. Then, the filter combinations $P_1+Z_2+P_2$ and $P_1+Z_1+P_2$ transmit maximally about 5% of incident light (when the directions of light vibration of these filters are parallel to each other). If there were such a considerable reduction of direct light only, this would be advantageous as the intensity of this light is normally reduced to a dozen or so percent for observation of the majority of common phase objects. In the situation under discussion, however, direct light should be additionally reduced relative to diffracted light by a number $N \approx 10$ (see Section 5.1), i.e., to about 0.5%, when a maximum image contrast of very thin phase objects is required. This reduction is achieved by rotating the analyser P_2 through a suitable angle α. The same is true of the Nikon device (Fig. 5.30). Accordingly, both systems need a high-power light source if one wishes to obtain microscopical images of satisfactory brightness.

A much more advantageous situation obtains in the KFV system (Fig. 5.31); here only direct light passes through the three polarizing filters, $P_1 + R + P_2$, whereas diffracted light traverses two polarizers, $P_1 + P_2$, and its intensity is reduced to about 9%. Consequently, if $N \approx 10$, the intensity of the direct light should be additionally reduced (by rotating the analyser P_2) to about 0.9%. This and the former value (9%) are two times as high as the corresponding values for the Polanret system. At any rate, the KFV device requires a two times weaker light source than either Osterberg's system, Nomarski's Polanret version, or the Nikon interference phase contrast device.

Like the Polanret system and Nikon device, the KFV attachment requires a microscope with a transferred-image unit (Figs. 5.27 and 5.30) for re-imaging the exit pupil of objectives in a easily accessible intermediate plane in which the polarizing phase plates Ph (Fig. 5.31) must be placed. Some modern inverted biological microscopes are, fortunately, constructed so that they posses such an intermediate exit pupil plane. The KFV device contains four polarizing phase plates (Ph) corresponding to 10, 20, 40, and $100\times$ standard objectives. The phase plates act jointly with four condenser diaphragms arranged in a rotating disc. While this book was being prepared for publication, this device become commercially available from the Central Optical Laboratory, Warsaw. The photomicrographs in Figs. XXII–XXV illustrate the general performance of this device.

5.8. Phase contrast microscopy by using interference systems

Both phase contrast and interference microscopy display phase objects and structures of different refractive indices or different thicknesses. The phase contrast method provides a higher sensitivity to minute changes in optical path in the specimen, but, on the other hand, produces disturbing halo and shading-off effects, which make it difficult to carry out accurate studies of large objects. This defect is not present in some types of interference microscopes, and a true interference method does not discriminate between large and small objects, so that all phase objects can be imaged with equal fidelity regardless of their shape and dimensions. Furthermore, a true interference microscope is capable of giving both qualitative and quantitative information on the distribution of optical path differences. It is true that one can measure optical path differences by using some variable phase contrast devices (see Section 5.7), but this applies only to very small objects and fine structures, whereas some interference microscopes allow one to measure optical path differences of large objects such as biological cells and tissues.

These examples show that a combination of phase contrast and interference microscopy could be useful to reveal more completely the nature and properties of the specimen. In this section, some systems for both phase contrast and interference microscopy are presented with a special emphasis to their use for phase contrast observation, whereas their interferometric possibilities will be discussed in the third volume of this book.

5.8.1. Interphako

The Interphako system offers an approach to microscopy with different phase contrast and interference methods. It was developed by Beyer and Schöppe [5.40, 5.41], and is commercially available from C. Zeiss, Jena, together with the Amplival or Peraval microscopes for transillumination microscopy, or the Epival microscope for the study of specimens in reflected light.

This system essentially consists of a compact Mach–Zehnder interferometer composed of two identical rhomboidal prisms M_1 and M_2 (Fig. 5.32) arranged behind the objective Ob. The skew surfaces S_1 and S_3 of these prisms constitute semitransparent mirrors, whereas surfaces S_2 and S_4 are fully-reflecting mirrors. An image-transfer system, T_1 and T_2, is inserted in front of the interferometer; it re-images the back focal point F' of the objective Ob to an easily accessible plane F'' inside the interferometer. For maintaining this plane in a fixed position, a reflecting prism RP is used, which is suitably translated in the direction p parallel to the optical axis of the re-imaging system T_1 when one objective is replaced by another of different magnification. The geometrical image of the object O is projected to infinity by a negative lens NL, and is then focused by the image-transfer system T_1 to an intermediate image plane O'. This latter is then transferred by lenses T_2 into the front focal plane of the ocular Oc. Before that, however, the light is divided into two beams by the semitransparent surface S_1 of the rhomboidal prism M_1. One beam follows path *1* and the other path *2*. They are then recombined by the semitransparent mirror S_3 of the rhomboidal prism M_2 and interfere with each other.

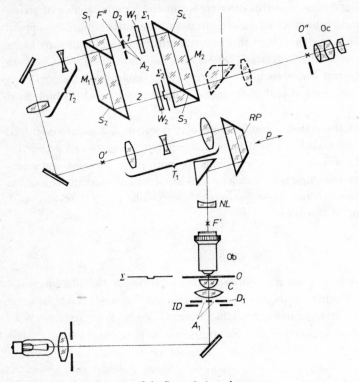

Fig. 5.32. Optical diagram of the Interphako microscope.

This system is capable of giving either an interference image, with uniform and fringed field of view and variable wavefront shear (see Chapter 16, Vol. 3), or three different kinds of phase contrast: "Interphako" (i.e., *interference-phase contrast*), normal positive and negative phase contrast, and alternating phase contrast after Beyer (see Subsection 5.6.1). Coloured phase contrast equipment is also provided with the Interphako system, although this kind of phase contrast has rather less practical advantage.

The most interesting aspect of this system is the interference-phase method. It is achieved by inserting an annular diaphragm D_2 between the rhomboidal prisms in the focal plane F''. This diaphragm is conjugate with another diaphragm D_1 located in the front focal plane of a condenser C. The direct light, therefore, whose primary source is the condenser annulus A_1, passes through the annular opening A_2, whereas the diffracted light is stopped. Thus an essential difference is established between the light beams 1 and 2. Beam 1 consists of the direct rays only, and can be regarded as a reference beam whose wavefront Σ_1 is not deformed by the object (this is true for fine objects only), whereas beam 2 includes both the direct and diffracted rays, and its wavefront Σ_2 is the same as the original object wavefront Σ. Recombining the wavefronts Σ_1 and Σ_2, produces a phase-interference image (interphako image) of the object under examination (see Figs. XXVa and b). Brightness of this image (compare Fig. XXVI) or colours (if white light is used) can be varied continuously by means of a phase compensator consisting of two glass wedges W_1 and W_2. By sliding along one of these wedges one can change the phase difference between the reference wavefront Σ_1 and deformed image wavefront Σ_2. When the wedges are appropriately calibrated, it is possible to measure the optical path difference. Exact measurements are, however, only possible in the case of very small objects because for large objects the diffracted rays are insufficiently deviated from the direct beam and cannot be effectively separated by the annular diaphragm D_2; moreover, the wavefront Σ_1 is deformed by the object. This deformation is the greater the larger the object, the smaller the effective focal length of the imaging system, and the greater the width of the annuli A_1 and A_2. This is, in fact, the usual problem of halo and shading-off effects.

Typical phase contrast and Beyer's alternating phase contrast (Fig. 5.24) are achieved by using turret-mounted phase rings instead of the annular dia-phragm D_2 and by blocking light path 2.

For coloured phase contrast, a highly dispersive phase plate is used. This produces greatly differing phase shift ψ between the direct and diffracted light; over the range of wavelengths of the visible spectrum this phase shift varies from about $360°$ to $540°$. According to their individual phase difference, phase details are thus variously coloured. A particularly striking colour phase contrast is obtained when the objects are embedded according to the so-called colour-dispersion method (dispersion staining, see Subsection 6.3.3).

The most popular microscope incorporating Interphako system is the Peraval-Interphako (Fig. 5.33). Its basic equipment consists of: a unit for phase-interference contrast and shearing interference, a unit for positive and negative phase contrast after Beyer (Fig. 5.24), a unit for negative phase contrast after Beyer and for central dark-ground observation, a condenser rotating disc with annular diaphragms for all kinds of phase contrast and dark-ground observations as mentioned above, planachromatic objectives $6.3 \times /0.12$, $12.5 \times /0.25$, $25 \times /0.50$, $63 \times /0.80$, and $100 \times /1.3$ corrected at infinite image distance, and different accessories for shearing interference microscopy.

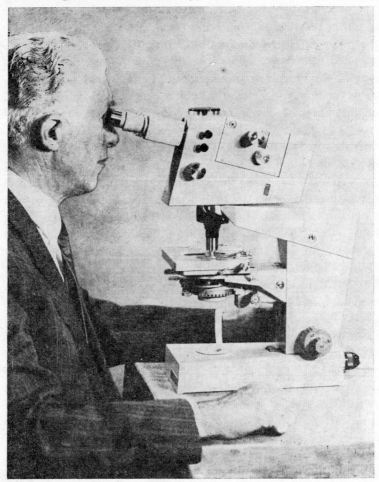

Fig. 5.33. Peraval-Interphako microscope.

Each type of phase contrast method uses two sets of phase plates with phase rings of two different dimensions. One set (phase rings of larger diameter) are for objectives 6.3, 12.5, and $25 \times$, and the other set (phase rings of smaller diameter) is for objectives 63 and $100 \times$. However, the low power objective ($6.3 \times$)

does not function in accordance with Beyer's method using two phase rings (Fig. 5.24) but with a single phase ring and, of course, a single conjugate annular diaphragm.

Returning to Fig. 5.32, it should be added that in reality the annular diaphragms D_1 and D_2 have three concentric annuli: two for phase-interference contrast and a third, which lies outside the former, for bright-field observation. By progressively closing the iris diaphragm ID, one can pass uninterruptedly from bright-field to phase contrast method, through less to more contrasty imaging. As can be seen, for this kind of phase contrast microscopy, Beyer's method, described earlier in Subsection 5.6.1, is also applied.

The above-mentioned units are interchangeable and it is not possible to pass immediately from phase-interference contrast to normal positive or negative phase contrast as one unit must first be replaced by another. Passing from phase-interference contrast to shearing interference microscopy and vice versa also requires an exchange of several accessories and some time-consuming manipulations. In spite of these drawbacks, the Peraval-Interphako is a very useful instrument.

5.8.2. *Variable phase contrast microscopy based on the Michelson interferometer*

This microscope is the outcome of work on simultaneous variable phase contrast and interference microscopy performed by the present writer in 1966–1968 and partly described in papers [5.42], [5.43], and [5.11].

The basic optical elements of the system finally developed are shown in Fig. 5.34. The most important is a polarizing interferometer of the Michelson type, which incorporates an interference polarizer IP used as the beam-splitter, two quarter-wave plates Q_1 and Q_2, two mirrors M_1 and M_2, and a Sénarmont compensator consisting of a quarter-wave plate Q_3 and linear polarizer (analyser) P_2. This interference system is located between the objective Ob and ocular head (RP and Oc). The lenses, L_1 and L_2, act as a re-imaging system and are in reality more complicated than shown.

The interference polarizer IP consists of two right-angle prisms, the hypotenuse surfaces of which are covered, by vacuum evaporation, with a stack of dielectric thin films TF alternately of high and low refractive index and cemented together (see Subsection 1.4.3). A light beam incident upon the interference polarizer IP is split into two parts: a reflected and a transmitted component polarized at right angles to each other. The reflected component vibrates perpendicularly to the plane of light incident on IP and the transmitted component vibrates in

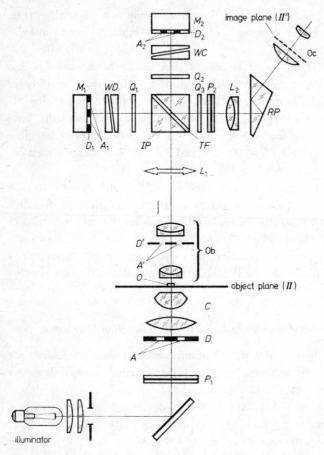

Fig. 5.34. Optical diagram of variable phase contrast and interference-phase microscope based on a polarization interferometer of the Michelson type [5.43].

this plane. The quarter-wave plates, Q_1 and Q_2, rotate the directions of light vibrations through 90° by the double passage of the light, so that the beam reflected by the mirror M_1 passes through the interference polarizer IP, whereas the other beam reflected by the mirror M_2 is totally reflected from the stack of thin films TF and recombines with the first beam. Both beams, after passing through the quarter-wave plate Q_3 and analyser P_2, vibrate in the same direction and thus can interfere with each other. The quarter-wave plates Q_1, Q_2, and Q_3 are set in such a manner that their axes (slow and fast) make an angle of 45° with the vibration directions of light beams split by the interference polarizer IP. This system has been used both for the continuously variable phase contrast method and for shearing interference microscopy with continuously variable wavefront shear. This latter technique will be dealt with in Chapter 16, so that here only the former will be described.

For variable phase contrast microscopy an annular diaphragm D is located in the front focal plane of the condenser C, and two diaphragms D_1 and D_2, one of which has an annular opening A_1 and the other an absorbing ring A_2, are located close to or in the mirrors M_1 and M_2 in such a manner that the ring A_2 and annulus A_1 are conjugate to the condenser annulus A. The image A' of this latter is formed in the back focal plane of the objective Ob and projected by means of the re-imaging lens L_1 into the planes of the diaphragms D_1 and D_2.

In these conditions, the direct light emerging from the condenser annulus A is totally absorbed by the ring A_2 but transmitted by the annular opening A_1; after being reflected by the mirror M_1, it passes through the interference polarizer IP to reach the ocular Oc. The diffracted light, on the other hand, is stopped by the diaphragm D_1 and reflected by the mirror M_2 only. It is then reflected by the interference polarizer IP, and on passing through the quarter-wave plate Q_3 and analyser P_2 can interfere with the direct light reflected from the mirror M_1. This situation is illustrated in detail in Fig. 5.35, which shows the paths of two rays: direct b and diffracted d, arbitrarily selected from the light beam leaving the object. Each of these rays is split into two components: reflected b_1, d_1 and transmitted b_2, d_2. The reflected component d_1 of the diffracted ray and the

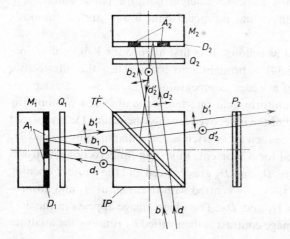

Fig. 5.35. A more detailed ray diagram of the top part of the interference system shown in Fig. 5.34.

transmitted component b_2 of the direct ray are absorbed by the diaphragm D_1 and D_2, respectively, whereas the reflected component b_1 of the direct ray and transmitted component d_2 of the diffracted ray are reflected by the mirrors M_1 and M_2, and directed, as rays b_1' and d_2', to the ocular.

Rotating the analyser P_2 varies the phase difference between the direct and diffracted light vibrations. This phase difference is $\psi = 2\theta$, where θ is the angle between the direction of light vibrations in the analyser P_2 and one of the axes, e.g., slow axis, of the quarter-wave plate Q_3 shown in Fig. 5.34. Thus it is possible to achieve a continuous variation of the phase shift ψ between the direct and diffracted light beams. The amplitude ratio of the direct and diffracted beams can be varied by rotating the quarter-wave plates Q_1 and Q_2 or by using an additional linear polarizer P_1 located before the interference polarizer IP, e.g., below the sub-stage condenser C. Rotating this polarizer alters the amplitude ratio of the light vibrations of the direct and diffracted beams, while their phase difference remains constant within the range of a given quadrant. When the vibration direction of the polarizer P_1 passes from the neutral to antineutral quadrants (see Subsection 5.7.1), the phase shift ψ jumps suddenly by 180°, as was the case in the Polanret system. If α denotes the angle between the direction of light vibration in the polarizer P_1 and principal section of the interference polarizing splitter IP (this section is the plane of the paper in Figs. 5.34 and 5.35), then for $\alpha = 0$ only the diffracted light reaches the ocular Oc, and for $\alpha = 90°$ only the direct light passes to the ocular. For intermediate positions, the intensity ratio τ_r varies as $\tan^2\alpha$. Thus, a phase contrast system is achieved wherein both the phase difference ψ and intensity ratio τ_r of the direct and diffracted light beams are continuously variable.

This microscope is operated as follows. It is first aligned for Köhler illumination and the object O (Fig. 5.34) is brought into focus. Then the interference system is adjusted by means of a wedge compensator WC and wedge diasporometer WD so that zero order uniform field interference is observed without any lateral shearing in the exit pupil of the objective (for this, the ocular Oc is replaced by an auxiliary telescope). To see such an interference, the diaphragms D_1 and D_2 should be decentred or removed for a moment. In setting for zero order uniform field interference, the diaphragms D_1 and D_2 are replaced and the annular opening A of the condenser diaphragm is exactly centred with the annulus A_1 and absorbing ring A_2 of the diaphragms D_1 and D_2. The object image appears in the field of view of the ocular Oc, and image contrast is then varied by rotating the analyser P_2 and polarizer P_1.

Excluding one of the diaphragms D_1 or D_2 yields other possibilities for

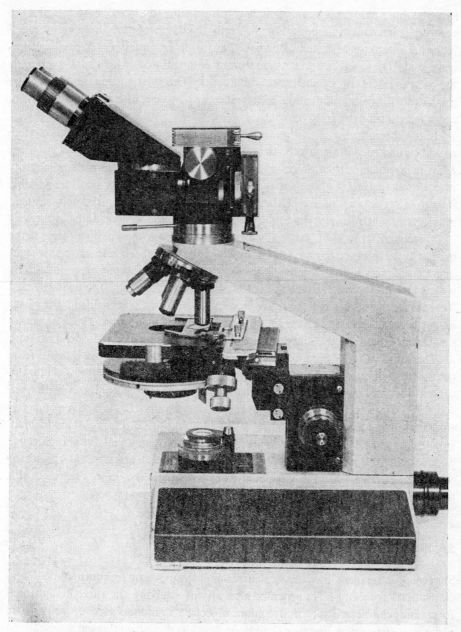

Fig. 5.36. A prototype microscope the optical system of which is shown in Fig. 5.34.

investigating phase objects. In particular, it is possible to obtain phase-interference contrast (Interphako method). For this, only the diaphragm D_2 need be excluded from the interference system shown in Fig. 5.34. The analyser P_2 and quarter-wave plate Q_3 together form a Sénarmont compensator, which makes possible the measurement of optical path differences. This system can also easily be transformed into a polarization interference microscope with variable wavefront shear (see Section 16.2).

It has been stated that the variable phase contrast version of this system requires great care in the production of glass prisms for the interference polarizer IP, with particular attention being paid to pyramidal error (non-perpendicularity of polished surfaces to base surfaces). This error should be as small as possible (less than 10 seconds of arc). Otherwise, it is impossible to achieve uniform field interference or a single (unduplicated) image of the object under study.

Moreover, stray light, especially that reflected from the back surfaces of the quarter-wave plates Q_1 and Q_2, is highly undesirable. To eliminate this light and increase image contrast, the plates Q_1 and Q_2 are set not quite perpendicularly with respect to the rays 1 and 2 (Fig. 5.34) but a little obliquely. The light reflected from the front surfaces of the quarter-wave plates and that reflected from the glass–air surfaces of the beam-splitter IP is of no importance as vibrates in such a manner that cannot pass to the image plane of the microscope.

The most valuable feature of this system is its ability to pass quickly from continuously variable phase contrast to phase-interference contrast and also to central dark-ground and bright-field microscopy. It is equipped with four standard objectives of magnifying power 10, 20, 40, and $100 \times$ and a condenser with a revolving disc in which four annular diaphragms (D, Fig. 5.34) are located. In all there are three pairs of diaphragms D_1 and D_2: one pair for the objective of $10 \times$ and $20 \times$ magnifying power, the second pair for objective $40 \times$, and the third for objective $100 \times$. A view of this microscope is shown in Fig. 5.36. This is a prototype built in the Central Optical Laboratory, Warsaw.

5.9. Stereoscopic phase contrast microscope

Stereoscopic imagery has become increasingly important for both optical and electron microscopy, as it provides a means of studying the three-dimensional form and characteristics of irregular objects. A monoobjective *stereoscopic phase contrast system* has therefore been developed with biological applications in mind [5.44].

5.9.1. Underlying principles and mode of operation

This microscope makes use of a well-known stereoscopic technique based on dividing the aperture of an optical instrument into two halves by means of crossed polarizers. Several variations of this technique have been investigated and as a result a versatile instrument has been constructed for the effective combination of phase contrast and stereoscopic observation. Its basic optical elements are shown in Fig. 5.37. The aperture diaphragm D, containing a transparent annulus A, is covered by semicircular polarizers P_1 and P_2. These polarizers are crossed and arranged in such a manner that the light passing through vibrates 45° from the bisecting line B–B, which is perpendicular to the line joining the eyepieces E_1 and E_2. The annular opening A of the condenser C is conjugate with the phase

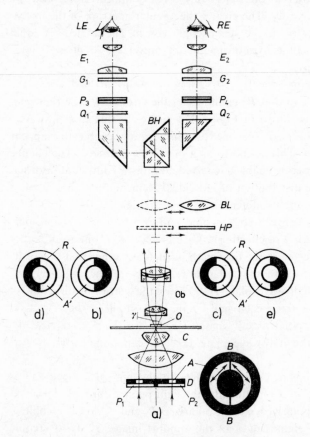

Fig. 5.37. Schematic diagram of stereoscopic phase contrast microscope [5.44].

ring of a phase objective Ob. Two other polarizers P_3 and P_4 are placed in the
ocular tubes of a typical binocular head *BH*. Rotating the polarizers P_3 and P_4
around the axis of eyepieces E_1 and E_2 extinguishes either the left or the right half
of the annulus image A' formed on the phase ring R. If we set the polarizer P_3
so as to extinguish the right half of the annulus image A' (Fig. 5.37b) and the
polarizer P_4 so as to extinguish the left half of this image (Fig. 5.37c), the left
eye *LE* of the observer receives only the light passing through the left half of the
exit pupil of the objective Ob and the right eye *RE* only the light passing through
the right half of this pupil. The object O then appears to be observed with the
left eye from the left side and with the right eye from the right side only. The
condition for stereoscopic observation is therefore satisfied. The phase details
of the specimen located closer to the objective Ob really appear closer and the
more distant details further away. The greater the angular aperture of the microscope system, the more effective is the stereoscopic vision. The angle γ is equal
to 15°, 22°, 31°, and 62° for objectives of magnifying power 10, 20, 40 and 100×
(oil immersion), respectively.

Conditions for three-dimensional observation are also satisfied when the
polarizers P_3 and P_4 are set so that P_3 extinguishes the left and P_4 the right half
of the annulus image A', as shown in Figs. 5.37d and e. In this case, however,
the stereoscopic effect is reversed; specimen details nearer the objective appear
further away and vice versa. This is known as a *pseudoscopic image*. The change
from stereoscopy to pseudoscopy enables one to obtain much additional information concerning the relative distribution of individual elements in complex specimens and to demonstrate small differences in their spatial location. It is very
useful to be able to switch between stereoscopic and pseudoscopic vision, particularly in biological research. For effecting a quick change from one to the other
a half-wave plate *HP* is inserted in the light path in such a manner that one of its
axes makes an angle of 45° with the polarization planes of the polarizers P_1
and P_2. In this case, the half-wave plate *HP* rotates the vibration direction of
the polarized light (coming from each half of the exit pupil of the objective)
through 90°. By setting the polarizers P_3 and P_4 in the position for stereoscopic
vision and then inserting the half-wave plate *HP*, we can immediately switch
over to a pseudoscopic image.

Light incident on the inclined prism surfaces in the binocular head *BH* is
subject to some depolarization. Consequently, the light passing through the prism
system becomes elliptically polarized, and the polarizers P_3 and P_4 do not completely extinguish the left and right halves of the annulus image A'. To overcome
this defect, quarter-wave plates Q_1 and Q_2 are placed before the polarizers P_3

and P_4. By suitable adjustment of these quarter-wave plates elliptically polarized light again becomes linearly polarized. A Bertrand lens BL is inserted in order to set the plates Q_1 and Q_2 and polarizers (P_3, P_4) precisely and also to match the annulus image A' and the phase ring R. The microscope can of course also provide a good stereoscopic image without the quarter-wave plates Q_1 and Q_2.

For the quantitative evaluation of depth in the specimen, two identical glass plates G_1 and G_2 with stereoscopic marks can be placed in the front focal planes of the eyepieces. If these plates are properly adjusted, an observer with normal vision sees a three-dimensional image of the specimen together with a single (stereoscopic) image of the stereometric marks. The latter image appears to be "immersed" in the specimen image. A series of stereometric marks can be arranged in such a manner as to produce a stereoscopic scale of depth. Another procedure for the stereoscopic measurement of depth is based on the method of "floating image" (see Subsection 13.3.3).

Highly sensitive negative phase objectives, as described in Subsection 5.5.2, are especially suitable for this stereoscopic phase contrast system. With these objectives, the stereoscopic effect and depth perception are, in general, better than with positive phase objectives. This has been shown with living cells cultured in vitro [5.45] and with many other phase specimens.

It is obvious that using ordinary rather than phase objectives permits amplitude specimens to be observed stereoscopically in bright field [5.46]. In this case, as found experimentally, annular condenser apertures rather than full cone illumination are advantageous; the stereoscopic effect and resolution are thereby much improved. The amplitude objectives described in Subsection 6.1.2 can be used in order to achieve a further increase in stereoscopic effect and image contrast. In practice, the higher the image contrast the greater the stereoscopic effect.

The stereoscopic phase contrast microscope as schematically presented in Fig. 5.37 (but without quarter-wave plates) is manufactured by PZO, Warsaw, under the name "stereophase" microscope. Its symbols are: MB3OS (older model) and Biolar S. Standard equipment are: KFA and KFS objectives (see Section 5.5) of magnifying power $10\times$, $20\times$, $40\times$, and $100\times$ (oil immersion), a turret condenser with four annular diaphragms D as shown in Fig. 5.37, a unit comprising a half-wave plate HP and Bertrand lens BL, and a binocular head with polarizers P_1 and P_2. The latter can be quickly removed, thus allowing an immediate switch from stereoscopic to normal phase contrast observation.

5.9.2. Lateral resolution

Stereoscopic vision is also obtained by placing the bisecting polarizers P_1 and P_2 (Fig. 5.37) behind the objective Ob. This position is, however, less satisfactory. Above all, the resolving power of the objective is halved in the direction perpendicular to the bisecting line, whereas in the case of the system presented in Fig. 5.37, where the aperture of the condenser is bisected, the resolving power of the objective is the same in all directions and, moreover, does not differ from that of a normal phase contrast microscope. This can be explained on the basis of the diffraction theory of image formation.

Suppose the object is a linear grating with a period only slightly greater than the resolution limit of the objective. The grating is oriented so that its grooves are parallel to the bisecting line of the condenser (or objective). Consequently, the diffraction pattern in the back focal plane of objective appears as illustrated in Fig. 5.38, where diagram a relates to the stereoscopic system presented in Fig. 5.37, and diagram b shows what happens when the bisecting polarizers P_1 and P_2 are placed behind the objective. In these diagrams, EP denotes the objective exit pupil, P_1 and P_2 are the directions of the light vibrations in the semicircular polarizers that bisect the condenser aperture (or exit pupil of objective), B–B is the bisecting line, A_0, $+A_1$ and $-A_1$ are the diffraction images of the condenser

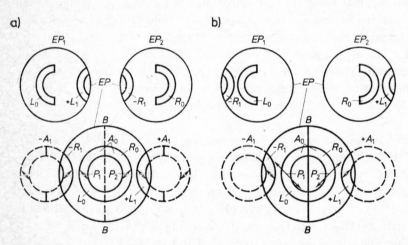

Fig. 5.38. Auxiliary diagrams for discussion of the lateral resolution of a stereoscopic phase contrast microscope whose optical diagram is shown in Fig. 5.37. A line grating used as a test object is oriented so that its grooves are parallel to the bisecting line BB of the substage condenser (diagram a) or of the objective (diagram b).

annulus produced by the object grating, EP_1 and EP_2 are the images of the exit pupil of objective observed through the left-hand and right-hand eyepiece, respectively, when the microscope is adjusted for stereoscopic vision. The eyepiece polarizers P_3 and P_4 (Fig. 5.37) are so set that they extinguish the right half R_0 (Fig. 5.38a) and left hand L_0 of the direct (zero diffraction order) image A_0 of the condenser annulus, respectively. In this arrangement, P_3 also extinguishes the right half $-R_1$ of the first order diffraction image $-A_1$, while P_4 extinguishes the left half $+L_1$ of $+A_1$. Thus, the observer views the left images L_0 and $+L_1$ through the left-hand eyepiece E_1 (Fig. 5.37) and the right images R_0 and $-R_1$ through the right-hand eyepiece E_2. The halves L_0 and $+L_1$ are mutually coherent, and light waves issuing from them can interfere with each other and produce a resolved image of the object grating (Fig. 5.39a), which is observed by the left-hand eyepiece.

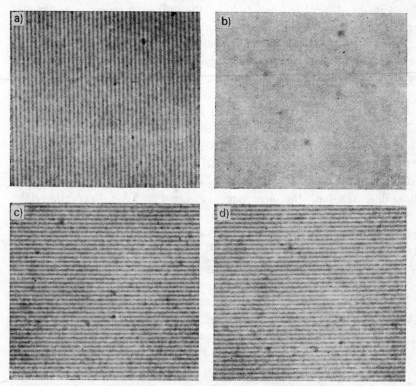

Fig. 5.39. Photomicrographs a and c show that the lateral resolving power of a stereoscopic microscope with bisected condenser aperture (Fig. 5.37) is the same in each direction, whereas lateral resolution of a similar microscope but with bisected exit pupil of the objective is two times reduced in direction perpendicular to the bisecting line (photomicrographs b and d).

At the same time, light waves emerging from the right halves R_0 and $-R_1$, which are also mutually coherent, give the resolved image of the grating in the right-hand eyepiece.

Quite a different situation occurs when P_1 and P_2 (Fig. 5.37) are inserted behind the objective Ob. If the polarizer P_3 of the left-hand eyepiece E_1 extinguishes the right half R_0 of the direct annulus image A_0 (Fig. 5.38b), the left half $+L_1$ of the first diffraction order image $+A_1$ is also extinguished, and, conversely, the polarizer P_4 of the right-hand eyepiece E_2 also extinguishes the left half L_0 and right half $-R_1$ of the zero and first diffraction order images A_0 and $-A_1$. In this case (if the Bertrand lens BL (Fig. 5.37) is inserted), the observer views the left half L_0 of the direct image A_0 and right half $-R_1$ of the diffracted image $-A_1$ through the left-hand eyepiece, and the right half R_0 of the direct image A_0 and left half $+L_1$ of the diffracted image $+A_1$ through the other eyepiece. The halves L_0 and $-R_1$ are opposed, their mutual coherence is poor, and light waves issuing from them cannot effectively interfere to produce a resolved grating image in the left-hand eyepiece (Fig. 5.39b). Nor can light waves emerging from the halves R_0 and $+L_1$ produce a resolved image of the object grating in the right-hand eyepiece since here the situation is analogical to that in the left-hand eyepiece. The result is the same as if the objective exit pupil had been half-obscured by a light absorbing diaphragm.

Suppose the grating is now oriented so that its grooves are perpendicular to the bisecting line $B-B$ (Fig. 5.37) of the condenser or objective. Then the diffraction pattern in the back focal plane of objective appears as shown in Fig. 5.40, whose diagrams have the same symbols as in Fig. 5.38; diagram a relates to the stereoscopic system presented in Fig. 5.37, while diagram b represents the version of this system with bisecting polarizers P_1 and P_2 placed behind the objective Ob. The diagrams of Fig. 5.40 show that the diffraction patterns observed by the left-hand and right-hand eyepieces are now identical for both versions of the stereoscopic system; hence, identically resolved grating images (Figs. 5.39c and d) are observed. We can therefore state that the resolving power of the stereoscopic phase contrast microscope built along the lines shown in Fig. 5.37 is not reduced and is the same in all directions, whereas the resolving power of a similar microscope but with a bisected objective exit pupil is two times degraded in the direction perpendicular to the bisecting line.

To illustrate the performance of the stereoscopic phase microscope (model MB3OS or Biolar S) presented here, some stereopairs are set out in Fig. XXVII. Each pair shows a three-dimensional image of some typical phase and amplitude objects. The reader can receive this image when his left eye looks only at the left

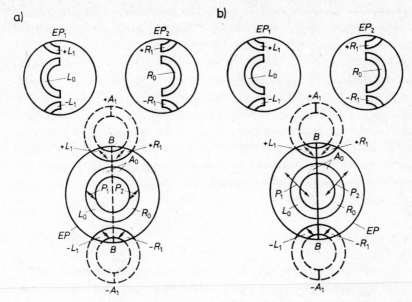

Fig. 5.40. Diagrams similar to those shown in Fig. 5.38 except that the object grating is oriented so that its grooves are perpendicular to the bisecting line *BB* of the condenser (diagram a) or of the objective (diagram b).

and his right eye only at the right hand picture of a given stereopair. For this, it is sufficient to put a vertical screen, e.g., a sheet of paper about 25 cm high, between the eyes and the stereopair pictures. After some practice, the left-hand and right-hand images coincide, and one obtains a single stereoscopic image.

5.9.3. Stereoscopic axial resolution

The stereoscopic axial resolution (Δz) of the microscope under discussion can be evaluated by using a formula resulting from the first order geometrical theory of the common stereoscopic microscope. This formula has the form

$$\Delta z = \frac{250 s_a}{M \tan \gamma} \text{ [mm]},\tag{5.22}$$

where Δz is the smallest difference in depth which can be perceived through a stereoscopic microscope by an experienced observer, M is the total magnifying power of the microscope, γ is the angle of convergence under which the left and right eyes of the observer look at an object under examination, and s_a

is the stereoscopic acuity of binocular vision. The latter quantity is found to be
6″ to 60″, and a mean value equal to 30″ (= 0.000145 rd) is usually accepted.

For typical stereoscopic microscopes, the angle of convergence $\gamma = 14°$,
and the total magnifying power M can be up to $200\times$. A value of Δz as small
as 1 μm then results from Eq. (5.22). In a monoobjective stereoscopic microscope,
the condenser aperture (or exit pupil of objective) of which is divided into two
halves, the angle of convergence γ can be as great as 60° (for a $100 \times /1.3$ immersion objective), and the total magnification can be up to 1,500. The minimum
detectable difference Δz in depth perceived through such a stereoscopic microscope
is then equal to about 0.03 μm. As can be seen, such a low value for stereoscopic
axial resolution allows accurate measurement of the depth in microscopic specimens (see Section 13.3).

5.10. Phase contrast for incident light

The general principles underlying the phase contrast method for reflected-light
microscopy are analogous to those holding for the transmitted light microscope,
and there are only certain technical differences between biological and metallographic phase contrast microscopes. There are three slightly differing ways of
transforming a normal incident light microscope into a phase contrast instrument.
These will be outlined briefly.

5.10.1. System with phase plate immediately behind the beam-splitter

This system (Fig. 5.41) was designed by Jupnik *et al.* [5.14]. A light source S
is imaged on an aperture diaphragm D by a collector (Col). The rays coming from
an opening A of the diaphragm D pass through a field lens FL preceded by a field
diaphragm FD, are reflected by the beam-splitter BS, pass through the objective
Ob, and reach the object O under examination. The diaphragm opening A is
imaged at A' in the interior of the object when the object reflecting surface is in
coincidence with the object plane Π of the objective Ob. This coincidence is,
of course, obligatory. The light reflected specularly by the object (this is the direct
light beam B_u) passes through the objective Ob again, partly (about 50%) traverses
the plate BS and forms a focused real image A'' of the aperture A. This image
coincides with a phase plate Ph. In practice, Ph is a ring (or narrow strip) and the
diaphragm opening A is therefore an annulus (or slit). In order not to overelab-
orate Fig. 5.41, a circular opening and disc-shaped phase plate are shown. The
light (B_d) diffracted by the object surface passes beside the phase plate Ph as an

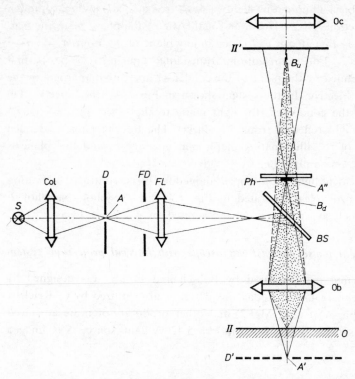

Fig. 5.41. Optical system of an incident light phase contrast microscope after Jupnik, Osterberg, and Pride [5.14].

extended beam, and is focused by the objective Ob in its image plane Π'. The direct and diffracted beams, B_u and B_d, interfere with each other and produce a phase contrast image of the object surface. The image is observed through an ocular Oc. The same situation therefore obtains as previously presented in Figs. 5.2 or 5.6 for transparent object microscopy.

But here an additional comment is required. In Fig. 5.41 the phase plate Ph is not placed in the back focal plane of objective but behind it, so the location of this plate is different from that defined earlier for transmitted-light phase contrast systems (see Fig. 5.2 or 5.6). In order to coincide the objective back focal plane with the phase plate Ph (Fig. 5.41), this latter should be located between the beam-splitter BS and objective Ob, or even inside the objective (the back focal plane of microscope objectives with magnifying power greater than about $5\times$ lies in the interior of their optical systems). Such a configuration is inconvenient (object illuminating light passes through phase plate), although possible (see Subsection 5.10.3). However, for the site of the phase plate the general condition holds that it can be located in any plane of the Fourier spectrum of spatial frequencies. For a conventional transmitted light microscope the first Fourier plane is defined by the plane of the condenser aperture diaphragm image produced by the objective. In the system shown in Fig. 5.41, the objective Ob functions as both the condenser (for light going to the object O) and image-forming lens (for light reflected from the object). The Fourier plane is defined by the image A'' of the illuminating diaphragm opening A, and this plane is coincident with the phase plate Ph.

Examples of incident light microscopes adapted for phase contrast observation according to the principles illustrated in Fig. 5.41 are Neophot, Neophot 2, Nu and Nu2, which are (or were) manufactured by C. Zeiss Jena.

5.10.2. Incident light phase contrast microscope with internal projection system

This microscope, first manufactured by Bausch and Lomb, was designed by Benford and Seidenberg [5.47]. Today this design is also utilized by C. Reichert Wien (metallographs MeF and MeF2) and other producers of more advanced microscopes. The scheme is illustrated in Fig. 5.42. A light source S is imaged

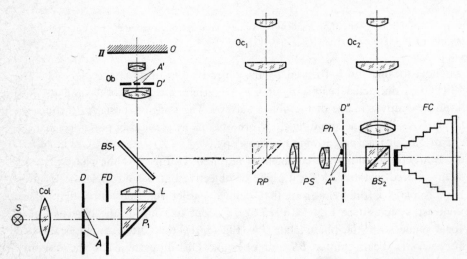

Fig. 5.42. Optical system of an incident light phase contrast microscope after Benford and Seidenberg [5.47].

on an annular diaphragm D, whose image D' is projected by the prism P_1 and lens L on the back focal plane of the objective Ob. For every point of the source S, the light leaving the objective thus forms a parallel beam. The light reflected specularly by an object O passes through the objective again and is reflected (about 50%) from a 45°-slanted beam-splitting plate BS_1. A removable prism RP reflects the light into an ocular Oc_1, which allows one to observe an object image without phase contrast (e.g., bright-field image). If the prism RP is excluded, the light passes through a projection optical system PS, which projects the second image D'' of the annular diaphragm D onto a phase ring Ph. Behind this ring there is a beam-splitting cube BS_2, which allows a phase contrast image of the object surface II to be directed to another ocular Oc_2 and/or to a photographic camera FC. The field diaphragm FD is, as usual, imaged on II.

The main advantage of this system is that the images A' and A'' of the condenser diaphragm annulus A have identical dimensions regardless of the focal length of the objective Ob, so that the same phase plate Ph can be used with objectives of different power and only one diaphragm D is required.

5.10.3. System with phase plate in objectives

In comparison with the systems described earlier this (Fig. 5.43) is far simple but less useful in more advanced research. The phase ring Ph lies inside the objective Ob in (or near by) its back focal plane. At the same plane the images of an annular diaphragm D and its opening A are projected by a lens L. The diaphragm D is intensively illuminated by the image of a light source S focused on D by means of a collector Col. The light directed by the beam-splitter BS to the objective Ob is absorbed and/or reflected by the phase ring Ph so that a high power light source is needed (e.g., a 12 V/100 W halogen lamp). That portion of light which leaves the objective is specularly reflected and diffracted by the surface of the object under study (O). Both light beams, direct and diffracted, enter the objective again; the direct beam produces the reflected image A'' of A on Ph, while the diffracted beam passes alongside it. Both beams interfere with each other giving a phase contrast image of the object surface observed through an ocular Oc. In front of the ocular it is possible, if necessary, to insert a tube lens system TS which corrects the mechanical tube length of the standard microscope since this length is extended by the beam-splitting plate BS. The field diaphragm FD is, of course, imaged on the object O.

Some simple incident-light phase contrast microscopes make use of this idea, for instance the MM5 model available from E. Leitz Wetzlar or the Standard

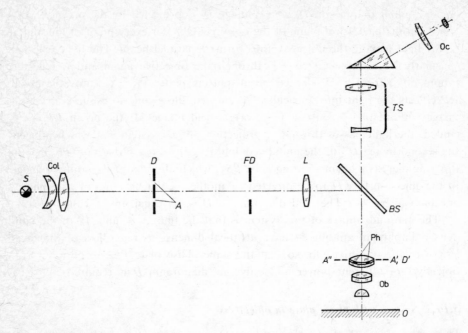

Fig. 5.43. Optical system of an incident light phase contrast microscope with a phase plate within the objective.

UM model manufactured by C. Zeiss Oberkochen. The phase rings of the MM5 microscope are covered with a special antireflecting film which extinguishes reflections from the beam-splitter BS to the objective Ob. The system shown in Fig. 5.43 has also been used successfully by Popielas [5.48] in a small metallurgical microscope MET3 produced by PZO, Warsaw. He has shown that KFS phase objectives with soot-dielectric phase rings (see Subsection 5.5.3) give incomparably better images of many typical light-reflecting specimens than do standard phase objectives with dielectric-metallic phase rings. This statement is self-evident since the phase rings of KFS objectives do not reflect light.

5.11. How and where to use the phase contrast microscope

To anyone considering the current status of phase contrast microscopy it is obvious that this technique is already widely used, especially in biology and medicine. For many years standard phase contrast devices have been in routine use in different fields of cell biology and clinical medicine. Phase contrast microscopes are part of the basic equipment of most microscopical laboratories and the technique constitutes one of the principal points in the programme of microscopy courses in biology, medicine, biophysics, veterinary medicine, and other branches of the life sciences.

Although phase contrast microscopy is widespread today, its general principles are nevertheless insufficiently well know in practice. It would therefore seem advisable here to provide some basic instructions relating to the correct use of a typical phase contrast microscope.

5.11.1. Adjustment of phase contrast microscopes

The preliminary adjustment of a phase contrast microscope is the same as for a bright-field microscope. However, special attention should be paid to correct illumination. Above all, the Köhler principle of illumination must be strictly complied with.

In this method of illumination, the condenser C (Fig. 5.6) is first adjusted (focused and centred) so as to obtain a sharp and central image of the field diaphragm FD in the object plane II. The diaphragm FD is open so that only the actual field of view of the microscope is illuminated. If the image of this diaphragm is greater than the real field of view of the microscope, useless light is transmitted within the microscope objective by specimen peripheries outside the visible field and image contrast is reduced. The larger the aperture and magnification of the objective, the greater the contrast reduction. Therefore, when one phase objective of low magnification is used as a screening objective for exact studies by using middle- and high-power phase objectives, it is necessary to correct systematically the opening of the field diaphragm FD according to the actual field of view of the objective to be used. The bulb B (Fig. 5.6), collector Col, and field diaphragm FD form either a separate or built-in illuminator of the microscope. The former must be matched with the optic axis of the microscope.

Next the coiled filament of a bulb B (Fig. 5.6) is imaged onto the annular diaphragm D of the condenser C by focusing the collector Col or bulb B. This adjustment is easily observable through the auxiliary telescope inserted into the ocular tube in place of an eyepiece and focused at the exit pupil of the objective (if the microscope is equipped with a Bertrand lens, the latter, in conjunction with the eyepiece, may be used as the auxiliary telescope). Images of the filament can be seen in Fig. 5.44, but only the first image (a) represents the correct illumination of the annular diaphragm. The second image (b) shows that the filament image is decentred and the alignment of the light source must be corrected, whereas the third picture (c) indicates that the bulb has a filament unsuitable for phase contrast microscopy. In the last case, the length of the filament is adequate, but its width is not; the clear annular area of the condenser diaphragm cannot therefore be completely illuminated, and some directional disturbances occur in phase contrast imaging. Phase contrast microscopy in any case requires bulbes with a closely coiled filament (more closely coiled than is shown, for the sake of simplicity, in Fig. 5.44), otherwise the field of view of the microscope, especially when low-power objectives are used, is not uniformly illuminated. In a common bright-field microscope, this defect is reduced by inserting a ground-glass filter below the condenser. This practice is, however, inadmissible in phase contrast microscopy.

Fig. 5.44. Correct (a) and incorrect (b, c) imaging of a light source (bulb filament) in the clear annulus of the annular condenser diaphragm.

The third very important step in the adjustment of the microscope consists in centring the image A' of the condenser diaphragm annulus A (Fig. 5.6) with the phase ring R. A correct coincidence of A' with R is shown in Fig. 5.45b. Any decoincidence of these elements, even as small as shown in Fig. 5.45a, effects the quality of phase contrast imaging adversely.

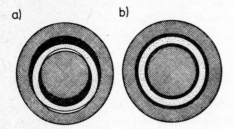

Fig. 5.45. Condenser annular diaphragm incorrectly (a) and correctly (b) centred with the phase ring.

All the steps described above are also obligatory for incident-light phase contrast microscopy. In the latter, however, an additional requirement is that the object surface under study must be perpendicular to the optic axis of the microscope objective; otherwise, it will be found difficult to centre the annular diaphragm with the phase ring. Some incident-light microscopes are therefore equipped with an additional object stage, which has levelling screws or other attachments facilitating the proper adjustment of the object surface to be examined.

In phase contrast microscopy it is also very important to maintain all optical elements in a state of absolute cleanliness. In particular, this applies to the object slide, cover slip, and front surfaces of the condenser and objective. Any contamination, and especially finger-marks on these elements and surfaces are liable to radically decrease image contrast.

5.11.2. General remarks on specimens for phase contrast microscopy

Ways of preparing specimens for phase contrast microscopy are described in detail in more specialized books, e.g., in [5.14], [5.26], [5.49], and [2.48]. Only some general remarks will therefore be given here.

Most biological objects may be examined quite well in water, in many isotonic solutions, and in the liquids normally accompaning them. Such *immersion specimens* should be completely dilute and as thin as possible. In general, the thiner the preparation, the better its image and the more easily the results can be interpretated. Immersion specimens include primarily prepared cells and tissues, bacteria, fungi, algae, and other animal and plant microorganisms. Numerous technical materials (microcrystals, fibres, powders, etc.) are also examined as immersion specimens.

Any phase contrast microscope displays the optical path differences (δ), or phase shift (φ), occurring in the specimen. For a given thickness t (Fig. 5.46) of the object under study (O), the intensity (or brightness) of its phase contrast

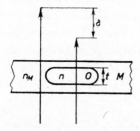

Fig. 5.46. Immersion phase specimen (illustrating formula (5.23)).

image depends on the difference between the refractive indices, n and n_M, of the object and its surrounding medium M. In general, it is desirable to select an immersion liquid which allows the optical path difference δ to be inside the range of unreversed contrast (see Section 5.1). Otherwise, interpretation of the object image is difficult or even impossible. Immersion specimens should therefore be prepared so as to satisfy the condition

$$\delta = (n - n_M)t < \delta_r, \tag{5.23}$$

where δ_r is the limit of unreversed contrast. In this case, the relations of intensity between the object image and background are as shown in Table 5.6. If, for instance, $t = 5$ μm and $\delta_r = 100$ nm, the refractive index difference $n - n_M$ should be smaller than 0.02.

TABLE 5.6

Brightness relations between phase contrast image of immersion objects and background when object phase shift φ is smaller than the range of unreversed contrast φ_r

Relation between refractive indices of object (n) and surrounding medium (n_M)	Relations of brightness between object image and background	
	Positive phase contrast	Negative phase contrast
$n > n_M$	Object image is darker than background	Object image is brighter than background
$n < n_M$	Object image is brighter than background	Object image is darker than background

Another class of specimens for phase contrast microscopy comprises thin sections of solid materials (polymers, glasses, ceramics, etc.) whose structure consists of a matrix and grains. Let us consider a small fragment of a section as shown in Fig. 5.47, where G is a single grain and M is its surrounding matrix. In general, the optical path difference δ between light rays passing through the matrix and grain is equal to

$$\delta = n_1 \Delta t_1 + n_M t_M + n_2 \Delta t_2 - nt, \tag{5.24}$$

where n, n_M, and t, t_M are the refractive indices and thicknesses of the grain and matrix, n_1 and n_2 are the refractive indices of the media on each side of the specimen, and Δt_1, Δt_2 are the elevations (or depressions) of the grain beyond the matrix surfaces. Assuming that the media are the same on both sides of the section, refractive indices $n_1 = n_2 = n_0$, and

$$\delta = t_M(n_M - n) + (\Delta t_1 + \Delta t_2)(n_0 - n). \tag{5.25}$$

In this case, the condition $\delta < \delta_r$ can be fulfilled by making the section as thin as possible and/or by matching refractive indices n and n_0. If the section is immersed in air, $n_0 = 1$, and only t_M, and sometimes also Δt, are at the microscopist's disposal. However, any manipulation with Δt is difficult because no section can be entirely flat due to the microrelief resulting from preparation techniques or to the diverse hardness of various grains.

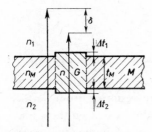

Fig. 5.47. Fragment of a thin section of a transparent solid specimen (illustrating formulae (5.24) and (5.25)).

A large class of phase specimens consists of objects whose structure is examined by observing their surfaces in reflected light or in transmitted light if the object is transparent. The structure displays as elevations and depressions (Fig. 5.48) and also as difference in *phase jumps* of reflected light (Fig. 5.49). Transmitted light creates the situation described previously (Eq. (5.24) or (5.25), in which

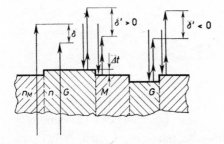

Fig. 5.48. Specimen with a surface phase relief (illustrating formulae (5.26) and (5.27)).

$\Delta t_1 + \Delta t_2 = \Delta t$). In reflected light, on the other hand, the optical path difference δ' is produced by elevation and depression, and, in general, also by the above-mentioned difference in phase jumps ($\Delta\varphi_j$). Then

$$\delta' = n_0 2\Delta t \pm \Delta\delta_j, \tag{5.26}$$

where $\Delta\delta_j$ is the optical path difference caused by $\Delta\varphi_j$. If φ_{jM} and φ_{jG} are phase jumps created by the matrix and grain, respectively, then $\Delta\delta_j = (\varphi_{jM} - \varphi_{jG})\lambda/2\pi$. For glass and other dielectric materials, φ_j is equal to about 180°, whereas for

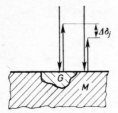

Fig. 5.49. Optical path difference caused by a difference of phase jumps of reflected rays.

metals and semiconductors, it is smaller than 180° (Table 5.7). It is often difficult to distinguish in the phase contrast image a structure resulting from surface relief and phase jump differences. In such cases, a procedure based on the replica technique is helpful because a replica does not record phase jumps differences. Plastic and other replicas belong to phase specimens which are very suitable for examination by phase contrast microscopy.

TABLE 5.7

Phase jumps of light reflected specularly from surfaces of different solid materials

Phase jump φ_j [°]	Material
140–150	Silver, gold, copper
150–155	Nickel, bismuth
155–160	Tin, aluminium, carbon, platinum, zinc, iron
160–170	Chromium, graphite
170–180	Silicon
180	Glass

Let us now return to Fig. 5.48 and Eq. (5.26) but ignoring $\Delta\delta_j$. The condition $\delta' < \delta_r$ is then fulfilled if $\Delta t < \delta_r/2n_0$. In this case, relations of intensity between the image of grains and matrix are as shown in Table 5.8. If, e.g., $\delta_r = 100$ nm

TABLE 5.8

Brightness relations between phase contrast image of light reflecting phase objects and background when object phase shift φ is smaller than the range of unreversed contrast φ_r

Relations between grain surface and matrix surface	Relations of brightness between grain image and its matrix	
	Positive phase contrast	Negative phase contrast
Grain surface is elevated above matrix surface (Fig. 3.48, left)	Grain image is brighter than matrix	Grain image is darker than matrix
Grain surface is lower than matrix surface (Fig. 5.48, right)	Grain image is darker than matrix	Grain image is brighter than matrix

and $n_0 = 1.515$ (immersion oil), then $\Delta t < 33$ nm, and for $n_0 = 1$ (air), $\Delta t < 50$ nm.

It is interesting to see what the relation is between δ and δ' (Fig. 5.48) when $n_0 = 1$, $n = n_M$ and $\Delta \delta_j = 0$. Now Eqs. (5.25) and (5.26) reduce simply to $\delta = \Delta t(1-n)$ and $\delta' = 2\Delta t$, respectively, and $\delta'/\delta = 2/(1-n)$ or

$$|\delta'| = \frac{2|\delta|}{n-1}. \tag{5.27}$$

As can be seen, one and the same elevation (or depression) produces an optical path difference that is several times as great in reflected light as in transmitted light. If, e.g., $n = 1.52$ (glass), δ' is four times as great as δ. Thus the condition of unreversed contrast is easier to fulfil for transmitted light phase contrast microscopy. But, on the other hand, the incident-light phase contrast microscope allows smaller elevations or depressions to be displayed (if stray light is properly extinguished).

5.11.3. Phase contrast microrefractometry

If the refractive index n_M of an immersion (mounting) medium is equal to that (n) of an object under study, the optical path difference $\delta = 0$ at any thickness t of the object (Fig. 5.46). As long as $n > n_M$, the object image is darker (within the range of unreversed contrast) than the background in positive phase contrast, and when $n < n_M$, the contrast changes its sign and the object image becomes brighter than the background. It is self-evident that the converse holds good when a negative phase contrast device is used. Knowing the refractive index n_M of an immersion liquid in which the object image completely disappears, one can determine the unknown refractive index n of the object. This procedure is a matching method. Like other procedure of this kind (e.g., Becke's method, see Subsection 6.3.3), it is accurate and well reproducible on condition that the optical properties of the object are not changed by matching liquids. The smaller an interval Δn there is between successive liquid samples, the easier it is to match and accurately determine an unknown refractive index. Typically, the interval Δn is equal to 0.001, but liquids with $\Delta n = 0.0005$ or even 0.0002 are also in use.

Before the advent of transmitted light interference microscopes, *phase contrast microrefractometry* was successfully applied in cell biology to determine the refractive index of living cells and concentration of solids in them [5.26, 5.50, 5.51]. The refractive index n of a living cell depends upon the concentration C of its solids, the majority of which are proteins, lipoproteins, and amino-acids. For practical purposes, the living cell can be considered as a water solution of these solids (mainly proteins). In general, a 1% increase in concentration of a protein solution causes its refractive index to rise by about 0.0018. Nearly all the substances that are commonly found disolved or finely dispersed in living protoplasm (see Section 16.12 and Table 16.9) have a very similar refractive increment. A general relation can thus be formulated as follows:

$$n = n_s + \sigma C, \tag{5.28}$$

where n is the refractive index of the solution, n_s that of the solvent (usually water), σ is the specific refraction increment ($\sigma \approx 0.0018$), and C is the concentration of solids in grams per 100 ml solution. Equation (5.28) may be written as

$$C = \frac{n - n_s}{\sigma} \quad \text{[g/100 ml]}, \tag{5.29a}$$

or

$$C = \frac{n - n_s}{100\sigma} \quad \text{[g/cm}^3\text{]}. \tag{5.29b}$$

Knowing the concentration of solids, the concentration of water (C_w) can also be determined. However, the latter is not simply equal to $100 - C\%$ because one gram of dry protein does not occupy 1 cm^3 but approximately 0.75 cm^3 [5.26]. Consequently,

$$C_w = 100 - 0.75C. \tag{5.30}$$

Phase contrast microrefractometry of individual living cells and cell populations (mainly blood corpuscles) is discussed in detail by Ross [5.26]. This technique has not, however, been used widely in cytological research, probably because trasmitted light interference microscopes are more practical and more versatile for this and other studies in cell biology (see Section 16.12).

The main disadvantage of phase contrast microrefractometry as presented above is the need for several changes of calibration liquid. The refractive index of an object (e.g. cell) can be determined if the liquids can be changed several times without destroying the object. This procedure is, of course, inconvenient and very time-consuming. Another phase contrast technique, which needs only a single high dispersion liquid, is therefore more practical. This is almost the reverse of the procedure presented in Subsection 5.3.3, concerning the measurement of the sensitivity of a phase contrast microscope. Suppose an object O (Fig. 5.46) is surrounded by a liquid M the dispersion curve of which is more abrupt than that of the object (see curves n' and n, respectively, in Fig. 5.16). If both curves intersect for a wavelength λ_0 within the visible spectrum, the optical path difference $\delta(\lambda_0) = 0$, and the object is invisible in monochromatic light of wavelength λ_0. For any other wavelength, $\delta(\lambda) \neq 0$ and the object is clearly observable. As long as $\lambda > \lambda_0$, the optical path difference $\delta < 0$, and the object image is darker than the background in positive phase contrast, and when $\lambda < \lambda_0$, the optical path difference $\delta > 0$, and the object image is brighter than the background. If negative phase contrast is used, the converse holds good. The matching wavelength λ_0 for which the object image disappears can be fixed very precisely when the wavelength λ of monochromatic light is continuously varied through the visible spectrum from red to blue, and vice versa. Satisfactory results can also be obtained if a set of narrow pass-band interference filters is used instead of the monochromator.

This technique is especially suitable for determining the refractive index of solid microobjects such as strips of thin films, microcrystals, or powders. Its sensitivity and accuracy depend on the magnitude of the intersection angle between the dispersion curves shown in Fig. 5.16 (the bigger the better) and the sensitivity of the phase contrast microscope. The use of highly sensitive phase contrast devices, e.g., the KFA device (see Subsection 5.5.2), allows the refractive index of such objects as strips of thin film to be determined with an accuracy equal even to 0.0002 if the matching wavelength λ_0 is fixed within ± 2 nm, and the dispersion curve of the immersion liquid is known with an accuracy better than 0.0002 (for wavelength λ_0 and its vicinity).

An interesting application of phase contrast microrefractometry has recently been described by Bożyk [5.53]. She used the KFA device (see Subsection 5.5.2) for measuring the refractive profile and core diameter of optical fibres. The measuring procedure depended on microdensitometry of phase contrast images recorded on a photographic film. The fibres under study were embedded in a high dispersion liquid.

5.11.4. Fields of application of phase contrast microscopy

Phase contrast microscopy has been in use for a long time, and its practical applications are so manifold that it is impossible to give any complete review in this book. Only some examples are mentioned here, and the reader who wishes to know more is referred to specialized monographs dealing with this and related fields of light microscopy [5.14, 5.26, 5.27, 5.56–5.62].

Phase contrast microscopy is, of course, most widely applied in biology and biomedicine. The bibliography in this field is so large that only the most representative papers and books will be cited in the following text.

It is in cytology and histology that phase contrast microscopy is most widely used [5.14, 5.22, 5.24–5.27, 5.50, 5.51, 5.56, 5.67]. Indeed, the phase contrast microscope is particularly well suited to the examination of living cells, tissues, and microorganisms too transparent in their natural state to be observed with a normal microscope. Cell membrane, nuclei, nucleoli, mitochondria, fibroplast spindles, chromosomes, Golgi apparatus, cytoplasmic granules can all be clearly identified. Bajer [5.67], as well as other researchers [5.68], has succeeded in obtaining excellent observations of mitosis. It should be emphasized that the phase contrast method can be applied equally well to the study of both animal and plant cells [5.69]. Extensive research has been undertaken in connection with the diagnosis of tumours and the growth and behaviour of mammalian cells cultured in vitro [5.24, 5.25, 5.70, 5.71]. Very valuable observations have also been made in haematology [5.59, 5.60]. Blood and medula cells are especially suited for phase contrast microscopy. Viruses [5.72, 5.73], bacteria, phagi, some fungi, and eggs of parasitic worms are also suited to this type of microscopy [5.14, 5.27]. Positive results are also obtained in the study of urine [5.74, 5.75]. Although of limited range, the phase contrast microscope is a valuable tool for the freshwater microscopist and paleontologist [5.76].

Mineralogy, crystallography, and chemistry are another field where phase contrast microscopy is usefully applied [5.14, 5.27, 5.77–5.79]. Colourless microcrystals, powders, and particles of solids, the refractive index of which differs only moderately from that of the surrounding medium (immersion liquid), are particularly well suited for examination by phase contrast microscopy. Here, phase contrast microrefractometry (see Subsection 5.11.3) is a useful technique for the quantitative or semiquantitative identifying analysis of microcrystals and various solid particles [5.78]. Clay and dust, fat, grease, oil and soap, paint and pigments, as well as foods, drugs, and pharmaceuticals have also been examined by some researchers who threw new light on the physics and chemistry of these materials. Textile fibres and such defects of theirs as cracks and fine inclusions are also well suited for phase contrast observation [5.80, 5.81].

The next vast field of application for phase contrast microscopy is metallography and, in general, the physics of surfaces [5.14, 5.27, 5.82–5.87]. Some minute structures, subtle defects, network dislocations, and the like can be displayed only by using an incident-light phase contrast microscope (or differential interference contrast microscope, see Subsection 7.2.4) for the study of properly prepared surfaces of solids (metals, alloys, semiconductors, etc.). For instance, stacking faults in silicon epitaxial wafers are among the minute but serious defects well revealed by the phase contrast method [5.48, 5.84–5.87].

To sum up, the potentialities of phase contrast microscopy are considerable. They have been made good use of in various fields of science and industry, in particular, in biology and medicine.

*

*   *

Although for routine microscopy the phase contrast method has already made considerable advances, nevertheless further refinements are desirable.

The major problem is how to eliminate halo and shading-off effects. These defects are, of course, inherent in the phase contrast technique and will never be solved in a satisfactory manner. However, continuously variable phase contrast systems allow these undesirable effects to be controlled and reduced to some extent, in order to optimize the quality and fidelity of phase contrast imaging. In the author's view, what is lacking is a cheap and widely available phase contrast microscope with continuously variable image contrast.

A useful development is the trend for combining the phase contrast with other techniques, in particular, with differential interference contrast microscopy (see Chapter 7). A versatile microscope comprising both variable phase contrast and a shearing interference system with continuously variable wavefront shear (see Section 7.4) offers the greatest advantages for the extensive study of phase specimens.

An interesting system capable of generating various contrast types has recently been described by Lang and Dekkers [5.88]. This system is suitable for imaging both phase and amplitude structures in either differential or non-differential contrast with adjustable magnitude. Like the Polanret system, it contains a polarizer, two birefringent plates, and an analyser. One of the birefringent plates is inserted in the entrance pupil of a condenser and the other in the exit pupil of the objective. Both plates, and thus both pupils, are divided into two conjugate semicircular areas with different anisotropic properties, so that this system may be regarded as a double Polanret system which gives two images incoherently superimposed. The resultant image can be either a phase contrast image with the amplitude contrast suppressed, or vice versa.

There has been also considerable interest in recent years in developing methods that can provide direct, accurate, and real time measurements of phase objects which produce even large optical path difference variations. Some approaches to this problem have been proposed by Thompson, Sommargren, and Sprague [5.89–5.91]. New potentialities for phase contrast microscopy lie also in scanning optical microscopy (see Subsection 12.2.1).

Finally, it is interesting to note that several authors have become interested in studying the problem of phase contrast imaging on the basis of the theory of image formation in partially coherent light. A number of theoretical works dealing with different aspects of this problem have been published in the last two decades. Most of these works are listed by Khan and Rao [5.92]. Results in this field are both interesting and stimulating for further progress in phase contrast microscopy.

6. Amplitude Contrast, Dark-Field, Optical Staining, Modulation Contrast, and Other Related Techniques

The phase contrast method dealt with in the preceding chapter is the most valuable among those based on a modification of the Fourier spectrum enabling us to observe phase specimens and phase-amplitude objects of low contrast. Similar procedures are also applied in other techniques known as *amplitude contrast*, *central dark-field*, *microstrioscopy* (or *schlieren microscopy*), and *dispersion staining*. These and some related techniques are discussed in this chapter.

6.1. Amplitude contrast microscopy

Phase objects which transmit practically the same amount of light as their surrounding medium were considered in Chapter 5. However, it appears that using a technique similar to that of standard phase contrast microscopy allows the image contrast of light absorbing object to be controlled (increased, reduced, or even reversed). These possibilities are predicted by both the vector representation and algebraic theory of phase contrast. They also result, of course, from the Fourier approach to microscopical imagery.

6.1.1. Theoretical considerations

As a strict theory of amplitude contrast is quite complicated, only some approximate considerations will be undertaken here, and mainly such problems will be stressed which are required for a general insight into the physical principles and some practical aspects of this technique.

Vectorial representation. Suppose a plane light wave strikes normally a small amplitude object (see Fig. 2.1a in Volume 1). Transmittance of this object is τ_0 and its refractive index does not differ from that of a transparent surrounding medium.

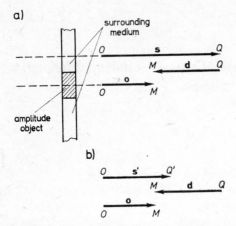

Fig. 6.1. Vector representation of the amplitude contrast method.

The situation presented in Fig. 2.1a can now be supplemented by a vectorial diagram consisting of three vectors (Fig. 6.1a): $\mathbf{s} = OQ$, $\mathbf{o} = OM$ and $\mathbf{d} = QM$, which represent, respectively, the direct light traversing only the surrounding medium, the light passing through the object, and the diffracted light. Vectors $\mathbf{s}$ and $\mathbf{o}$ are parallel to each other because the object does not change the phase of the transmitted light wave but only reduces its amplitude; hence vector $\mathbf{o}$ is shorter than $\mathbf{s}$, and their lengths are $|\mathbf{o}| = a' = \sqrt{\tau_0}\,a$ and $|\mathbf{s}| = a$, where a' and a are the amplitudes of the light wave exiting the object and its surround.

By analogy with phase contrast, the vector $\mathbf{o}$ may be treated as the sum of vectors $\mathbf{s}$ and $\mathbf{d}$. Since $\mathbf{s}$ and $\mathbf{o}$ are parallel to each other and $|\mathbf{o}| < |\mathbf{s}|$, the vector $\mathbf{d}$ must therefore be antiparallel to $\mathbf{s}$. Hence it appears that an amplitude object produces diffracted light which is shifted in phase by an angle of $180°$ with respect to the direct light.

According to the diffraction theory of image formation, the object image is a result of interference of the direct and diffracted light waves. Thus, in accordance with Eq. (1.110b), the image intensity $I'_0 = (|\mathbf{s}| - |\mathbf{d}|)^2$, whereas the background intensity $I'_b = |\mathbf{s}|^2$; hence the image contrast is given by

$$C' = \frac{I'_b - I'_0}{I'_b} = \frac{2|\mathbf{s}||\mathbf{d}| - |\mathbf{d}|^2}{|\mathbf{s}|^2}. \tag{6.1a}$$

As the intensity of diffracted light is normally small relative to that of direct light, i.e., $|\mathbf{d}|^2 \ll |\mathbf{s}|^2$, Eq. (6.1a) takes the form

$$C' = \frac{2|\mathbf{d}|}{|\mathbf{s}|}, \tag{6.1b}$$

which shows that the image contrast C' may be increased if the amplitude $|s|$ of the direct light is reduced. Consequently, if N is a number by which the intensity of this light is divided, then the vector s transforms into $s' = OQ'$ (Fig. 6.1b). The length of this new vector is $|s'| = |s|/\sqrt{N}$. For the sake of simplicity, let the primary vector s be a unity vector; then $|s'| = 1/\sqrt{N}$ and Eq. (6.1b) takes the form

$$C' = 2\sqrt{N}|d| \tag{6.2}$$

which shows that as N increases so does image contrast C'.

Algebraic basis. When phase contrast theory is extended over phase-amplitude objects, the Richter's condition (Eqs. (5.10)–(5.12)) for optimum image contrast takes the following, more general form [5.11]:

$$\tau_p = \frac{1}{N} = 1 - 2\sqrt{\tau_0}\cos\varphi + \tau_0, \tag{6.3}$$

$$\sin^2\psi_{op} = \frac{\tau_0}{\tau_p}\sin^2\varphi. \tag{6.4}$$

Here τ_p is the ratio of the light transmittance of the conjugate area to that of the complementary area of the phase plate, ψ_{op} is the optimum phase difference between these two areas, τ_0 is the ratio of the light transmittance of the object to that of an equal thickness of the surrounding medium, and φ is the phase shift produced by the object.

Considering amplitude objects for which $\varphi = 0$, one can easily show that Eqs. (6.3) and (6.4) become

$$\tau_p = (1 - \sqrt{\tau_0})^2, \tag{6.5}$$

$$\psi_{op} = 0 \quad \text{for } \tau_0 < 1, \tag{6.6a}$$

or

$$\psi_{op} = 180° \quad \text{for } \tau_0 > 1. \tag{6.6b}$$

The condition $\tau_0 > 1$ signifies that the object absorbs less than an equal thickness of its surrounding medium. This case is neglected, and only objects which absorb more light than their surround ($\tau_0 < 1$) will be considered. When $\tau_0 \to 1$, the light transmittance of the object and its surround becomes alike, and the required value τ_p for optimum positive contrast approaches zero.

From what was said above, it is clear that the image contrast of weakly absorbing amplitude objects can be optimally increased by using a ring (R, Fig. 5.6) with zero phase shift and transmittance $\tau_p = (1 - \sqrt{\tau_0})^2$. Such a ring will be termed an *amplitude ring*.

Fourier transform approach. Let us now consider a line grating and assume that its amplitude transmittance distribution is described by the following function:

$$f(x) = 1 + b\cos 2\pi x, \tag{6.7}$$

where x is the coordinate perpendicular to the grating lines and parallel to the object plane of microscope objective, whereas b is the amplitude of function $f(x)$. Let b be much smaller than 1 ($0 < b \ll 1$).

When a parallel beam of monochromatic light strikes the object plane normally, the intensity distribution $I(x)$ just behind the grating is described by $[f(x)]^2$, i.e.,

$$I(x) = 1 + 2b\cos 2\pi x + b^2\cos^2 2\pi x. \tag{6.8a}$$

The quadratic term of this equation is small relative to the remaining terms and can be neglected. Thus, one can write

$$I(x) = 1 + 2b\cos 2\pi x. \tag{6.8b}$$

The plots of Eqs. (6.7) and (6.8b) are represented, respectively, by the solid and dashed curves in Fig. 6.2a.

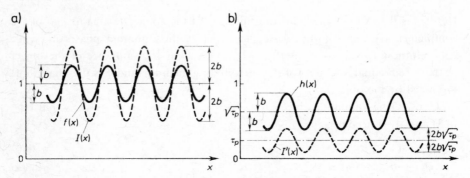

Fig. 6.2. Illustrating the Fourier transform approach to the amplitude contrast method: a) plots of Eqs. (6.7) and (6.8b), b) plots of Eqs. (6.11) and (6.12b).

The amplitude distribution of light in the back focal plane of the objective is given by the Fourier transform $F(u)$ of $f(x)$, i.e.,

$$F(u) = \int_{-\infty}^{+\infty} (1 + b\cos 2\pi x)\exp(-2\pi iux)\,dx, \tag{6.9a}$$

or

$$F(u) = \int_{-\infty}^{+\infty} \exp(-2\pi iux)\,dx + \int_{-\infty}^{+\infty} b\cos 2\pi x\exp(-2\pi iux)\,dx, \tag{6.9b}$$

where u is the spatial frequency related to the coordinate x ($u = 1/p$; p is the grating period). The second integral in Eq. (6.9b) expresses the diffracted light, whereas the first integral expresses the direct light. Now let the intensity of the latter be attenuated by a plate of transmittance τ_p ($0 < \tau_p < 1$). Returning to Fig. 5.2 it is clear that this plate (Ph) must be located in the back focus (F'_{ob}) of the objective. The attenuation mentioned above is equivalent to the multiplication of the first integral in Eq. (6.9b) by $\sqrt{\tau_p}$ and so the Fourier transform $F(u)$ takes the form

$$F'(u) = \int_{-\infty}^{+\infty} \sqrt{\tau_p}\exp(-2\pi iux)\,dx + \int_{+\infty}^{-\infty} b\cos 2\pi x \exp(-2\pi iux)\,dx. \tag{6.10}$$

The intensity distribution $I'(x)$ in the image plane[1] of the microscope objective is given by $[h(x)]^2$, where $h(x)$ is the Fourier transform of $F'(u)$, i.e.,

$$h(x) = \sqrt{\tau_p} + b\cos 2\pi x, \tag{6.11}$$

and

$$I'(x) = [h(x)]^2 = \tau_p + 2\sqrt{\tau_p}\,b\cos 2\pi x + b^2\cos^2 2\pi x. \tag{6.12a}$$

The quadratic term of the above equation is very small relative to the remaining terms and it can be neglected. Thus Eq. (6.12a) takes the form

$$I'(x) = \tau_p + 2\sqrt{\tau_p}\,b\cos 2\pi x. \tag{6.12b}$$

The plots of Eqs. (6.11) and (6.12b) are represented by the solid and dashed sinusoids in Fig. 6.2b. As can be seen, the distribution of the function $h(x)$ is the same as that of the function $f(x)$, but the former has a lower background level. However, the intensity distribution $I'(x)$ compared with $I(x)$ has both a lower background level and lower intensity variation. The result is a higher image contrast C' compared with the object contrast C. From Eqs. (2.4b) and (3.64b) it results that

$$C = \frac{4b}{1+2b}, \tag{6.13}$$

and

$$C' = \frac{4\sqrt{\tau_p}\,b}{\tau_p + 2\sqrt{\tau_p}\,b}. \tag{6.14}$$

[1] For the sake of simplicity, it is assumed that the coordinate x' in the image plane is the same as in the object plane, and the image magnification is equal to 1.

Since $\sqrt{\tau_p} > \tau_p$, Eqs. (6.13) and (6.14) lead to

$$\frac{C'}{C} = \frac{\sqrt{\tau_p} + 2\sqrt{\tau_p}\,b}{\tau_p + 2\sqrt{\tau_p}\,b} > 1. \tag{6.15}$$

Figure 6.2b refers to a situation when $\sqrt{\tau_p} > b$. If, however, $\sqrt{\tau_p} < b$ (Fig. 6.3), the fidelity in intensity distribution $I'(x)$ is disturbed by secondary maxima that appear between the main maxima.

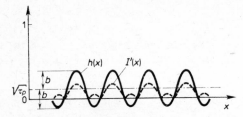

Fig. 6.3. As in Fig. 6.2b, but $\sqrt{\tau_p} < b$.

6.1.2. Practical implementation and properties

From what has been said here, it is evident that a device similar to the standard phase contrast system (Fig. 5.6) can be used for observing amplitude objects with low contrast (C). As a matter of fact, such a device was developed by the author of this book and has long been available from the Polish Optical Works (PZO). It is known as *amplitude contrast device KA* [6.1, 6.2]. Other researchers have also devised systems similar to the KA device. Their properties will be presented below.

KA device. One reason for making this device was that in stained biological specimens there are always details and structures insufficiently stained and therefore difficult to distinguish by bright-field microscopy. It is true that such details can be observed well enough by the phase contrast method, but this distorts the image of more intensely stained objects.

The construction of the KA device is similar to that of the positive phase contrast equipment KFS described in Subsection 5.5.3. The amplitude rings are made by the same process and consist of a soot layer and a dielectric layer. The latter cancels the negative phase shift of the soot layer (see Fig. 5.23). For this, the thickness t of the dielectric layer is given by

$$t = \frac{|\delta|}{n_M - n}, \tag{6.16}$$

where δ, n, and n_M are as in Eq. (5.18).

In initial experiments with the KA system typical achromatic objectives $10 \times /0.25$, $20 \times /0.40$, $40 \times /0.65$ and $100 \times /1.25$ (oil immersion) were used. These were changed into amplitude-contrast objectives by placing, according to Fig. 5.23 and Eq. (6.16), an amplitude ring between the two lenses nearest the objective back focus. Amplitude rings were prepared with light transmittance $\tau_p = 0.25$, 0.20, 0.15, and 0.10, and geometrical parameters $d/d_{E'_{ob}} = 0.5$, $w/d_{E'_{ob}} = 0.1$, and $a/a_{E'_{ob}} = 0.17$. Here d, w, and a are, respectively, the diameter, width, and area of the amplitude rings, $d_{E'_{ob}}$ and $a_{E'_{ob}}$ are the diameter and area of the exit pupil of objectives.

The same achromatic microscope objectives but without the amplitude ring, as well as those fitted with positive phase rings with a phase shift of $+90°$ and transmittance of about 0.2, were employed for comparative studies. The quality of each objective was tested by means of a shearing interferometer (see Section 4.3.1 and Fig. 4.11). Only objectives with the same degree of aberration correction were selected for the comparative study of amplitude contrast imagery. A condenser with annular aperture diaphragms conjugate with the amplitude and phase rings was used for all observations. Thus, the illumination of all test specimens was identical for amplitude contrast, phase contrast, and ordinary bright-field observation. Some results are shown in Figs. XXVIII–XXXI.

The experiments have shown that the best results are obtained with stained biological specimens when the soot amplitude ring absorbs 75–80% of the direct light ($\tau_p = 0.25$–0.20). This value was selected for the final version of the KA device manufactured by PZO, Warsaw. When the amplitude ring absorbs more than 85%, a decrease of image contrast is observed. A halo also appears, similar to the phase contrast halo.

The basic features and advantages of this device can be seen by studying photomicrographs of some fairly lightly stained specimens in Figs. XXVIII–XXXI. Each of the amplitude contrast photomicrographs (marked "a") is confronted with the bright-field image of the same specimen taken under comparable conditions. The exposure time was controlled automatically, using a photomicrographic attachment with an automatic exposure-meter.

It can be seen that in comparison with the ordinary bright-field microscope the KA device obtains greater image contrast for insufficiently or moderately stained specimens. This is achieved both by using ring-shaped illumination and by reducing the light intensity of the zero order diffraction maximum according to the theoretical prediction discussed above. Another reason for the higher contrast is the elimination of the diffuse (stray) light from the field of view and the delineation of fine details. The last feature is a consequence of the improvement of the optical transfer function for partially coherent illumination (compare Fig. 6.5). This advantage is particularly useful in the investigation of minute components and structures of biological cells. When a specimen includes both large heavily stained objects and some fine low-contrast details, amplitude contrast enhances the latter more than the former. This effect permits small details on a background of large objects or in their vicinity to be observed better than in the bright-field microscope.

It is true that weakly stained objects can be observed quite well with a standard positive phase contrast microscope, but some undesirable phase artefacts may then efface interesting amplitude details. Amplitude contrast, on the other hand, tends to emphasize amplitude details and minimize phase details. However, the amplitude contrast device is useless for objects which are intensely stained. It tends to minimize or even reverse the contrast of the image of objects absorbing more than about 80%. These properties, after all, results from the Fourier transform theory of amplitude contrast (see Fig. 6.3). For further information the reader is referred to the source references [6.1, 6.2] and to articles by Göke [6.3, 6.4].

The KA device was originally developed for transmitted-light microscopy. Next, it was also successfully adapted to reflected-light microscopy by Popielas [5.48, 6.5], who demonstrated that the amplitude contrast method also has many interesting advantages for studies of reflecting specimens. Using this method, some finely etched structures appear more clearly than in standard metallurgical microscopes.

Nomarski amplitude contrast device. Like the KA device, this contains an annular diaphragm D (Fig. 6.4) in the front focal plane of a condenser C and an amplitude ring R in the back focal plane of an objective Ob. An essential difference between both devices lies in the size of the amplitude ring R at which the real image of the condenser annular opening A is formed. Namely, the ring R extends up to the edge of the exit pupil of objective and its inner diameter d_i lies between $0.7d_{E'_{ob}}$ and $0.9d_{E'_{ob}}$ ($d_{E'_{ob}}$ is the diameter of the exit pupil of objective). The transmittance

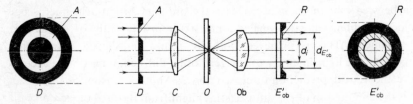

Fig. 6.4. Schematic diagram of the Nomarski amplitude contrast device.

τ_p of the amplitude ring is determined as a function of the relative diameter d_r by the relation [6.6]

$$\tau_p = \frac{d_r^2}{(1+d_r)^2},\qquad\qquad (6.17)$$

where $d_r = d_i/d_{E'_{ob}}$. This formula gives $\tau_p = 0.25$ for very narrow rings ($d_r \approx 1$) and $\tau_p = 0.20$ for $d_r = 0.8$.

The annular opening A of the condenser C should be exactly adjusted to the size of the amplitude ring R; the last element must be filled completely by the image of A. If a high-power immersion objective is used, the condenser must also be immersed; hence the latter must have a better optical quality than that required in conventional phase contrast microscopy. It is moreover advantageous to employ high-power oculars (20 to 25 ×) in order to bring out fully the quality of images of object details which are close to the resolution limit defined by Eq. (3.61).

The amplitude ring of large diameter d_r acts primarily as a selective filter favouring diffracted light created by the fine structures of the object under examination. Calculations based on an approximate theory show that by such a filtering process the quality of images can be partially improved by enhancing the contrast of periodic objects whose periods lie between 2ϱ and ϱ (ϱ is the resolution limit defin-

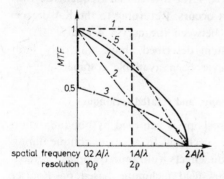

Fig. 6.5. Plots of the modulation transfer function (*MTF*) corresponding to different modes of illumination [6.6].

ed by Eq. (3.61)). The curves shown in Fig. 6.5 illustrate the modulation transfer functions (*MTF*) corresponding to different modes of illumination. Curve *1* characterizes the coherent imagery (compare curve *e* in Fig. 3.59) when a circular condenser diaphragm is closed up to minimum, whereas curve *2* corresponds to the incoherent imagery (compare curve *a* in Fig. 3.69) when the condenser aperture is equal to that of the objective. In the first case, the fidelity of imagery is perfect, but the resolution is limited to 2ϱ, whereas in the second case, the resolution is improved to the theoretical limit ϱ, but the contrast of high spatial frequencies is very small. Quite a different situation obtains if illumination is produced by a very narrow annular opening (curve *3*). The image contrast of fine objects is improved, but that of middle and low frequencies is considerably diminished. Curve *4* corresponds to the situation represented by curve *3* but with an amplitude ring of $\tau_p = 0.25$ placed in the back focal plane of objective and conjugated with the annular opening of the condenser as shown in Fig. 6.4. In this case, there is an excellent resolution of the finest structures and the useful magnification can be doubled. Thanks to the amplitude ring, the *MTF* represented by curve *3* is greatly improved and the general image contrast is increased. As the inner diameter of the amplitude ring and, of course, that of the condensers annulus diminish, so does the *MTF* for high spatial frequencies. Curve *5* illustrates the situation when the amplitude ring of transmittance 20% is as large as 0.2 times the pupil radius $r_{E'_{ob}}$, i.e., the width of the ring $r_{E'_{ob}} - d_i/2 = 0.2 r_{E'_{ob}}$. Although the contrast is generally better than usual (curves *2* and *3*), it is nevertheless somewhat lower than with a narrow amplitude ring, especially in the range of the highest spatial frequencies accepted by the microscope objective. On the other hand, curve *5* is more horizontal than curve *4* for low spatial frequencies, thus improved fidelity between image and object occurs. Returning to the KA device, its *MTF* may be said to be intermediate between the curves *4* and *5*.

The Nomarski amplitude contrast system described above was only built as an experimental prototype and has never been available commercially.

6.2. Oblique illumination, dark-field microscopy, and related techniques

Dark-field or dark-ground imagery in microscopy is achieved by oblique illumination or by spatial filtering of the Fourier spectrum. In practice, oblique illumination that produces some superperception effects (ultramicroscopical perception)[2] is more popular. However, the dark-field technique based on Fourier filtering has recently attracted more attention (see Ref. [6.7] and the papers cited there).

The properties and interpretation of dark-field images are more complicated [6.8] than those of bright-field images. Some authors merely use a phase grating

---

[2] It is sometimes stated that the resolving power of the microscope is better under dark-field than under brigh-field illumination. In fact, it is only the perception limit that is better and image contrast of small objects that is increased.

[5.3, 4.8] for explaining the principles of dark-field imagery, whereas others take into consideration only small particles. Practices of this kind frequently lead to some confusion because theoretical consideration which hold good for a phase grating, for instance, do not apply to amplitude gratings or small particles, and vice versa. In other words, there is no generalization in dark-field imagery and any particular class of objects must be considered individually. Such an individual approach falls outside the scope of this chapter and is not, in fact, necessary. Although conventional dark-field microscopy is well established in practice, nevertheless it is currently restricted to certain technological domains. In biology and biomedicine, on the other hand, little interest has been shown in this technique during the last few decades (an exception is fluorescence microscopy with dark-field illumination). The lack of interest has mainly been attributed to the development of phase contrast microscopy.

Under these circumstances, the following discussion will deal with some basic instrumental aspects of dark-filed microscopy, whereas for more detailed theoretical information the reader is referred to the cited literature [2.13, 5.3, 4.8].

6.2.1. Oblique illumination

There are some procedures which belong partly to bright-field and partly to dark-field microscopy. These procedures are known as *oblique* or *skew illumination*. For trans-illuminated specimens, this illumination is achieved by decentring the bright-field condenser or its properly closed aperture diaphragm. Another procedure is to permit light to enter the condenser from only one side or through asymmetric openings of various shape [2.12, 6.9, 6.10].

A versatile condenser for both dark-field illumination and different modes of oblique illumination was developed by Zselyonka and Kiss [6.11]. The functional principle of this condenser is that the illumination of the specimen under examination is due to the common effect of rays characterized by different directions, different intensities, different spectral range as well as some other features. Among other things, it produces asymmetric shading effects and gives an impression of three-dimensional images.

In the past, procedures of oblique illumination were often used in transmitted light microscopy of phase objects. Oblique illumination was also valuable in one of the immersion methods for evaluating the refractive index of transparent solids [6.12]. In incident-light microscopy, oblique illumination is sometimes used to observe specimens whose characteristic details suffer from lack of intensity contrast. In this case, oblique illumination produces asymmetric shading-off effects, which accentuate any slight relief or difference in light of the specimen surface. A result is an impression of a three-dimensional image. Moreover, oblique illumination in reflected-light microscopes enables one to detect surface unevenness and surface roughness. The best known instrument for this purpose is the *Schmaltz microscope* (or *light-cut microscope*) with a slit epi-illuminator (see Ref. [2.44]) and its improved version developed by Menzel (see Refs. [6.8] and [1.31] for details).

6.2.2. Oblique dark-field illumination

In *dark-field* or *dark-ground illumination* the rays strike the specimen at such an angle that the direct light is not accepted by the microscope objective; only the diffracted and scattered light from the specimen enters the objective. Oblique illuminating rays may come from all sides as well as from one side with respect to the optic axis of objective. One-sided oblique illumination is only used in some special cases, whereas symmetrical illumination from all sides is that which is normally in use. This latter is accomplished by means of *special dark-field condensers*. The simplest of these is based on a bright-field condenser (Fig. 6.6a) whose aperture is partially blocked by a circular stop, and a hallow cone of light emerging from the outer annular opening of the condenser passes through the specimen in such a manner that the direct light misses the microscope objective when both the condenser and the objective are focused on the object under observation. This method of obtaining dark-field illumination is applicable only to low- and middle-power objectives whose numerical aperture is smaller than 0.65. For high-power objective (numerical aperture $A \geqslant 0.65$), more specialized dark-field condensers are necessary to supply an intense hallow cone of light of greater aperture angle σ than results from $A = n\sin\sigma$, where n is the refractive index of the immersion medium. There are different types of such condensers (see Ref. [2.12] for details), but for transmitted-light microscopy the *cardioid condenser* (Fig. 6.6b) and the *paraboloid condenser* (Fig. 6.6c) are the most popular. The former consists of two glass mirrors the first of which is spherical and the second is cardioidal surface. This system satisfies the condition of aplanatism, is thus free from spherical aberration and coma, and does not, of course, suffer from chromatic aberration. Normally it is used with oil immersion; in this case its typical numerical aperture is 1.2–1.3 (sometimes up to 1.4). The paraboloid condenser makes use of reflections on a single surface which is made from a paraboloid truncated by a plane surface perpendicular to the condenser axis. The

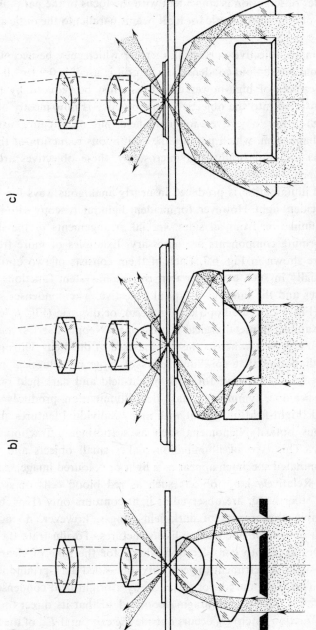

Fig. 6.6. Dark-field condensers: a) conventional with annular diaphragm, b) cardioid, c) paraboloid.

specimen under observation is coincident with the focus of the paraboloid mirror. This focus is, of course, stigmatic for light beams parallel to the optic axis of paraboloid.

The maximum objective numerical aperture which may be accepted by both the cardioid and paraboloid condensers, when oil is used, is 0.9 to 1.0. The larger numerical apertures of high-power objectives must be reduced by means of a funnel stop inserted into the objective. This practice is now mostly obsolete and currently high-power objectives for dark-field microscopy are usually fitted with an iris diaphragm, which permits the continuous reduction of the objective numerical aperture. For bright-field microscopy, these objectives are employed with a fully open iris diaphragm.

Dark-field illumination is produced in nearly analogous ways for both transmitted and incident light. However, for incident-light microscopy with symmetrical dark-field illumination from all sides, special arrangements of the illuminating and image-forming components are necessary. Examples of more frequent configurations are shown in Fig. 6.7. Each of them consists of two optical systems coupled coaxially in such a manner that the outer system functions as a dark-field condenser and the inner as a typical objective. The condenser system may be catoptric (Fig. 6.7a), catadioptric (Fig. 6.7b), or dioptric (Fig. 6.7c). The first is used, for example, in metallurgical microscopes (Neophot series) manufactured by VEB Carl Zeiss Jena, whereas the third is used in the Leitz Ultropak system.

It is worth noting that epi-illuminators of large metallurgical microscopes are constructed so as to serve for both bright-field and dark-field observations.

Roughly speaking, symmetrical dark-field illumination produces the reverse of the normal bright-field image but with many individual features and artefacts due to various optical phenomena such as scattering, diffraction, refraction, reflection, etc. This type of illumination makes small objects and fine details of a transilluminated specimen appear as a light or coloured image against a dark background. Relatively large objects such as red blood cells or strips of thin films, on the other hand, are observed as light contours only (Figs. 6.8b and c).

These simple appearances of dark-field images, however, do not apply to objects which consist of fine grating-like structures. To illustrate the difference, let Fig. 6.9 be taken into consideration where, for the sake of clearness of the drawings but without detriment to generality, one-sided dark-ground illumination is presented. This is produced, for instance, by a bright-field condenser (see Fig. 6.6a) fitted with a circular diaphragm decentred so that its direct image (or the zero order diffraction patch) Q_0 occurs outside the exit pupil E'_{ob} of the microscope objective. A grating-like object produces diffractions patches of the first (Q_{-1},

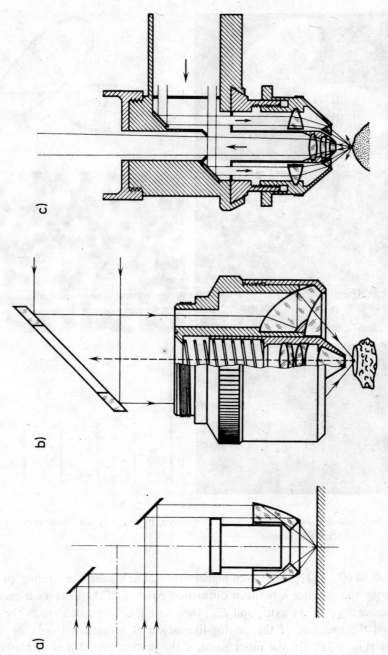

Fig. 6.7. Dark-field condensers for reflected-light microscopy: a) catoptric, b) catadioptric, c) dioptric (Leitz Ultropak system).

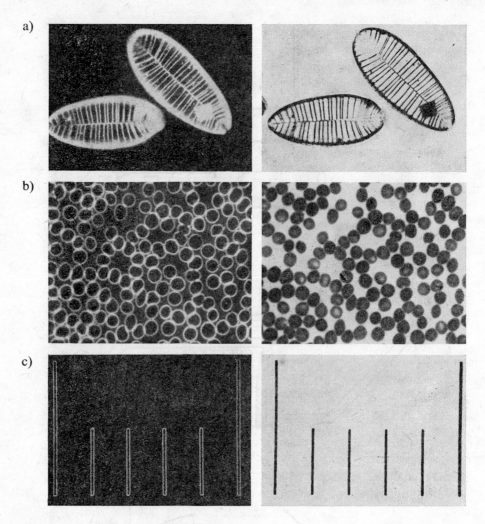

Fig. 6.8. Dark-field and bright-field images: a) diatoms, b) smear of red blood cells, c) fragment of an object stage micrometer.

Q_{+1}), second (Q_{-2}, Q_{+2}), and even higher orders. The smaller the grating period p the larger the distance h between diffraction patches. If the distance h exceeds the diameter $2r_{E'_{ob}}$ of the exit pupil E'_{ob}, then no diffraction patch enters the exit pupil and the presence of the grating-like object is accentuated only by light contours (Fig. 6.9a). On the other hand, if the grating period p is increased so that the distance h becomes as small as to permit the diffraction patch Q_{-1} to enter the exit pupil E'_{ob}, then the whole object appears as a bright image on a dark background (Fig. 6.9b); however, the grating structure is not observed. According to the Abbe theory (see Section 3.3), the structure can be resolved if at least two diffraction patches (Q_{-1} and Q_{-2}) are accepted by the exit pupil of the objective (Fig. 6.9c).

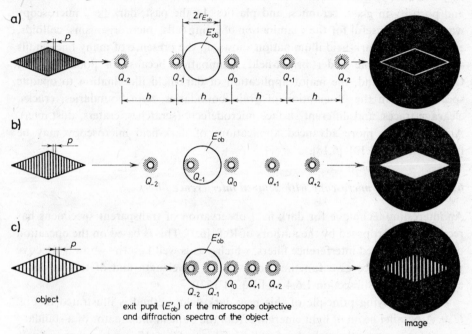

object

exit pupil (E'_{ob}) of the microscope objective and diffraction spectra of the object

image

Fig. 6.9. Appearance of dark-field images of fine grating-like structures (explanation in the text).

To resolve a grating structure with so small periods p as shown in Figs. 6.9a and b, a microscope objective with greater numerical aperture must be used to permit us to bring at least two diffraction patches (Q_{-1} and Q_{-2}) into the exit pupil of the objective.

The situation shown in Fig. 6.9 is similar if an annular dark-field diaphragm is used (Fig. 6.6) instead of a decentred circular diaphragm.

The dark-field images shown in Fig. 6.9 are illustrated in more detail by photomicrographs in Fig. XXXII, where photomicrograph a corresponds to Fig. 6.9a, c to Fig. 6.9b, and e to Fig. 6.9c.

In reference to Fig. 6.9, it is worth while to note that a sinusoidal amplitude grating, which produces only three diffraction patches (Q_0, Q_{-1}, and Q_{+1}; see Figs. 1.92 and 1.93), cannot be resolved if oblique (one-sided or symmetrical) dark-field illumination is used.

The major application of dark-field illumination to transparent specimens is in the examination of materials of low contrast that scatter and refract light. In particular, this technique is suitable for observing small particles, inclusions, and porosity in glass, ceramics, and plastics. In the past, dark-field microscopy was extremely useful for the examination of living cells, microorganisms, colloids, etc. In general, dark-field illumination shows up the presence of many fine details which are invisible under bright-field illumination because of poor contrast. On the other hand, the major application of dark-field illumination to opaque specimens is in the observation of grain boundaries, twin boundaries, cracks, cleavage traces, and different surface microdefects (scratches, craters, dust, etc.). Areas of some more advanced applications of dark-field microscopy may be found in Refs. [6.13]–[6.18].

6.2.3. Dark-field microscopy with detuned interference filters

An interesting technique for dark-field observation of transparent specimens has recently been proposed by the authors of Ref. [6.7]. This is based on the operation of suitably detuned interference filters, which—as is well known—show a selective peak transmittance as a function both of the wavelength and of the incidence angle (compare Subsection 1.6.4 and Eq. (1.152)).

The underlying principle of this new dark-field method is illustrated in Fig. 6.10. A parallel beam of light emerging from the built-in illuminator of a standard microscope passes through the first interference filter IF_1 and strikes a finely ground glass GG. Characteristics of the incident beam are the peak wavelength λ_1 and the half bandwidth $\Delta\lambda_1$ of the filter. Beyond the ground glass there is a second interference filter IF_2. Its peak wavelength is λ_2 and half bandwidth $\Delta\lambda_2$. Typically, $\lambda_2 > \lambda_1$ and $\Delta\lambda_2 > \Delta\lambda_1$, e.g., $\lambda_1 = 488$ nm, $\lambda_2 = 514$ nm, $\Delta\lambda_1 = 4$ nm, and $\Delta\lambda_2 = 8$ nm. The object to be examined (O) is placed just behind the interference filter in the object plane Π of the microscope objective. The ground glass GG acts as a diffuser which spread the incident light over a large solid angle.

The peak wavelengths of the filters IF_1 and IF_2 are mutually separated by $\lambda_2 - \lambda_1$, and light of wavelength λ_1 that normally strikes the second filter cannot pass through it. However, as follows from Eq. (1.152), this filter can transmit rays of wavelength λ_1 that fall on it obliquely at an angle of incidence θ. This angle depends on the amount of the detuning $\lambda_2 - \lambda_1$. The last quantity is chosen so that θ is larger than the objective aperture angle σ. Consequently, only light scattered and diffracted by the object O can enter the microscope objective Ob, while the direct light misses the image-forming system; hence the same situation obtains as when the dark-field condensers described previously are used. Rays of different angles θ are produced by the diffuser GG.

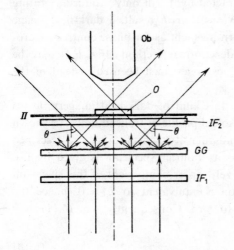

Fig. 6.10. Principle of dark-field microscopy with detuned interference filters [6.7].

As can be seen, the method is very simple. Its only defect is that by filtering the conventional light source to quasi-monochromatic light its intensity is greatly reduced, but the authors maintain that illumination is generally sufficient for visual observation. It is suggested that more intense images can be obtained if a laser source is employed instead of the conventional illuminator coupled with the interference filter IF_1. The problem of speckles (see Subsection 11.5.1), which is an attribute of highly coherent laser light, can then be overcome by continuously moving the diffuser in its plane and detecting the temporally integrated images. In the present writer's opinion, this arrangement with laser light is too expensive and would not, in general, be acceptable to professional microscopists.

The technique presented above can be classed among procedures generally known in optical literature as Fourier filtering in the object plane.

6.2.4. Central dark-field microscopy

Dark-field imagery is also achieved by spatial filtering in the Fourier plane. As was pointed out earlier (see Section 3.5), this plane is normally in the exit pupil (or back focal plane) of the microscope objective. Returning to Fig. 3.17 we can see that the Fourier spectrum of a grating consists of a series of diffraction spots ... $Q_{-2}, Q_{-1}, Q_0, Q_{+1}, Q_{+2}, ...$ The central spot or zero diffraction maximum is formed by direct (undiffracted) light. If this spot is completely screened off by means of an opaque stop, the diffracted light will only produce a grating image as in the situations described previously; as a result, a dark-field image will appear. For such imagery, the same arrangement as for phase contrast microscopy (see Figs. 5.2 or 5.6) or amplitude contrast method (Fig. 6.4) may be employed except that a completely opaque screen must be used instead of the phase or amplitude plate.

To provide a theoretical approach to this kind of dark-field imagery, let us consider an amplitude grating with a sinusoidal profile of transmittance, as shown in Fig. 6.2a. Its amplitude transmittance and intensity transmittance are described by Eqs. (6.7) and (6.8), respectively, and its Fourier spectrum is given by Eq. (6.9). The opaque screen means that the zero order component of this spectrum is completely extinguished. This situation is equivalent to $\tau_p = 0$, hence the first integral in Eq. (6.10) is also equal to zero. Consequently, Eq. (6.11) now takes the form

$$h(x) = b\cos 2\pi x, \tag{6.18}$$

and the intensity distribution in the image plane is described by

$$I'(x) = [h(x)]^2 = b^2 \cos^2 2\pi x. \tag{6.19}$$

The plots of Eqs. (6.18) and (6.19) are shown in Fig. 6.11. As can be seen, the spatial frequency in the image $I'(x)$ is doubled in comparison with that in the object (Fig. 6.2a). The fidelity between the grating and its image is thoroughly

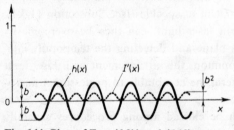

Fig. 6.11. Plots of Eqs. (6.8b) and (6.19).

upset. This situation was discussed in Subsection 3.3.1 and shown in Fig. 3.24 and is typical of sinusoidal amplitude gratings which produce only three diffraction maxima: Q_0, Q_{-1} and Q_{+1}. An entirely different and more complicated dark-field image occurs if the grating produces a greater number of diffraction maxima than three. An example is shown in Fig. Vf, where the basic frequency and the second frequency are visible in the grating image. The next interesting dark-field image is shown in Fig. VId, where only basic frequencies along x- and y-directions are visible as in the bright-field image (Fig. VIa). From these examples it reappears that the dark-field images of gratings or grating-like objects depend in the first place on the type of grating. If, however, the object is extended, i.e., has the form of a disc or strip, its central dark-field image is first of all represented by edge enhancement as shown in Figs. 6.8b and c.

The central dark-field technique is almost automatically implemented by means of variable phase contrast devices described in Section 5.7.

6.2.5. Microstrioscopy

Microstrioscopy or schlieren microscopy is achieved when all the diffraction maxima on one side of the zero order maximum are excluded from the image formation. In reference to Fig. 3.17, this microscopical technique occurs if the left-hand diffraction spots $\ldots Q_{-2}$, Q_{-1} or the right-hand spots Q_{+1}, Q_{+2}, $\ldots$ are masked. A basic consequence of such an operation is that the intensity distribution in the strioscopic image of a phase object is proportional to the gradient of the refractive index or optical path. As a matter of fact, the schlieren technique has been little used in microscopy, but it serves its purpose for macroscopical examination of different transparent media in which refractive index changes occur [6.8, 6.19–6.26]. For example, this technique is often employed to detect striae in optical glass and other homogeneous materials [6.26]. These disturbances result from local variations in the refractive index. Usually such objects do not produce diffraction spectra whose maxima of the first and higher orders are distinctly separated from the zero order maximum. Strioscopic images are therefore difficult to interpret, and there is equivocal correspondence between optical path variation and intensity distribution in the image plane.

For the interested reader, more details regarding microstrioscopy may be found in Refs. [2.13], [5.3], [6.23]–[6.26]. Some authors, however, confuse the strioscopic method with central dark-ground microscopy. In the present writer's opinion, the most appropriate nomenclature for these and other microscopical techniques is given by Zernike in Ref. [5.3].

6.2.6. *Ultramicroscopy*

Ultramicroscopy is a particular case of oblique dark-field microscopy when a specimen is illuminated by a powerful beam of light in a direction at right angles to the optic axis of the microscope objective. Ultramicroscopical techniques is directed towards the detection and observation of very small particles (< 0.5 μm) by means of the Tyndall scattering and of their diffraction images (Airy patterns). The apparatus used for this purpose is typically the *slit ultramicroscope* of Siedentopf and Zsigmondy (Fig. 6.12).

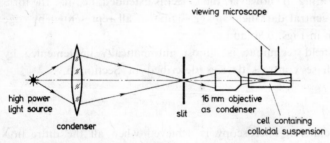

Fig. 6.12. Slit ultramicroscope.

Currently, the use of ultramicroscopy is rare and restricted to highly specialized laboratories [2.89] although in the past the technique was more popular, especially in the study of colloids in suspension, of hydrosols, and of dispersions in an oil base. Ultramicroscopy does, however, still have application in the field of microelectrophoresis. Colloidal particles differ largely in their ability to scatter light and require proper setting and focusing of the illumination beam. Under favourable conditions, particles down to 0.01 μm in size may be observed and studied by means of the slit ultramicroscope [2.12]. As was pointed out earlier (see Subsection 3.6.7), on dark ground the visibility of a small particle that scatters light is merely a matter of luminous flux emitted by the particle. Therefore, the smaller the particle size, the more intense must be the light beam focused on it.

It is worth recalling that in the early twentieth century the ultramicroscopical technique was successfully applied to the study of such phenomena as Brownian movement.

It is also worth while to note that illumination typical for ultramicroscopy was applied to long-exposure photomicrography of living biological microobjects and their movements. This procedure was first used by S. Dryl, a Polish cell biologist (see: *Bull. Acad. Polon. Sci., Ser. Biol. Sci.*, Section 2, **6** (1958), 429–430).

Figure 6.13 shows an arrangement of this kind improved recently by Hausmann [6.18], who used optical fibres for illuminating the specimen at right angles to the optic axis of the microscope objective.

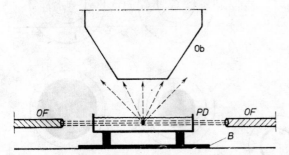

Fig. 6.13. Dark-field arrangement used by Hausmann [6.18] for taking long-exposure photomicrographs of moving biological objects; Ob—low-power objective, *PD*—Petri dish, *B*—dark background, *OF*—optical (light guiding) fibres.

6.3. Optical staining

In cytology and histology, biochemical stains are widely used for examining cells and tissues by means of bright-field microscopes. The purpose of staining is to create one or more colours to differentiate between chemically different parts of a biological object as well as between the object and its surrounding medium. The kind of staining falls outside the scope of this book and therefore its discussion will completely be omitted.[3]

There is, however, a class of methods which permit the direct examination of specimens by using merely optical means to colour their images. These will be discussed below; however, methods that use polarized light and interference phenomena of white light are excluded from this chapter.

6.3.1. Rheinberg illumination

A clever modification and extension of oblique dark-field illumination was suggested by Rheinberg in 1896, and is called *Rheinberg illumination* or *differential colour illumination*. In this method, a filter consisting of two complementary colours, C_1 and C_2 (Fig. 6.14), is placed in the first focal plane of a microscope

---

[3] For the interested reader, a comprehensive review of biochemical staining may be found in Ref. [2.12].

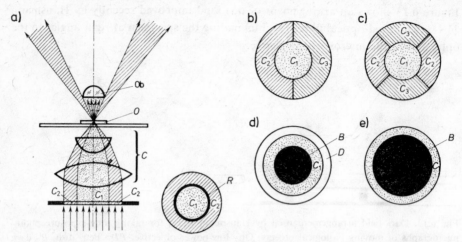

Fig. 6.14. Principle of Rheinberg illumination (a) and filters employed in substage condenser for modifying this type of illumination: b) and c)—multisector filters, d) and e)—annular filters for spectral-Rheinberg contrast according to Taylor [6.28]; C_1, C_2, and C_3—areas of different colours, B—opaque circle, D—clear annulus transmitting white light. The diagrams b to e show the condenser aperture seen in the objective exit pupil.

condenser C.[4] The component C_1 is a central circle, which is surrounded by the annular-shaped component C_2. The inner diameter of the latter is somewhat greater than $2f_c' A$, where f_c' is the focal length of condenser and A is the numerical aperture of the objective. Usually one takes

$$2r_2 = 1.1 \times 2f_c' A \tag{6.20}$$

as in oblique dark-ground illumination obtained by means of a bright-field condenser equipped with a central stop. The diameter $2r_1$ of the central circle C_1 can be varied by placing narrow opaque rings R to separate this circle from the outer annulus C_2. From C_2 the microscope objective Ob accepts only the light diffracted by the object under examination, while from C_1 it accepts both direct rays and diffracted light. However, the intensity of the latter is weak because the central circle C_1 is chosen so as to be darker than the outer annulus C_2. Thus the object O will have primarily the colour C_2 on a background of colour C_1. Effective combinations of colours include: violet circle—red annulus, blue circle—yellow-orange annulus, green circle—red annulus, and vice versa. Visually effective

---

[4] Originally, Rheinberg had used coloured gels so that object and background could be observed in different colours.

images are also obtained by using a clear uncoloured annulus with a coloured circle. In this case, a transparent object will be seen as a white image on a coloured background. Particularly striking and aesthetically beautiful images can be produced when diatoms are used as objects [6.27].

The arrangement as shown in Fig. 6.14 is the standard one, but several modifications are also in use. For instance, the outer annulus can be divided into two or even four sectors of alternating colours as shown in Figs. 6.14b and c. The sector colour filters are especially useful for the study of diatoms, crystal faces, warp-woof fabrics, and wood sections whose length-width dimensions may be displayed in contrasting colours. Another modification—the spectral Rheinberg system developed by Taylor [6.28]—is shown in Figs. 6.14c and d, which represent the portion of condenser aperture covered by the objective aperture, while the condenser aperture zone outside the objective aperture may be left clear or coloured.

During the late 1930s, a special condenser for Rheinberg illumination, the *Mikropolychromar*, was commercially available from Carl Zeiss Jena. At the present time, this condenser or other similar devices are no longer obtainable, but microscopists interested in Rheinberg illumination can quite easily adapt any bright-field condenser to this technique. Circular and annular or even sector filters can be constructed quite simply by using coloured acetate foil. In order to determine the correct inner diameter $2r_2$ of the annulus C_2 (Fig. 6.14a), it is sufficient to set up the microscope properly by focusing on a transparent specimen, remove the ocular, and, by observing the objective exit pupil, set the condenser iris diaphragm so as to make its image coincide with the edge of the pupil; the diameter of the iris opening is the minimum value of the diameter $2r_2$ for the annulus C_2.

Rheinberg illumination is suitable for objectives of magnification from $5\times$ to $100\times$. However, the numerical aperture of high-power objectives must be reduced via an objective iris or funnel stop.

In general, Rheinberg illumination is a useful accessory for those who are interested in studying the warp and weft of textiles, fibres, protozoa, insects, crystals, wood sections, and other low-contrast objects.

6.3.2. Double illumination

Another old and useful method of two-colour illumination, similar to the Rheinberg technique, is the double illumination method proposed by MacConail [6.28], who used both transmitted and reflected light of different colours, preferably

complementary, for producing optical staining. The light transmitted by a specimen under examination determines the background colour of the field of view,
while the colour of the object image is mainly produced by the reflected light.
In general, it is helpful to fix the intensity of the transmitted light at a level comparable with or slightly lower than that of the reflected light. The advantage
of this technique is that the condenser can be diaphragmed to satisfy Eq. (3.103),
so that colour contrast may be increased by reducing the glare associated with
the Rheinberg illumination.

A modification of the MacConail method, suggested by Wilson [6.29], is
shown in Fig. 6.15. A typical microscope incorporates additionally a sub-condenser disc D with an off-centred hole, an objective mirror (reflector) M, and
a semicircular stop S placed behind the objective Ob. The main functions of the
sub-condenser disc are: (1) to reduce the amount of light entering into the con-

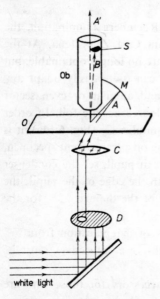

Fig. 6.15. Double-illumination microscopy [6.29].

denser C, and (2) to guide this quantity of light at such an angle so that oblique
dark-field illumination results (without the mirror M and semicircular stop S
in place). The only function of the reflector M, which is attached to the objective
Ob, is to reflect the direct rays passing through the specimen O and missing the
objective. These rays are denoted by A; they reflect at an angle that allows them
to illuminate the object under observation. A part of these rays is diffracted by

the object and enters the objective. Other rays, denoted by B, have an incidence angle different from that of rays A and these produce an image like that observed in the oblique dark-field method. The direct (undiffracted) rays from the beam B are blocked by an opaque stop S. This arrangement is particularly suitable for observing diatoms in colour.

6.3.3. Dispersion staining

The method now known as dispersion staining was first described by Crossmon [6.30], refined by Schmidt [6.31] and Cherkasov [6.32], and led to a routine technique by McCrone and his co-workers [6.33–6.46]. This consisted in providing a simple attachment [6.43] for the conventional polarizing microscope[5] and in publishing accurate dispersion staining data on a large number of inorganic substances in a useable form, and in offering courses of instruction in this technique.

Roughly speaking, the dispersion staining method is based on the *Christiansen effect*, which reveals itself as characteristic colours on the borders of colourless microscopic particles illuminated with white light. The colouration is caused by the different dispersion curves (see Fig. 1.39) of the particle and its surrounding medium. The variation of the refractive index according to the wavelength of light is a specific identifying feature of objects and is being increasingly used in a great variety of problems concerning the identification of microscopic particles. Thus dispersion staining is not only a qualitative but also quantitative or, at least, semiquantitative technique of light microscopy.

Normally, the Christiansen effects is invisible and in order to make use of it some microscopical accessories are required; the most useful of them will be presented below.

Oblique dark-field dispersion staining. This is the oldest method of producing dispersion staining [6.30]. A specimen to be examined is mounted in a high dispersion liquid whose refractive index is equal to that of the specimen only for a particular, preferably yellow or green wavelength in the visible spectrum (Fig. 6.16). This wavelength will be called the *matching wavelength* and denoted by λ_m. When white light of a hollow cone of oblique dark-field illumination contains rays of wavelength λ_m, then these rays, as well as those of wavelengths slightly different from λ_m, will pass undeviated through the mount and will miss the

---

[5] This attachment is commercially available from McCrone Research Associates Ltd., 2 McCrone Mews, Belsize Lane, London NW3 5BG, England.

microscope objective (see Fig. 6.6). Conversely, rays of wavelength distinctly greater and smaller than λ_m will be deviated (refracted) and on entering the objective will produce a coloured dark-field image of the specimen. Similar effects are obtained with both symmetrical and one-sided oblique dark-field illumination.

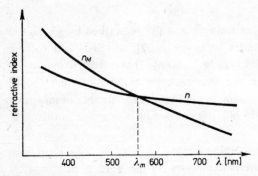

Fig. 6.16. Refractive index dispersion curves for a solid particle (graph n) and ambient liquid (graph n_M) showing a match in refractive index at light wavelength λ_m.

The dark-field system with a hollow illumination cone of light aperture produces dispersion colours which do not respond to a narrow band of wavelength around the matching wavelength λ_m corresponding to the refractive index match, and the colours are therefore less well defined and fainter. This defect is reduced in an alternative dark-field dispersion staining technique, which uses an ordinary bright-field condenser (Abbe type of numerical aperture 1.2) equipped with an annular diaphragm [6.47]. The dimensions of the clear annulus are selected so that the angular aperture of the hollow illumination cone is just larger than that of a low-power microscope objective ($10 \times /0.17$). Below the annular diaphragm there are three properly spaced circular apertures, one of which (the nearest to the light source) is that of an ordinary iris diaphragm. This three-aperture system produces a well-collimated beam, which impinges upon a clear annulus of 4 mm internal diameter and 5 mm external diameter located immediately below the condenser lens. This dispersion staining arrangement is mainly suitable for the microscopical identification of small particles of one particular type in the presence of other particles.

When correctly adjusted, this system gives excellent intensity contrast of images of small particles against a very dark background. There is an additional advantage for the annular illumination system as opposed to axial illumination with a circular condenser aperture: light passing through a small aperture allows diffraction fringes to appear. This effect cannot occur when an annular aperture is employed. Moreover, an aperture of this kind allow the microscope objective to retain the same resolving power normally obtained with a full objective aperture and with oblique illumination.

Standard technique of dispersion staining. Let us now consider Fig. 6.17, which illustrates the dispersion staining mechanism of a colourless and prismatically truncated particle. The particle is immersed in a layer of high-dispersion liquid enclosed between a microscope slide and a cover slip. A parallel beam of white light from a bright-field condenser is directed upwards to the specimen. The

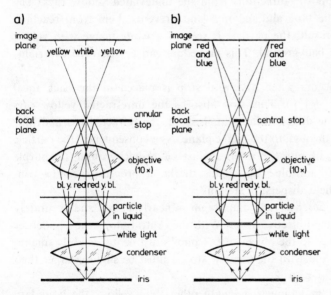

Fig. 6.17. Schematic representation of dispersion staining microscopy (according to Ref. [6.44]); three colours only are taken into consideration (yellow, red, and blue): a) with annular stop, b) with central stop.

condenser is fitted with an iris whose the most convenient position is at the front focal plane of condenser. The objective, preferably of magnification 10×, incorporates an opaque screen with central hole (Fig. 6.17a), or an opaque central disc (Fig. 6.17b). For the sake of brevity, these screens will be called *annular stop* and *central stop*, respectively. They are placed alternately in the back focal plane of objective. The particle-liquid mount is optically homogeneous only for λ_m, whereas for wavelengths on either side of λ_m optical discontinuities occur between the particle and liquid. Suppose λ_m is a yellow wavelength. At shorter wavelengths (blue), the refractive index of the liquid exceeds that of the particle; at longer wavelengths (red), the converse is true. As a result, light of wavelength λ_m passes through the specimen undeviated, whereas light of blue and red wavelengths, λ_b and λ_r, incident at any angle other than normal to a particle surface, is deviated in proportion to the magnitude of the refractive index difference between the particle and the liquid at λ_b or λ_r.

Under these circumstances, the annular stop (Fig. 6.17a) allows the yellow unrefracted rays to pass through to the image plane of microscope objective. Because of the prism-like shape of the particle, the mismatching blue and red rays are refracted in opposite directions from the undeviated yellow rays. The annular stop absorbs the blue and red rays and prevents them from reaching the image plane. As a result, the particle is seen as a white image with yellow borders against a white background. This procedure can be regarded as brightfield dispersion staining.

Another situation occurs when a central stop is placed in the back focal plane of objective (Fig. 6.17b). This stop absorbs the unrefracted yellow rays from the particle and the direct white light. Instead, the refracted red and blue rays are allowed to pass through to the image plane. As a consequence, the particle is seen as a dark image with purple borders against a dark background (the purple colour is a mixture of red and blue rays). Thus, the latter procedure can be considered as central dark-field dispersion staining.

In both cases, the condenser iris diaphragm is nearly closed and accurately centred with respect to the objective stops. With the central stop, the condenser diaphragm image observed in the objective exit pupil must be just slightly smaller than the central stop, while with the annular stop it must be slightly larger than the opening of the annular stop.

It is clear that for a matching wavelength other than yellow, the boundary colours of the particle image are different from those mentioned above (see Table 6.1). If the refractive index values of the particle and liquid are too far apart to match for a wavelength from the middle of the visible spectrum, no dispersion colours will be seen. If the particle is anisotropic and transilluminated with polarized light, different dispersion colours are obtained, depending on the orientation of the particle.

The most suitable liquids for typical dispersion staining procedures and for general identification of microscopic particles are the Cargille sets of certified high dispersion refractive index liquids (HD-liquids). For most work, the HD-1/2

TABLE 6.1

Dispersion staining colours [6.44]

Spectral colour (according to [6.48])	λ range [nm]	λ mean* $= \lambda_m$ [nm]	Central stop colours "A" light source (2856 K)	
			λ [nm]	Colour
Violet	400–450	440	579	Yellow
Blue	450–480	470	581	Yellow
Blue-green	480–510	495	594	Orange
Green	510–550	530	530	Red-purple
Green-yellow	550–570	560	560	Purple
Yellow	570–590	580	460	Blue
Orange	590–630	605	500	Blue-green
Red	630–700	650	503	Blue-green

* Corresponds to λ_m in the visible spectrum for the annular stop dispersion staining colours.

set is sufficient. It comprises 31 liquids ranging from 1.500 to 1.800 at intervals of 0.010 [6.44].

If reproducible dispersion colours are required, the colour temperature of the light source used must be defined. A low-voltage microscope lamp (15 to 20 W), operated slightly above the rated voltage, is preferable; it corresponds fairly well with the CIE[6] light source "A" with a colour temperature of 2854 K [1.19, 6.48].

The dispersion colours observed are most easily represented by the *CIE chromaticity diagram* (Fig. 6.18).[7] This diagram takes the form of a triangle inside which all possible chromaticities (*x*-, *y*-coordinates) are to be found, while the pure spectral colours lie on a curving boundary. The corresponding spectral wavelengths are distributed on this boundary, terminating in extreme violet ($\lambda = 380$–400 nm) and extreme red ($\lambda = 700$–780 nm). These pure (saturated) colours correspond to the matching wavelength λ_m of the annular stop dispersion staining colours. The other boundary of the chromaticity triangle is a straight line which joins the extreme violet and red spectral colours. This line is the locus of a family of pure purple colours. As these last colours are a mixture of the extreme violet and red, they cannot be described by wavelength values similarly to the pure spectral colours. Instead, they are represented as complementary wavelengths by the symbol $\overline{\lambda}$. The purple line is divided into three equal-length

[6] Commission Internationale de l'Éclairage.

[7] Please refer to [1.19], [6.48] for details.

parts, viz. violet-purple, purple, and red-purple. By joining the two dividing points to the locus of chromacity of the CIE light source "A" and projecting them on the curving boundary of the chromaticity triangle, their identity is determined by the complementary wavelengths $\bar{\lambda} = \overline{575}$ and $\lambda = \overline{555}$.

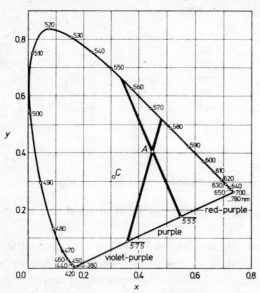

Fig. 6.18. The CIE chromaticity diagram.

The presentation of dispersion staining data by plotting direct dispersion curves, as shown in Fig. 6.16, has certain shortcomings. A more useful way of presenting these data is by the *dispersion staining graph* (Fig. 6.19), in which the refractive indices n_D of the matching liquids and the matching wavelengths are plotted, respectively, along the vertical (y) and horizontal (x) coordinates, while the wavelength scale is that used in the Hartmann dispersion formula (see Eq. (1.71b)):

$$\lambda = 200 + \frac{A}{H}, \tag{6.21}$$

where λ is expressed in nm, A is a constant, and H is the distance (in mm) along the scale corresponding to a given value of λ. For example, a value $A = 40,000$ is accepted for presenting the dispersion staining data in the McCrone Dispersion Staining Charts for Commercial Asbestos Minerals.[8]

[8] These charts are available from McCrone Research Associates Ltd., Belsize Lane, London.

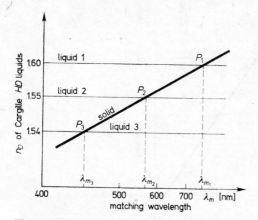

Fig. 6.19. Points P_1, P_2, and P_3 resulting from intersections of the skew and horizontal lines indicate the match wavelengths λ_{m_1}, λ_{m_2}, and λ_{m_3}, for which refractive indices of the Cargille liquids and that of a solid particle are identical.

In this way, the dispersion staining curves of the majority of solid materials are almost invariably straight inclined lines, with the Cargille liquids plotting as horizontal straight lines. The intersection points of the horizontal and inclined lines directly determine the matching wavelengths in each liquid (identified by its n_D as y-ordinate). In illustration, Fig. 6.20 shows Hartmann plots for one of the fibres measured by the dispersion staining technique.

There are numerous practical applications of the dispersion staining technique, especially that developed by McCrone. They include industrial hygiene and air pollution [6.38, 6.40, 6.46, 6.49–6.54], natural and man-made fibres and polymer films [6.55, 6.39, 6.56, 6.63], biological tissues [6.35, 6.57–6.59], and forensic examinations [6.45, 6.54, 6.60–6.64]. The main uses of this technique are in the determination of refractive indices, identification of solids, and optical staining of sections of fixed tissues to give selective colouration.

Microstrioscope as dispersion staining microscope. In white light, each microstrioscope displays an optical path gradient as non-symmetrical variations of colours in the image plane. Returning to Fig. 6.17, strioscopic dispersion staining is achieved when instead of an annular or central stop an opaque screen is used, which masks only one side of the objective exit pupil. In this case, unrefracted yellow rays and refracted rays—blue from one side of the particle and red from the other—pass through to the image plane. As a consequence, the particle is perceived as a white image whose opposite borders are non-symmetrically coloured

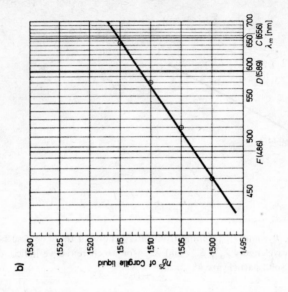

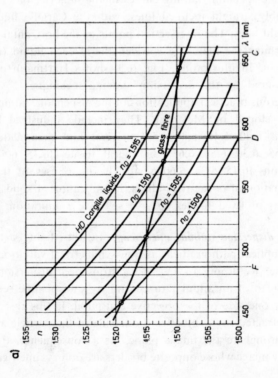

Fig. 6.20. Normal (a) and Hartman (b) plots for a glass fibre.

and seen against a white background. This is *strioscopic dispersion staining* with bright-field background, while strioscopic dispersion staining with dark-field background occurs when the screen prevents unrefracted rays from reaching the image plane of the microscope. In the latter instance, one should rather speak of pseudostrioscopy; the particle appears as a dark image with red and blue opposite borders.

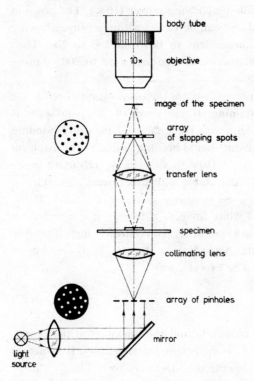

Fig. 6.21. Schematic diagram of a dispersion staining and strioscopic microscope [6.24].

An interesting system for both microstrioscopy and dispersion staining was proposed by Dodd [6.23] and next improved by McCrone [6.24]. This is a typical microscope whose usual substage condenser has been replaced by a strioscopy assembly (Fig. 6.21). A mirror reflects light from any convenient light source to illuminate a random array of about a hundred pinholes of diameter $d \approx 10$ μm. These pinholes are imaged with a collimating lens, in combination with a transfer lens, in the plane of the microscope stage. In this plane, is placed an array of stopping spots. The last array is a negative of the pinhole array. Their preparation is a sophisticated procedure described in detail by Dodd and McCrone [6.23, 6.24]. Above all, it is important to obtain stopping stops of proper diameter, which must be only slightly larger than the diameter of the images of the corresponding pinholes. The specimen under study is placed between the collimating lens and the transfer lens. The latter forms the specimen image in the object plane of a microscope objective of low magnifying power ($10\times$). The final image is observed through an ocular of middle magnifying power ($10\times$). The position of the specimen is not rigorously fixed and can be varied within a range allowing the magnification achieved with a transfer lens to be from $1\times$ to $2\times$. Thus the total magnifying power of the microscope can be increased to $200\times$, using a $10\times$ objective and $10\times$ ocular.

This arrangement can be used simultaneously for the observation of refractive index gradients and for dispersion staining. If the array of stopping spots is adjusted so that the spots are coincident with the images of the corresponding pinholes, then any refractive index gradient should become visible by the displacement of the images of the pinholes. In McCrone's view [6.24], refractive index variations in the 10^{-6} range should be detectable with this system although the step variation is predominant if the highest sensitivity is to be realized. Usually, striae are expected to give black and white images; it is only if the refractive index gradient (in general, optical path gradient) is relatively high that the strioscopic image will show colour tints. According to Ref. [6.23], the minimum detectable refractive index gradient can be expressed as

$$\left(\frac{\mathrm{d}n}{\mathrm{d}x}\right)_{\min} = 1.22\lambda\,\frac{ne}{at}, \tag{6.22}$$

where n and t are the local index of refraction and the thickness of the object, λ is the wavelength of the light used, a the transfer lens diameter, and e the fraction of the obscured light spot (pinhole image) that is necessary to produce a visible field. Taking, for example, a transfer lens of diameter $a = 2.5$ cm, and putting $n = 1.52$, $t = 0.1$ cm, $\lambda = 5.5\times10^{-5}$ cm, and $e = 0.1$, the minimum detectable refractive index gradient will be found to be $(\mathrm{d}n/\mathrm{d}x)_{\min} = 4\times10^{-5}/$cm.

Compared to the standard dispersion staining technique, annular stop dispersion colours are observed with unregistered pinholes and stopping spots, whereas central stop dispersion colours occur when the pinholes and stopping spots are mutually coincident. The standard dispersion staining technique utilizes a single aperture in the front focal plane of the microscope condenser and a single central stop or annular stop in the objective back focal plane. The smaller the condenser aperture and objective stop, the more sensitive the method. The system with an array of very small apertures (pinholes) is thus a very sensitive dispersion staining instrument. An unusual feature of this system is the presence of additional colouring in the centres of thicker particles. In McCrone's view, these additional colours are due neither to strioscopic effect nor to dispersion staining but must be considered as interference colours. It is, however, doubtful whether these colours play any useful role in dispersion staining procedure.

Dispersion staining systems with easily accessible Fourier plane. The dispersion staining technique and microstrioscopy both depend on filtering the Fourier spectrum. The filtering operation takes place in the Fourier plane, which is normally found at the back focal plane of the microscope objective. This is an awkward place to insert central or annular stops and other spatial filters. In consequence, only low-power objectives are commercially available for dispersion staining; high-power objectives cannot easily be adapted to this purpose because their back focal plane is inside their lenses. This limitation is unfortunate since it precludes the full potentiality of the microscope as far resolving power is concerned.

However, no problem exists if the microscope contains a relay system for the intermediate imaging of the objective exit pupils. Such an instrument is quite optimal for dispersion staining as well as for other techniques based on filtering the Fourier spectrum.

Today microscopes with an easily accessible intermadiate exit pupil of objectives are more and more widely available. However, these instrument are expensive and therefore an alternative system proposed by Dodd and McCrone [6.25] may also be found useful. The system operates by filtering the Fourier spectrum in the microscope exit pupil (this pupil is also known as the Ramsden disc or eyepoint of ocular) so that the original ocular functions as a transfer (relay) lens. However, two accommodations must be made to the new location of the dispersion staining stops: an auxiliary ocular system must be used to generate a secondary eyepoint and because of the demagnification of the Fourier spectrum in the primary eyepoint the size of the dispersion staining stops must be proportionally reduced. This reduction is, indeed, a defect of the system described by Dodd and McCrone in Ref. [6.25].

Dispersion staining combined with phase contrast. It was stated earlier that conventional dark-field illumination gives coloured images due to dispersion staining. Similarly, phase contrast systems are able to produce dispersion colours when the relation between the refractive index dispersions of a phase object and its surrounding medium is as shown in Fig. 6.16. This fact was first described by Schmidt [6.31]. Recently, the present writer noticed that the highly sensitive phase contrast device KFA (see Subsection 5.5.2) was capable of giving effective dispersion colours due to the high absorptance (96–97%) of its phase ring. Such an opaque phase ring acts simultaneously as a central stop in the standard dispersion staining system. Some examples of phase contrast images coloured due to the dispersion staining capability of this device are shown in Fig. VIII.

Variable phase contrast systems presented in Sections 5.7 and 5.8 are also suitable for dispersion staining when their optical components are adjusted so that a central dark field (or very near dark field) appears.

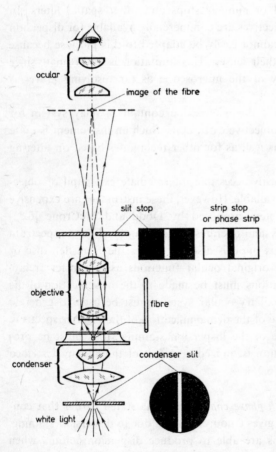

Fig. 6.22. Illustrating the dispersion staining of fibres using slit illumination.

Dispersion staining system with slit condenser aperture. Fibres and other elongated objects are particularly suitable for study by dispersion staining techniques. Staining efficiency is however improved if a slit condenser diaphragm is used in the front focal plane of the bright-field condenser, instead of a circular or annular aperture (Fig. 6.22). Consequently, a slit stop or opaque strip must be employed in the back focal plane of an objective. The condenser slit, fibre under examination, and objective stops should obviously be parallel to each other. If the condenser slit is sufficiently narrow, the sensitivity of the method is high. Moreover, it is easy to apply condenser and objective slits and objective opaque stops with variable width, which can be properly adjusted to the actual object.

The arrangement shown in Fig. 6.22 was attached to an inverted research microscope developed in the author's laboratory. A highly sensitive negative phase strip was additionally employed. As this microscope has a relay system, which projects the exit pupil of objectives at an easily accessible plane, all objective spatial filters (slit stop, strip stop, and phase strip) can be quickly exchanged and accurately adjusted to the condenser slit. It was stated earlier that dispersion colours depend on the position of the fibre under examination. An off-axis position produces asymmetrical dispersion colours although the condenser slit and objective filter are on the optic axis of the microscope. Almost identical phenomena occur when the fibre is axially positioned while the condenser slit and objective filter are conjugately decentred.

Dispersion staining as a refractometric tool. It was stated earlier that dispersion staining allows the refractive index of solid materials to be determined by matching the refraction of an unknown particle with that of a known immersion liquid. Thus, this technique must be classed with immersion methods among which the most popular is the *Becke method* (Fig. 6.23). Determination of the refractive index by this method is based on the fact that a colourless particle immersed in a liquid of the same refractive index is invisible. Otherwise, if the refractive indices of the particle and of the liquid are different, a bright *Becke line* surrounds the particle observed in monochromatic *axial illumination*.[9] The Becke line arises from refraction and total reflection of light at the boundary of the denser and rarer medium. This line is a narrow bright halo which borders the object image, and is usually more distinctly noticeable when the object is limited by approximately vertical planes. The main feature of the Becke line

---

[9] Ilumination with a full cone of light of low convergence is often called *axial illumination*. Usually it is obtained by closing the iris diaphragm aperture of the condenser.

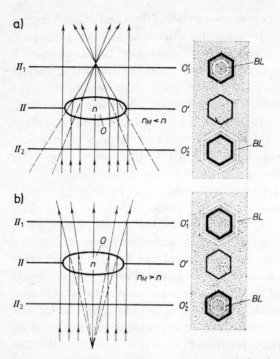

Fig. 6.23. Illustrating the Becke method: a) the refractive index n of a microobject O is higher than that (n_M) of a surrounding medium, b) $n < n_M$; Π, Π_1, and Π_2 are the planes on which the microscope is focused and O', O'_1, and O'_2 are the respective images of the object O observed through the microscope, BL is the Becke line.

is that it moves towards the optically less dense medium when the microscope is focused downwards (the distance between the specimen and the microscope objective decreases), and conversely, that it moves towards the optically denser medium when the microscope is focused upwards (the distance between the specimen and the microscope objective increases). As the refractive indices of the two media approximate, the Becke lines become less distinct until they completely disappear when both refractive indices are identical. This phenomenon provides a very sensitive gauge and in favourable circumstances difference of refractive indices of the order ± 0.001 to ± 0.0001 can be perceived [2.89, 5.80]. A drawback is the fact that a series of neutral liquids of known refractive index is required. A range of pure liquids such as listed in Table 2.5 is valuable for making different compositions of desired differences in refractive indices between successive mixtures whose components are of equal volatility. The refractive index (n_{mix}) of the mixture is approximately a linear function of composition, i.e.,

$$n_{mix} = \frac{n_1 V_1 + n_2 V_2}{V_1 + V_2}, \tag{6.23}$$

where n_1, n_2 and V_1, V_2 are the refractive indices and the volumes of the components, respectively. To obtain accuracy to the fourth decimal place for the refractive index of liquids, a temperature control of within 0.2°C is necessary. A large selection of liquids especially intended for immersion microrefractometry is offered by Cargille Laboratories, Cedar Grove, USA.

It is important to note that the Becke lines are mainly produced by the surface layer of the object examined. Many objects, e.g., textile fibres, possess a skin whose thickness is of the order of 1 μm. In such cases, the refractive index of this skin is measured and not that of the whole fibre.

If an object and an immersion liquid have the same refractive indices for yellow light but different dispersion curves (Fig. 6.16), then in polychromatic illumination the Christiansen effect occurs and the Becke lines appear red and blue. This colouration cannot be considered as a pure dispersion staining effect discussed previously. The Becke line is a halo, whereas the colouration produced by a dispersion staining system belongs to the object image itself.

A disadvantage of the Becke method is the repeated intermittent movement of the fine-focusing control. This procedure is necessary in order to estimate the refractive index match point. On the other hand, no defocusing movements are required if any dispersion staining technique is used, and the object under examination remains visible at the match point. The other advantage of the dispersion staining method is that a preparation of several immersion liquids imparts diagnostic colours only to particles of the specific substances to be examined. Other particles are either not optically stained or they appear in other colours. Sometimes values of the refractive index for different wavelengths, e.g., n_D, n_F, and n_C, are necessary to yield dispersion curves $dn/d\lambda$. The Becke method requires the use of a monochromator, whereas the dispersion staining technique yields the same values without this complication. In conclusion, the dispersion staining technique is more practical and more universal refractometric tool than the classical immersion method based on Becke lines, while the degree of accuracy offered by both methods is almost identical (routinely ± 0.001 for refractive index measurements).

In so far as the subject matter discussed above is important to microscopists, it is also worth mentioning the Jones procedure, which makes use of the white light spectrum projected into the object plane (or image plane) of the microscope

[6.65, 6.66]. The procedure depends on finding the wavelength within the visible spectrum where the Becke line disappears and requires the use of a wedge interference filter, or a grating or prism monochromator, with a short spectrum. In general, this procedure is more complicated than the classical Becke method or dispersion staining technique, but it may occasionally facilitate the refractive index determination by immersion methods.

It must be mentioned here that phase contrast microrefractometry as described in the preceding chapter (see Subsection 5.11.3) also belongs among immersion matching methods. It yields information on differences in refractive indices via the image contrast, which is a measure of the optical path difference between a given object and the adjacent medium. Phase contrast microscopes and especially the KFA device (see Subsection 5.5.2) are suitable for very precise miceorefractometry and additionally yield information on the distribution of the refractive index, e.g., across an optical fibre [5.52–5.54].

6.3.4. Other optically stained images

In the late 1940s and early 1950s, the successful application of Zernike phase contrast microscopy was responsible for certain conceptions of colour phase contrast imagery [5.9, 5.11, 5.27]. This was achieved by using special (e.g., birefringent) annular phase plates and special devices (e.g., *Varicolor*) capable of transforming the varying optical path differences produced by phase objects into visible coloured images. Nowadays those conceptions have been almost completely forgotten. Only the Nikon interference phase contrast device is still occasionally in use (see Subsection 5.7.3).

A great variety of coloured images of unstained objects are able to produce interference systems (see Chapters 7 and 16) when one of the interference fringes is spread over the field of view and when the specimen is illuminated with white light.

As regards colour contrasting via white light interference phenomena, it is worth mentioning the Pepperhoff procedure [6.67, 6.68]. This depends on the vacuum deposition of thin interference films for contrasting the phases (grains) of metal and alloy specimens. However, this procedure has not achieved wide application as working with high vacuum evaporation chambers is a very clumsy procedure.[10] An alternative procedure was devised by Bartz[11] at E. Leitz Wetzlar.

---

[10] According to a personal letter from H.-J. Preuss (E. Leitz Wetzlar).
[11] Bartz also contributed greatly to work on the metallographic interference films [6.69].

He developed a contrasting chamber that uses a sputtering technique in controlled low pressure gas atmosphere. This chamber to colour contrasting is used with reflected-light microscopes and is particularly valuable when identifying low-contrast polished metals, ores, coal, and ceramic materials.

Important contributions to colour contrasting techniques in metallurgical microscopy have also been made by Hasson [6.70–6.72] who developed, among other things, a reagent for the colour differentiation of molybdenum grains as a function of crystal orientation. The reagent is capable of superficially oxidizing the surface to form interference films [6.70]. The interference colours permit one to characterize the crystallographic orientation and, as such, have fruitful application in the metallographic study of molybdenum and of the combinations of this metal with other refractory metals as well as of various alloys.

Contrasting colours are also produced by anisotropic objects observed in polarized white light (see Chapter 15). One type of colouration (polarization interference colours or *Michel-Lêvy colours* as listed in Table 1.8) is seen when birefringent substances are observed between crossed or parallel polarizers. If these colours are plotted as individual points on the CIE chromaticity diagram (Fig. 6.18), they are all seen to be highly unsaturated and can be regarded as various mixtures of pure spectral colours [6.73]. Another type of colouration, known as *pleochroism*, is observed when certain more or less naturally stained crystalline substances change their colour if they are variously oriented with respect to the vibration direction of incident polarized white light.

Quite different coloured images occur in fluorescence microscopy (see Chapter 9). Specimens—mainly biological organisms, cells, and tissues—are stained with fluorescent dyes (*fluorochromes*) and irradiated with short-wavelength light (ultraviolet or violet-blue). The incident light is then converted into long-wavelength light (green, yellow, or red) by the fluorescent dye that has been absorbed by the specimen.

Recently, a new technique for producing coloured images of black-and-white or colourless structures, known as *pseudocolouring* [6.74–6.76], has gained popularity. That technique depends on Fourier spectrum filtering by means of spatial colour filters. If, for instance, the Fourier plane is covered by a spatial filter consisting of several annular sectors of different colours, then the colour variation in the filtered image will be determined by the directionality of the object microstructures. Conversely, if the filter consists of several circular zones of various colours, the individual spatial frequencies will give rise to different contributions to the image colouration. Actually, this type of image colouration applies rather to primary images which have been recorded on a black-and-white transparencies suitable for spatial filtering by means of an optical diffractometer (see Chapter 17).

However, it is worth noting that in microscopy the pseudocolouring technique was preceded by a very simple arrangement devised by Levkovich and Lozovskii [6.77]. This is similar to the typical phase contrast device (compare Fig. 5.6), but contains a two-colour filter instead of the phase plate. The ring-shaped conjugate area of the exit pupil of objectives is covered by a dense red filter and the complementary area by a less dense blue filter. No phase change is introduced into the direct and diffracted rays, which are only variously spectrally filtered. The image of a colourless object is formed additively by these two-colour beams.

In their article [6.77], the authors describe in detail how they prepared the filter combination mentioned above in order to examine certain biological objects (nerve fibres) by using a colour film and a simple photographic procedure. At the beginning of the experiment, the object image was blue against a bright red background. In time the background remained the same, while the object image changed its tint according to the variation of the diffraction properties of the object under examination.

6.4. Modulation contrast microscopy

In the early 1970s, Nomarski suggested that optical differentiation could be achieved by using oblique coherent illumination and a pure amplitude spatial filter located in the exit pupil of the microscope objective [6.78]. Usually, the amplitude transmittance of this filter is an one-dimensional function such as results from the interference of two slightly tilted plane wavefronts. However, when strongly oblique illumination is used, a sine-like distribution of the filter transmittance can also be divided into several discrete levels without important loss of image quality. This latter possibility was exploited by Hoffman and Gross, who developed a very simple device—the *modulation contrast system*—which converts phase gradients into intensity variations: one gradient is seen as dark, the opposite gradient as light, whereas all non-gradient regions (flat or of uniform refractive index) appear as grey. Light intensities in the image are modulated above and below an average grey value, thus name "modulation contrast" [6.79].

6.4.1. Theoretical principles

Any conventional transmitted-light microscope for bright-field observation can be easily converted into a modulation contrast microscope (MCM) by the addition of a slit diaphragm D (Fig. 6.24) in front of the condenser C and an amplitude filter (*modulator*) M after the objective Ob in a plane optically conjugate to the condenser slit S. In this case, this conjugate plane is the Fourier plane.

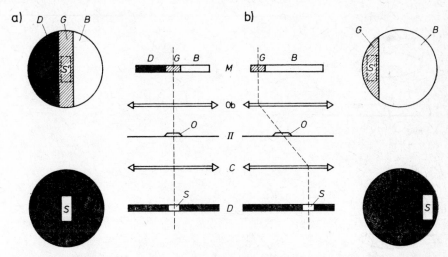

Fig. 6.24. Schematic diagram illustrating the principle of modulation contrast microscopy: a) with centred gray region of the objective exit pupil, axial illumination, b) with offset gray region, oblique illumination [6.79].

The most important element of this system is the modulator M. Typically, it consists of three areas: dark D, grey G, and bright B, of transmittance $\tau_D < \tau_G < \tau_B$. For most objects, transmittance values $\tau_B = 100\%$, $\tau_G = 15\%$, and $\tau_D = 1\%$ provide satisfactory images of a wide range of microscopic objects producing phase gradients. The luminous image S' of the slit S falls on and coincides with the grey area of the modulator. Light striking the object O (Figs. 6.24 and 6.25), where there are no phase gradients, remains undeviated and passes through the grey stripe of the modulator. This direct light is reduced to 15% of its former intensity and reaches the image plane producing a grey background and the image (O') of the zero-gradient features of the object O. Conversely, light striking the object where there are phase gradients is refracted and must pass, in whole or in part, through the dark or bright area of the modulator. In the situation as shown in Figs. 6.24 and 6.25, light reflected by the positive phase gradient passes through the dark area D, whereas that refracted by the negative phase gradient passes through the bright area B. As a consequence, the positive phase gradients are perceived as dark images and the negative ones as bright images against a grey background. This type of modulation leads to an illusion of three-dimensionality, which does nor represent the object geometry but the variation in thickness and/or refractive index of the object under examination.

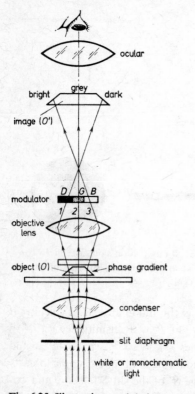

Fig. 6.25. Illustrating modulation contrast microscopy [6.79, 6.80].

The theory underlying the MCM shows that phase gradients, like spatial frequencies, are distributed over the entire exit pupil of the microscope objective. This fact is quite obvious (Fig. 6.26), but before the advent of the MCM was not previously recognized as a very useful natural agent permitting the visualization of invisible phase gradients by the use of a simple amplitude modulator. If, let us say, a plane wavefront Σ, propagating along the optic axis of objective, passes through a gradientless object O (Fig. 6.26a), the diffraction pattern occurring in the Fourier plane FP (Fig. 6.26b) has an axial maximum of light intensity Q_0. When, however, the same plane wavefront passes through a positive or negative

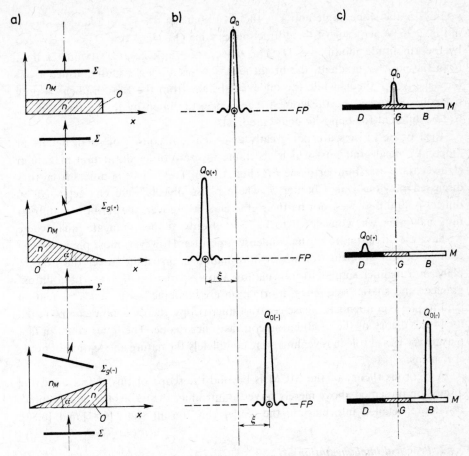

Fig. 6.26. Phase gradients (a), their localization (b), and modulation (c) in the Fourier plane.

phase gradient, it becomes a refracted (oblique) wavefront $\Sigma_{g(+)}$ or $\Sigma_{g(-)}$, which produces the same diffraction patterns $Q_{0(+)}$ or $Q_{0(-)}$ in the plane FP as does the wavefront Σ, except that this pattern is now shifted from the optic axis by a distance ζ proportional to the phase gradient γ. It is self-evident that opposite gradients shift their Fourier spectra Q_0 in opposite directions as well. Suppose the optical gradient is produced by the slope of a transparent object O as shown in Fig. 6.25. If n and n_M are, respectively, the refractive indices of the object and its ambient medium, then the phase gradient γ is defined as

$$\gamma = \frac{2\pi}{\lambda}\,(n_M - n)\,x\tan\alpha, \tag{6.24}$$

where α is the slope angle and x is the coordinate in the object plane as shown in Fig. 6.26. When each of the diffraction maxima Q_0, $Q_{0(+)}$ and $Q_{0(-)}$ is covered by the amplitude modulator M (Fig. 6.26c), the dark area D attenuates light from the positive gradient, the bright area B transmits light undiminished from the negative gradient, while the undeviated light from the zero gradient is only partially attenuated by the grey stripe. Hence, the intensity in the image plane will be different for opposite phase gradients.

Real phase objects are not greatly extended, but consist of small details of different optical gradients, which distribute the zero-order diffraction maximum Q_0 over the entire Fourier plane FP (Fig. 2.26b). This situation differs from that discussed previously in Chapter 5, where phase objects were characterized by optical path differences, not by their gradients. Moreover, the zero-order diffraction maximum was considered to be located only at the conjugate area, where we have the direct image of the condenser aperture. However, most phase objects contain optical gradient, for which information is distributed over the Fourier plane in a manner similar to information about spatial frequencies. In addition, gradient and spatial frequency information are combined. The lack of attention paid to phase gradients in phase contrast microscopy theory is now compensated for in the theory of the modulation contrast microscope. The latter could in fact play a very useful role in revealing more completely the nature and feature of phase specimens.

A rigorous theory of the MCM is beyond the scope of this chapter and the discussion presented above merely serves to introduce the results that are required. For more detailed information, the reader can consult Refs. [6.81] and [6.82].

6.4.2. Practical implementation

Returning to Fig. 6.24, it is important to note that the modulator M affects only light amplitude and does not cause phase changes. Normally, the bright B, grey G, and dark D areas of this modulator are neutral. If, however, they are differently coloured but with transmittance ratios as for the neutral modulator, then each phase gradient is imaged as before but in colours: similar gradients will have the same tint.

In general, two basic types of MCM are possible: a symmetrical system (Fig. 6.24a), where the condenser slit S and the conjugate grey stripe G are on the optic axis, and an asymmetrical system (Fig. 6.24b), where S and G are offset and the dark area of the modulator is outside the exit pupil of objective. In the former, resolution is approximately equal to λ/A (A is the numerical aperture of the objective), while in the latter it is twice as good and approaches $\lambda/2A$, the value usually obtained for oblique illumination; the asymmetrical system is thus privileged in practice.

In the symmetrical system, the dark and bright areas of the modulator are identical in size, while the grey region (G) is in the form of a narrow stripe equal in width to 0.1 of the diameter of the exit pupil of the objective. In the asymmetrical system, there is no dark area and the modulator consists of a large bright region and a small grey segment of circle. The bright region fills approximately 0.9 of the exit pupil. This system is commercially available from Modulation Optics Inc., Greenvale, N.Y., USA, as an attachment to conventional microscopes manufactured by other firms.

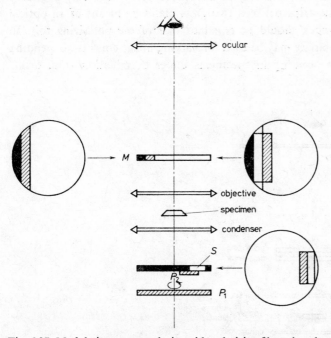

Fig. 6.27. Modulation contrast device with polarizing filters (see the text and Fig. 6.24 for explanation).

Another version of a modulation contrast attachment is schematically presented in Fig. 6.27. This is a system similar to that shown in Fig. 6.24b, but incorporates additionally two polarizers, P_1 and P_2, below the slit aperture of the condenser. The polarizer P_2 partially covers the slit S, while P_1 is rotatable round the optic axis of the condenser. When both polarizers are crossed, the effective condenser slit is narrow and its image is formed within the grey stripe of the modulator M. Otherwise, the portion of the slit S covered by the polarizer P_2 is imaged at the bright area of the modulator. Rotating the polarizer P_1 varies the spatial coherence of illumination and image contrast.

When the condenser slit is removed, all modulation contrast systems permit the use bright-field, dark-field, polarization, and fluorescence techniques [6.81]. They can be also used in combination with reflected-light microscopes.

In the present writer's opinion, it should be possible to obtain a modulation contrast device in which the grey stripe (Fig. 6.24a) or grey segment of cricle is made of a polarizing foil. The procedure for preparing such a modulator can be identical with that described previously in Subsection 5.7.4; it is only necessary to cement the polarizing stripe between two glass plates by means of an optical glue whose refractive index should be equal to that of the polarizing foil. In addition, a rotatable polarizer may be used for varying image contrast, depending on the phase gradients occuring in specimens under examination (Fig. 6.28).

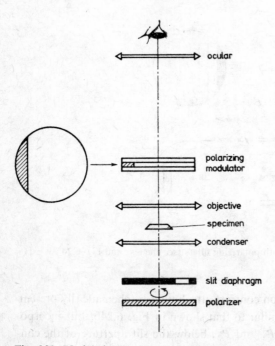

Fig. 6.28. Modulation contrast technique using polarizing foil as a modulation filter.

6.4.3. Properties and applications

The modulation contrast system enables phase gradients to be converted into intensity variations in the image. If one gradient, positive or negative, appears as darker, then the opposite gradient is brighter than an averaging background value. This kind of optical shadowing produces illusion of three-dimensionality (compare Fig. XXXIII), which represents the variations of the optical path difference but not geometric changes. Over a certain range of values the image intensity is a function of the first derivative of the optical path difference of a transparent phase object or of a geometric relief (profile) of an opaque object observed in reflected light [6.81]. The modulation contrast technique has a very short depth of field, providing optical sectioning of transparent objects wherein details above or below the plane of focus do not obscure or blur the image (see Fig. XXXIII). If stained specimens are examined, this technique can reveal more information by its gradient detection ability.

The modulation contrast microscope is extremely sensitive to optical gradients in a direction perpendicular to the condenser slit. If β denote the angle between the gradient and the normal to the edge of the grey stripe G (Fig. 6.24a), then the gradient detection ability is proportional to $\cos\beta$. Thus the user has the advantage of being able to select an orientation of the specimen to maximize the visibility of a specific gradient.

The modulation contrast microscope offers good visibility of fine details, small particles, and round edges of relatively large objects. With the grey stripe of the modulator and the condenser slit offset (Figs. 6.24b and 6.27), maximum resolution and maximum image contrast occur in directions nearly normal to the length of the condenser slit, whereas in the direction parallel to slit length the resolution is less and image contrast is minimal. Images of high contrast are produced by modifying the intensity of the direct and deflected rays[12] as they pass through the modulator. For a given magnification, the change in image intensity is directly proportional to the optical gradient and inversely proportional to

---

[12] Hoffman uses the terms "undeflected" or "undeviated" rays instead of direct or undiffracted rays [6.81] to refer to the condenser slit image which approximates to the zero order diffraction maximum, whose position in the Fourier plane of an objective is not displaced when the object is placed in the light path. However, according to Hoffman, deflected and deviated rays do not have the same meaning. In modulation contrast, "deflected" refers to the shift of the zero order intensity or slit image, while "deviated" refers only to the first and higher orders of diffracted rays; according to Hoffman, optical gradients do not deviate but deflect rays. As in general usage the words "deflected" and "deviated" are felt to be almost identical, it would appear better to stick to the term "diffracted" for diffracted rays.

slit image width and the transmittance of the modulator's grey region [6.79]. The image contrast is defined by Eq. (3.64b) and is inversely proportional to the background intensity (I_b'). Uncrossing polarizers P_1 and P_2 of the system shown in Fig. 6.27 increases the light intensity passing the bright region of the modulator M; in this case, I_b' is the sum of intensities from areas defined by the slit image on different parts of the modulator. Optimal images of gradient phase objects occur when the background intensity is approximately grey.

It is important to note that the properties discussed above are similar to those of an existing differential interference contrast (DIC) system to be described in the next chapter. The modulation contrast system may often be recommended as a substitute for the more expensive DIC system, especially in non-professional and less advanced professional microscopy.

By comparison with phase contrast microscopes, the modulation contrast microscope is generally less sensitive, but does not suffer from a halo around object images. This latter characteristic is a great advantage of the modulation contrast system. The appearance of images produced by these two systems is very different and a direct comparison of modulation contrast and phase contrast images is inadequate. Each system processes different information; the modulation contrast system above all converts phase gradients, while the phase contrast system converts optical path difference to intensity variations in the image plane.

In general, modulation contrast and phase contrast are complementary. The latter lends itself best to the observation relatively flat and thin phase objects, whereas the former is more suitable for viewing rounded phase objects.

The applications of modulation contrast microscopy are virtually unlimited and include: (1) all types of cells and tissues, and other biological objects, in live, stained, or unstained states, (2) surface and subsurfaces details of transparent materials such as crystals, plastics, and glass, (3) grain contours of opaque materials such as metallurgical and geological specimens, (4) structures of complex microelectronic circuits.

For further information concerning the properties and applications of MCM, the reader should consult Hoffman [6.81].

6.5. Monoobjective stereoscopic microscopy

A number of devices for monoobjective stereoscopic microscopy of high magnifying power should also be included among instruments discussed in this chapter. One of them was presented earlier in Section 5.9; another can be found in Ref.

[6.83]. The third, to be discussed below, consists of a versatile viewing attachment developed by Sojecki [6.84], which permits stereoscopic observation using only a typical microscope with a binocular tube. This attachment consists of a pair of special viewing systems (Fig. 6.29) positioned at the site of normal eyepieces (Fig. 6.30). They include relay lenses, L_1 and L_2, which project the primary exit pupil of microscope objectives into a secondary easily accessible plane, where a movable opaque screen S is located. This screen obstructs the left-hand or right-hand half of the secondary exit pupil. Simultaneously, the primary image

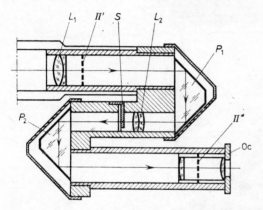

Fig. 6.29. Cross-section through one of two oculars ONS for monoobjective steroescopic observation.

plane II' of the objective is deplaced onto plane II'', which is coincident with the first focus of the ocular Oc. Two prisms, P_1 and P_2, change the direction of the light beam in order to diminish the overal size of the attachment and to preserve the original distance between the base and the exit pupil (Ramsden circle) of the microscope. The attachment does not introduce any additional magnification (its magnifying factor is $1 \times$), thus the size relations between the objective exit pupils and image planes are as $1:1$.

This stereoscopic attachment is commercially available from PZO, Warsaw. Its factory symbol is ONS. It may cooperate with any transmitted-light or reflected-light microscope equipped with a binocular tube. Depending on the screen position, a stereoscopic or pseudostereoscopic image can be obtained independently of the microscope magnification and of the technique of observation used (bright-field, dark-field, phase contrast, amplitude contrast, etc.) without change in magnification, field of view, or height of the Ramsden circle. Moreover, the ONS

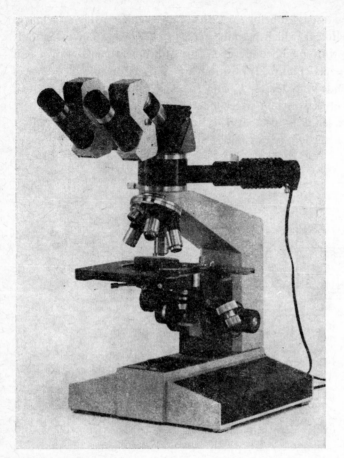

Fig. 6.30. Microscope Biolar (PZO, Warsaw) equipped with an epi-illuminator and the stereo-scopic oculars ONS.

attachment converts the inverted image in an ordinary microscope into an erect image.

The use of the ONS attachment is illustated in Fig. 6.30. A stereoscopic image is obtained when the obscuring screens (S, Fig. 6.29) mask, respectively, the left-hand and right-hand halves of the objective exit pupil imaged in the right-hand and left-hand ocular tubes. If the opposite halves are masked, a pseudostereoscopic image is obtained.

A more detailed description of the ONS system is given in an article by Göke [6.85].

One part of the ONS attachment only can, of course, be used with any monocular tube. In this case, any other diaphragm or filter (dispersion staining stop, filter for pseudocolouring, modulator for modulation contrast microscopy, etc.) can be used as required instead of the original screen for stereoscopic observation. This system is preferable to that proposed by Dodd and McCrone in Ref. [6.25] for microstrioscopy and dispersion staining microscopy since there is no demagnification of the secondary, easily accessible, exit pupil of objective.

*

* *

Finally, it is worth noting here that thanks to the immense progress that has taken place in fibre optics during the last decade, special illumination techniques have recently been extended by the introduction fibre illuminators for use in oblique dark-field microscopy [6.13, 6.18, 6.86, 6.87], Rheinberg illumination [6.86], and dispersion staining [6.88–6.90]. A fibre optics illuminator is capable, in particular, of supplying the intense narrow axial beam of Köhler illumination required for optimum dispersion staining. A fibre illumination system permits the microscopist to obtain bright, easily detectable near monochromatic dispersion staining colours of small colourless particles. As regards dark-field microscopy, fibre optics is capable of giving ring illumination without an annular diaphragm so that the maximum quantity of light energy is conveyed from the light source to the specimen. Fibre optics certainly represents an important step forward in light microscopy.

For the interested reader further details regarding the techniques of dark-field microscopy may be found in an article by P. Sebastian and R. Bock published quite recently in *Leitz-Mitt. Wiss. u. Techn.*, **9** (1987), 53–58. These authors showed that a dark-field microscope is also a useful tool for observation of the granular and fibrillar structures of specifically stained tissue sections.

7. Differential Interference Contrast

Differential interference contrast (DIC) method was introduced to microscopy by G. Nomarski (Fig. 7.1) over three decades ago [1.36, 7.1, 7.2], and today is classified among basic techniques of advanced light microscopy.

An important feature of DIC images is a relief appearance or shadow-cast effect. With transparent specimens this effect results both from the optical path gradient in the interior of objects and from their surface irregularities. To avoid misinterpretation, it should be established whether the relief appearance represents object surface topography or not. For this purpose, a combination of DIC and phase contrast is useful if unknown specimens are to be examined.

Another interesting feature of the DIC method is that it produces effective optical sectioning. This is particularly obvious when high-aperture objectives are used together with high condenser illumination apertures. The depth of field is then extremely shallow, and it is possible to obtain sharp and contrasting images of structures in a focused plane without appreciable image impairment by out-of-focus details.

The properties mentioned above resemble those of the modulation contrast system described in the preceding chapter. The DIC method was, however, invented earlier than the modulation contrast; thus the latter must be considered as a derivative of the DIC method. In general, the DIC system has better resolving power than the modulation contrast system; this is especially true of Nomarski's original system.

7.1. Background and principles of DIC microscopy

In order to understand well the DIC method, a recapitulation of basic properties of classical interferometry, with comparison to wavefront shear interferometry, will be useful. In principle, the DIC method is a particular version of shearing interferometry and can be realized using different classical interferometers.

Fig. 7.1. Georges Nomarski.

7.1.1. On the typical use of the Mach–Zehnder interferometer

Let us consider a classical Mach–Zehnder interference system (Fig. 7.2). This consists of two semi-silvered plates (S_1, S_2) and two fully-silvered mirrors (M_1, M_2), arranged symmetrically at 45° at the corners of a rectangle. The plate S_1 acts as a beam-splitter and S_2 as a beam-recombiner. A parallel light beam (B) with plane wavefront (Σ) incident upon S_1 is split into two parts: a transmitted component (B_1) and a reflected component (B_2). The former is then totally reflected from the mirror M_2 and recombined with the latter which is reflected from the mirror M_1 and passes through the plate S_2. Both beams can interfere together,

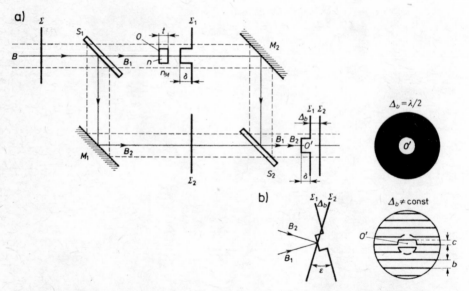

Fig. 7.2. Principle of the Mach–Zehnder interferometer (see text for explanation).

and an interference pattern is observed from E. It is self-evident that the beams B_1 and B_2 are widely separated within the interference system and this has proved a useful feature in the study of large phase objects. One beam, B_1, is made to pass through the object O to be investigated, while the other, the *reference beam B_2*, is passed through an empty space. In general, the object deforms the wavefront

Σ_1 passing through it, and the extent of deformation increases together with increases in the optical path difference δ introduced by the object. The latter quantity is defined by Eq. (2.2), in which n denotes the refractive index of the object, t is the object thickness, and n_M is the refractive index of the surrounding medium.

It can be shown that when the paths $S_1 M_2 S_2$ and $S_1 M_1 S_2$ are almost identical, a suitable adjustment of the mirrors produces a localized interference pattern, which can be uniform or fringed. The uniform pattern occurs when the *reference wavefront* Σ_2 and undeformed parts of the *object wavefront* Σ_1 are mutually parallel (Fig. 7.2a), if these wavefronts are inclined to each other at an angle ε (Fig. 7.2b), a fringe interference pattern appears.

To begin with, let us consider uniform interference and suppose, for a moment, that the object O is removed. Now, the beams B_1 and B_2 produce an interference field, whose intensity I is defined (according to Eq. (1.106)) by

$$I = a_1^2 + a_2^2 + 2a_1 a_2 \cos \psi_b, \tag{7.1}$$

where a_1 and a_2 are the amplitude of interfering light waves and ψ_b is their phase difference given by

$$\psi_b = \frac{2\pi}{\lambda} \Delta_b, \tag{7.2}$$

where λ is the wavelength of used light and Δ_b is the optical path difference between the wavefronts Σ_1 and Σ_2 (Fig. 7.2a). This last quantity is called the *bias difference of the optical path*; consequently, ψ_b is said to be the *bias phase difference*. A common name for Δ_b and ψ_b is the *bias retardation*; Δ_b is, however, expressed in length units, while ψ_b in radians or degrees of arc. If the amplitudes of two waves are equal ($a_1 = a_2 = a$), as is usually the case, Eq. (7.1) takes the form of Eq. (1.112), or

$$I = I_{max} \cos^2(\tfrac{1}{2}\psi_b), \tag{7.3}$$

where I_{max} is the maximum value of I ($I_{max} = 4a^2$). This value occurs for $\psi_b = 0$, $\pm 2\pi$, $\pm 4\pi$, ..., whereas for $\psi_b = \pm \pi$, $\pm 3\pi$, ..., the intensity $I = 0$. The solid curve in Fig. 7.3 represents the plot of Eq. (7.3).

Now, let the object O again be placed in the path of the beam B_1 (Fig. 7.2a). If $n > n_M$, the phase difference φ or optical path difference δ introduced by the object is negative ($\varphi < 0$ and $\delta < 0$); otherwise, φ and δ are positive. Figure 7.2 illustrates the situation where $\varphi < 0$. In this case, the intensity $I_{o'}$ within the interference image O' of the object O is defined by

$$I_{o'} = a_{1o'}^2 + a_2^2 + 2a_{1o'} a_2 \cos (\psi_b - \varphi). \tag{7.4}$$

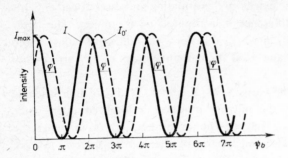

Fig. 7.3. Plots of Eqs. (7.3) and (7.5b).

If $a_1 = a_2 = a$ and the light is not attenuated by the object, then $a_{10} = a_1 = a$ and Eq. (7.4) takes the form

$$I_{o'} = 2a^2[1 + \cos(\psi_b - \varphi)] = 4a^2\cos^2[\tfrac{1}{2}(\psi_b - \varphi)], \tag{7.5a}$$

or

$$I_{o'} = I_{max}\cos^2[\tfrac{1}{2}(\psi_b - \varphi)]. \tag{7.5b}$$

The contrast C' of the interference image O' depends simultaneously on ψ_b and φ. According to Eq. (3.64b),

$$C' = \frac{I - I_{o'}}{I} = 1 - \frac{\cos^2[\tfrac{1}{2}(\psi_b - \varphi)]}{\cos^2(\tfrac{1}{2}\psi_b)}. \tag{7.6}$$

Maximum values of C' equal to unity occur when the second term of the above equation becomes zero, that means $\psi_b - \varphi = \pi, 3\pi, 5\pi, ...$; whereas minimum values of C' occur if $C' = 0$, hence the following equation must be fulfilled:

$$\cos^2[\tfrac{1}{2}(\psi_b - \varphi)] = \cos^2(\tfrac{1}{2}\psi_b).$$

A general solution of this equation is $\psi_{b(0)} = \varphi/2 + m\pi$, where $m = 0, 1, 2, 3, ...$ Note that the values $\psi_{b(0)}$ for the zero image contrast are determined by intersection points of the curves I and $I_{o'}$ in Fig. 7.3. In particular, the interference field is black if $\psi_b = \pi$ or $\Delta_b = \lambda/2$ and the image O' is bright when φ is different from zero and from $(2m+1)\pi$, as shown in Fig. 7.2a. By using a calibrated phase compensator, e.g., a two-wedge compensator, it will obviously be possible to measure the optical path difference δ. The easiest way is first to set the compensator so that the bias $\Delta_b = 0$ (interference field is brightest), and then to shift it until the bias $\Delta_b = \delta$ (image O' becomes brightest).

So far we have confined ourselves to monochromatic light, but interference also occurs when white light is used provided that the bias Δ_b is small—not larger than a few wavelengths. Varying Δ_b reveals different interference colours according to Table 1.8. Uniform or *even field interference* in white light tends to be referred to as *flat-tint interference*.

Let us now consider the situation shown in Fig. 7.2b. This comes about when one of the interferometer mirrors is set non-parallel with respect to the others. As can be seen, the bias Δ_b is not longer constant across interference field beyond the object image O', but is continuously variable from zero to a border value, which increases in direct proportion to the size of the angle ε at which the wavefronts Σ_1 and Σ_2 intersect. For these loci of the interference field where $\Delta_b = 0$, λ, 2λ, ..., bright interference fringes occur, while dark fringes pass through loci where $\Delta_b = \lambda/2$, $3\lambda/2$, $5\lambda/2$, ... The interfringe spacing b is defined by Eq. (1.114a), where $\gamma = \varepsilon$. In practice, ε is small and now the mentioned equation reduces simply to

$$b = \frac{\lambda}{\varepsilon}. \tag{7.7}$$

It is clear that b corresponds to an optical path difference $\Delta_b = \lambda$. The object image O' is seen as a local displacement (c) of the interference fringes. Measuring the interfringe spacing b and the fringe displacement c, yields the optical path difference δ from an obvious relation

$$\frac{c}{b} = \frac{\delta}{\lambda}. \tag{7.8}$$

The optical path difference $\Delta_b = 0$ determines the centre of the zero order interference fringe. This fringe is bright. It continues to be bright (achromatic) when white light is used, but fringes of the first and higher orders become coloured according to the scale of interference colours (see Table 1.8).

7.1.2. Mach-Zehnder interference system used as a wavefront shear interferometer

The design of the Leitz double-objective interference microscope (see Chapter 16) is based on principles illustrated in Fig. 7.2 and represents a classical concept of interference microscopy. A more modern configuration, known as a *wavefront shearing interferometer*, is shown in Fig. 7.4. This is, of course, the Mach–Zehnder system discussed above; the only difference being that the object O under examination is no longer in the path of one of two split light beams, but is located in the original beam (B) not yet duplicated into two components (B_1 and B_2). This apparently simple modification has some very significant consequences, especially in respect to differential interference contrast, which is the subject of the present chapter.

The wavefront shearing interferometer (in short, *shearing interferometer* or *shear interferometer*) does not produce a reference wavefront but two identical object wavefronts, Σ_1 and Σ_2, originating from a single wave surface Σ_p incident upon the object and passing through it as a locally deformed wavefront Σ. There are different types of wavefront shears: *lateral* or *transverse*, *longitudinal* or *axial*, *angular*, *radial*, etc. The most useful is lateral shearing, as shown in Fig. 7.4, and only this will be discussed here. Its basic feature is that the wavefronts Σ_1 and Σ_2 are recombined in such a way that one is laterally sheared a distance s relative to the other; thus different parts of these two wavefronts may be regarded as mutual references. Depending on the ratio of the shear (s) to the lateral size (l) of the object under examination, one obtains *complete shearing* ($s > l$), *partial shearing* ($s < l$), or *differential shearing* ($s \ll l$).

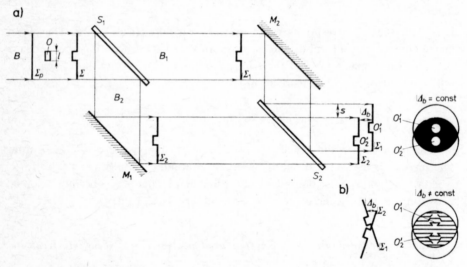

Fig. 7.4. Mach–Zehnder interferometer used as a wavefront shear interference system (see text for explanation and compare Fig. 7.2).

If the sheared wavefronts Σ_1 and Σ_2 are parallel to each other and $s > l$ (Fig. 7.4a), then two fully duplicated images, O_1' and O_2', of the object O occur against uniform field interference. Within one image, e.g., O_2', the optical path difference δ produced by the object O adds to the bias Δ_b, while within the other image (O_1') the optical path difference δ subtracts from Δ_b. Thus, the resultant optical path differences, Δ_1 and Δ_2, at O'_1 and O'_2 are respectively $\Delta_b - \delta$ and $\Delta_b + \delta$. Consequently, Eq. (7.5b) must now be written as two equations:

$$I_{o'_1} = I_{max} \cos^2 [\tfrac{1}{2}(\psi_b - \varphi)], \tag{7.9a}$$

$$I_{o'_2} = I_{max} \cos^2 [\tfrac{1}{2}(\psi_b + \varphi)], \tag{7.9b}$$

Note that this situation may be interpreted as if there were a plane reference wavefront Σ'_2 and one object wavefront Σ'_1 with two local phase deformations originating from two separated objects, of which one produces the optical path difference $+ \delta$ and the other $- \delta$ (Fig. 7.5a). When white light is used, the discrepancy in Δ_1 and Δ_2 appears as different interference colours of O'_1 and O'_2, symmetrical in relation to the colouration of the background.

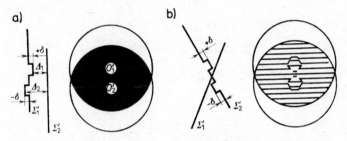

Fig. 7.5. Wavefronts producing the same interference patterns as those shown in Figs. 7.4a and b.

If the wavefronts Σ_1 and Σ_2 are inclined, a fringe interference pattern occurs (Fig. 7.4b), which also contains two object images (O'_1 and O'_2) accentuated by a displacement of fringes in opposite directions. In this case, too, the interference pattern may be considered as the result of interference of a plane reference wavefront and one object wavefront with two opposite phase deformations (Fig. 7.5b).

The set-up shown in Fig. 7.4 forms the basis of the Interphako microscope (see Subsection 5.8.1), which is manufactured by VEB Carl Zeiss Jena.

7.1.3. DIC microscopy as a method for displaying optical gradients

Let us now consider in greater detail a situation where the wavefront shear is very small (differential) relative to the lateral size of the object under examination ($s \ll l$), and assume that the interference system is adjusted at uniform-field interference. As a necessary precondition of DIC microscopy, this assumption will hold good throughout the rest of this chapter.

The most interesting DIC image appears when the object under examination has phase gradients. A small drop of immersion oil on a microscope slide constitutes such an object and resembles a planoconvex lens (Fig. 7.6). After passing through this object, a plane wavefront Σ_p takes the form of a locally curved surface Σ, which is then slightly split into two wavefronts Σ_1 and Σ_2. Let the Cartesian coordinates x, t and x, δ be associated with the same axial section of the object O and the wavefront Σ. The horizontal coordinate x is parallel to the shearing direction and expresses the lateral size of the object, while the

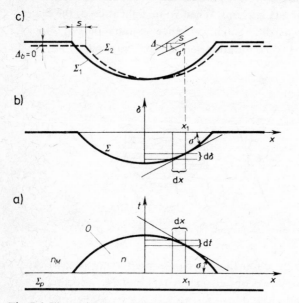

Fig. 7.6. Illustrating Eqs. (7.10)–(7.18).

vertical coordinates t and δ pass through the vertex of the object and express the object thickness and optical path difference, respectively. The last two quantities are not constant but are a function of x. A variation of x by an infinitesimal amount dx causes a change of δ by

$$d\delta = (n_M - n)\,dt, \tag{7.10}$$

where n is the refractive index of the object, n_M is that of the surrounding medium, and dt denotes the object thickness change corresponding to dx. From the geometry of Fig. 7.6a, it follows that

$$\frac{dt}{dx} = \tan\sigma. \tag{7.11}$$

Here σ is the slope of the object surface at the point x_1, where the variation $d\delta$ was considered above. By combining Eqs. (7.10) and (7.11), one obtains

$$\frac{d\delta}{dx} = (n_M - n)\tan\sigma. \tag{7.12}$$

On the other hand, from the geometry of Figs. 7.6b and 7.6c one has

$$\frac{d\delta}{dx} = \tan\sigma', \tag{7.13}$$

and

$$\tan\sigma' = \frac{\Delta}{s}, \tag{7.14}$$

where σ' is the slope of the wavefronts Σ, Σ_1, and Σ_2 at the region concerned (x_1), s is the wavefront shear, and Δ is the optical path difference between interfering wavefronts Σ_1 and Σ_2 at the region x_1 (it is assumed that the bias $\Delta_b = 0$). By combining Eqs. (7.13) and (7.14), one has

$$\Delta = s\frac{d\delta}{dx}. \tag{7.15}$$

This is a very important relation which shows that the optical path difference Δ between two infinitesimally sheared wavefronts does not express the optical path difference δ in the specimen, but is directly proportional to the derivative $d\delta/dx$, i.e., to the gradient of δ in the direction of wavefront shearing.

If the bias $\Delta_b \neq 0$, Eq. (7.15) takes the form

$$\Delta = \Delta_b + s\frac{d\delta}{dx}. \tag{7.16}$$

When the maximum value of the interference field intensity is assumed to be unity, the intensity in the region concerned is given by

$$I_{o'} = \cos^2\left[\frac{\pi}{\lambda}\left(\Delta_b + s\frac{d\delta}{dx}\right)\right]. \tag{7.17}$$

For $\Delta_b = \lambda/2$ (background is maximally dark), the image intensity $I_{o'}$ changes proportionally to the square of the derivative $d\delta/dx$ (provided $sd\delta/dx$ is small). However, for $\Delta_b = \lambda/4$, the intensity $I_{o'}$ is linearly related to $d\delta/dx$.

As in Eqs. (7.15)–(7.17) there is a differential of the optical path $(d\delta/dx)$, hence the name "differential interference" or "differential interference contrast". Note that $d\delta/dx$ can be both positive and negative. At one side (to the right) $d\delta/dx$ is positive, but at the other side of the object the derivative $d\delta/dx$ is negative. Hence in white light the interference colours of both object slopes will be different if $\Delta_b \neq 0$.

From Eqs. (7.12) and (7.15) it follows that

$$\tan\sigma = \frac{\Delta}{s(n_M - n)}. \tag{7.18}$$

By measuring Δ one can estimate the slope σ of an object if s and $n_M - n$ are known. Conversely, if σ is known, the refractive index difference $n_M - n$ may be determined. These and other measuring possibilities of differential microinterferometry will be discussed in more detail in Chapter 16.

7.1.4. DIC microscopy based on the double refracting interference system

DIC microscopy is particularly easy to undertake with the aid of a polarizing interferometer such as shown in Fig. 1.82. It is enough to place a Wollaston prism W in the back focal plane of a microscope objective Ob and a slit diaphragm D_c in the front focal plane of a condenser (Fig. 7.7a). The diaphragm D_c is preceded by a polarizer P, while the birefringent prism W is followed by an analyser A. The slit diaphragm and polarizer as well as the analyser are rotatable around the optic axis of the condenser C and objective Ob. The initial or zero orientation of the condenser slit, polarizer, and analyser with respect to the birefringent prism is defined as shown in Fig. 7.7d, where PP and AA denote the directions of light vibration in the polarizer P and analyser A, E denotes the edge of the apex angle α of the upper wedge of the Wollaston prism W, and SS is the direction of the slit S of the condenser diaphragm D_c. As can be seen, the polarizer P and analyser A are crossed and their directions of light vibration form an angle of $45°$ with the wedge edge E of the prism W, while the slit S is oriented parallel to this edge.

In these conditions, the general working principle of the DIC microscope is as follows. A plane light wave Σ_p, polarized linearly by the polarizer P, leaves the condenser C. When passing through the object under examination (O), the wave is subjected to a local phase shift corresponding to the optical path difference δ occurring in the object. The distorted wavefront Σ enters the objective Ob, and is sheared by the birefringent prism W into two wavefronts Σ_e and Σ_o, polarized at right angles. When passing through the analyser A, both wavefronts interfere with each other and make visible the transparent object O in the form of two images laterally duplicated to a greater or lesser degree, depending on the apex angle α of the prism W. The image O' is observed by means of an ocular Oc,

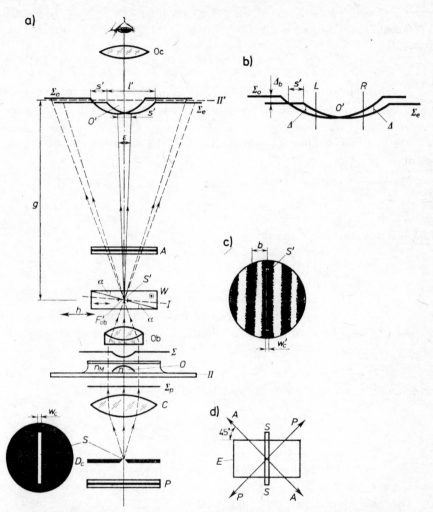

Fig. 7.7. Schematic diagram and operating principle of a transmitted-light differential interference contrast microscope using a typical Wollaston prism (W), inserted between two crossed polarizers (P and A), and a slit condenser diaphragm (D_c). The pattern shown in figure c is seen in the objective exit pupil when the ocular Oc is removed or supplemented by a Bertrand lens.

adjusted so that its first focus is coincident with the image plane Π' of the objective Ob. Uniform-field interference arises when the prism W is adjusted so as to bring its own interference fringes I in coincidence with the back focus F'_{ob} of the objective Ob (compare Fig. 1.82a and corresponding text in Subsection 1.6.5). A DIC image is obtained if the wavefront shear s' in the image plane Π' is very small with respect to the size l' of the object image O'. From the geometry of Fig. 7.7a, it results that

$$s' = \varepsilon g, \tag{7.19}$$

where g is the optical tube length of the microscope and ε is the angular shear of wavefronts Σ_e and Σ_o. The last quantity is very small (see Eq. (1.155)) and one can take $\tan \varepsilon \approx \varepsilon$.

For the sake of generality, it is convenient to use the wavefront shear s referred to the object plane Π, i.e.,

$$s = \frac{s'}{M_{ob}} = \varepsilon \frac{g}{M_{ob}}, \tag{7.20a}$$

where M_{ob} is the magnifying power of objective. By combining Eqs. (7.20a) and (2.44) and ignoring a negative sign for M_{ob}, we obtain

$$s = \varepsilon f'_{ob}. \tag{2.20b}$$

Here f'_{ob} is the focal length of objective.

For standard DIC procedure, s should be not greater than or comparable with the resolving power ϱ of the microscope objective, defined by Eq. (3.53). Thus, the following relation should be fulfilled:

$$\varepsilon f'_{ob} = \frac{0.82 \lambda}{A}. \tag{7.21}$$

By combining the above equation and Eq. (1.155), one obtains

$$\tan \alpha = \frac{0.82 \lambda}{2(n_e - n_o) A f'_{ob}}. \tag{7.22}$$

Here A denotes the numerical aperture of objective and $n_e - n_o$ is the birefringence of the material of which the Wollaston prism is made.

Equation (7.22) defines the apex angle α of the birefringent prism required in the DIC procedure. The greater the product $A f'_{ob}$, the smaller the angle α. For low-power objectives, $A f'_{ob} \approx 5$ mm, hence $\alpha = 20'$ if the Wollaston prism is made of quartz crystal; whereas for high-power objectives, $A f'_{ob} \approx 2.5$ mm, hence $\alpha = 40'$.[1] In practice, it is sufficient to use a single birefringent prism the apex angle of which is optimally adjusted to the most frequently used objective.

---

[1] Strictly speaking, standard high-power and middle-power objectives cannot be used with the interference system shown in Fig. 7.7a since the back focus F'_{ob} of these objectives is inside their optical system, and a typical Wollaston prism cannot be adjusted so as to bring its own interference fringes I into coincidence with F'_{ob}.

To obtain a good interference effect in the image plane Π' (Fig. 7.7a) and a well contrasting object image O', light beams emerging from the condenser C should be sufficiently coherent, i.e., the width w_c of the condenser slit S must be adequately narrow. For slit illumination, the coherence degree γ_{12} in the object plane Π is given by Eq. (3.11), where d is now equal to s. The modulus of γ_{12} is a number varying between 0 and 1. When it is equal to 1, light rays emerging from the condenser are perfectly coherent and give rise to ideal interference phenomenon. But satisfactory interference also occurs when $|\gamma_{12}| = 0.88$ (see Subsection 1.6.1). For this value of $|\gamma_{12}|$, the argument of function (3.11) is approximately equal to unity; hence we obtain the following condition for the width w_c of the condenser slit:

$$w_c = \frac{\lambda f_c'}{\pi s}. \tag{7.23a}$$

Taking into account Eqs. (7.20b) and (7.21), we have

$$w_c = \frac{f_c' A}{0.82\,\pi}. \tag{7.23b}$$

As can be seen, w_c increases in proportion to the condenser focal length f_c' and objective numerical aperture A. If, e.g., $A = 0.15$ and $f_c' = 10$ mm, then $w_c = 0.6$ mm.

In practice, the width w_c of the condenser slit should be adjusted so that its image S' (Fig. 7.7c) in the back focal plane of the objective Ob is narrower than one quarter of the spacing b between the interference fringes of the Wollaston prism observed in the objective exit pupil. The interfringe spacing b is defined by Eq. (1.156). Assuming that the width w_c' is equal to $b/4$ and taking into account Eqs. (1.156) and (7.20b), one obtains

$$w_c' = \frac{b}{4} = \frac{\lambda}{4\varepsilon} = \frac{\lambda f_{ob}'}{4s}. \tag{7.24}$$

Since $w_c' = w_c f_{ob}'/f_c'$, the above relation can be rewritten as

$$w_c = \frac{\lambda f_c'}{4s}. \tag{7.25}$$

The last expression resembles Eq. (7.23a) except for a slight difference in the denominator, which in Eq. (7.25) is 4 instead of $\pi = 3.14$.

By sliding the birefringent prism W in the transverse direction h as shown in Fig. 7.7a, the phase difference between sheared light waves (the bias Δ_b) changes. For $\Delta_b = 0$, the background and the middle of the object image O' are dark, but at slopes of O' there is a continuously variable optical path difference Δ between the sheared wavefronts Σ_e and Σ_o, defined by Eq. (7.15), and both sides of the object O are visible as bright regions (compare Fig. IXa). The intensity $I_{o'}$ of these regions, L and R (Figs. 7.7b and 7.8a), is defined by

$$I_{o'} = \sin^2 \frac{\pi \Delta}{\lambda} = \sin^2 \frac{\pi s \dfrac{d\delta}{dx}}{\lambda}. \tag{7.26}$$

This situation occurs when the interference system is adjusted so that the image S' of the condenser slit coincides with the centre of the zero order interference fringe of the Wollaston prism. In white light, this fringe is black if the polarizer P and analyser A are crossed, as shown in Fig. 7.7d. The image S' must, of course, be oriented parallel to the interference fringes of the Wollaston prism (Fig. 7.7c).

If, however, the prism W is slightly shifted from its zero position (Fig. 7.8a) to the left, then the wavefront Σ_o is advanced with respect to the wavefront Σ_e; hence a bias Δ_b is introduced and the right-hand region R of the object image O'

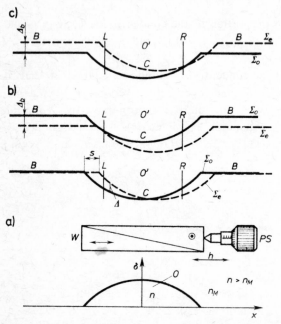

Fig. 7.8. Function of the transverse movement (h) of the Wollaston prism W in the DIC microscope, whose optical system is shown in Fig. 7.7a; PS—micrometer screw by means of which the Wollaston prism W is transversely slid (this screw will be called the phase screw).

becomes brighter (Fig. 7.8b), while the left-hand region L appears darker than the background B (compare Fig. IXb). Now, Eq. (7.26) must be rewritten as

$$I_{o'} = \sin^2 \frac{\pi\left(\varDelta_b \pm s\, \dfrac{d\delta}{dx}\right)}{\lambda}. \tag{7.27}$$

Here the signs "$+$" and "$-$" refer, respectively, to the right-hand region (R) and left-hand region (L). The intensity of the centre C of the lens-like phase object under consideration is the same as that of the background B. In general, these parts of the image where there are no optical gradients always have the same intensity as the background.

Note that the last expression is similar to Eq. (7.17) except that a cosine is now replaced by a sine function. If, however, the polarizer P is oriented parallel with respect to the analyser A (Fig. 7.7), no difference will occur between these two equations. As a parallel orientation of these two elements is less useful than a crossed position, it will not be considered in the following text.

If, conversely, the Wollaston prism W is shifted from its zero position (Fig. 7.8a) to the right, then the left-hand area L of the object image O' becomes brighter than the right-hand area (Fig. 7.8c). A further lateral movement of this prism in one or other direction causes the bias $\varDelta_b$ to increase, hence both the object image and the background become increasingly brighter, next their intensities diminish, increase, diminish, and so on, according to the square sine dependence on $\varDelta_b$. Usually, in DIC microscopy a bias $\varDelta_b$ not greater than 1.5λ is generally accepted, but for the majority of phase objects the optimal value is within the range $\lambda/30 - \lambda/10$.

In practice, especially for non-critical qualitative and descriptive studies of phase objects, the DIC method uses not monochromatic but white light, which causes all optical gradients to appear as coloured images. For a given bias retardation, the interference colours depend both on the magnitude and on the sign of the phase gradient, whereas for a given gradient, they depend on the bias retardation (compare Figs. IXb–e). When $\varDelta_b \neq 0$, the image centre of a lens-like or round object always has the same colour as the background,[2] whereas the tints of the edge regions, L and R (Figs. 7.8b, c), are symmetrical with respect to that of the background (B), i.e., the colour of one edge region is of a higher

[2] For anisotropic phase objects, the situation is different, but such objects are beyond the scope of this chapter. However, it is worth noting here that no difference occurs when the polarizer P (Fig. 7.7a) is placed behind the microscope objective Ob but ahead of the Wollaston prism W.

interference order while that of the other edge region is of a lower interference order, symmetrically to the background colour (compare Fig. IX and Table 1.8). This colour differentiation means that wavefront shapes can be estimated, which allows optical gradients to be evaluated. To arive at such an estimation, it is convenient to make the background purple by choosing the bias retardation $\Delta_b \approx 560$ nm. In this case, a small variation in optical gradients causes a rapid

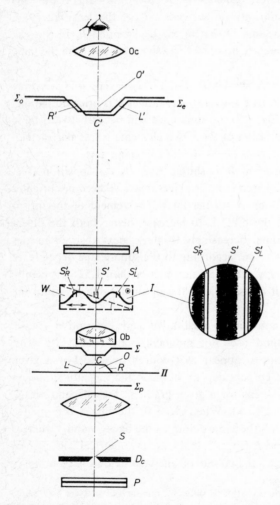

Fig. 7.9. Illustrating the interpretation of the DIC image of a gradient-like object observed through the same double refracting differential interference microscope whose optical system is shown in Fig. 7.7a (see the text and Fig. 7.7 for explanation).

change of interference colours from purple into red and blue (see Fig. IXd). For this reason, the purple colour is called the *sensitive tint*.

For most typical work, the bias retardation Δ_b is, however, maintained at low levels (about $\lambda/10$), which results in a dark grey background with a characteristic greenish tint (see Fig. IXb). It is important to note that the colour of the background and of the images of gradientless regions of the object is always that cut out by the image S' of the condenser slit S from the interference fringes I of the Wollaston prism W (Figs. 7.7a and c); opposite optical gradients, on the other hand, produce tints which are in I at opposite sides of S'. This general rule is more clearly expressed in Fig. 7.9. The central part C of a truncated phase object O and its surround produce a direct image S' of the condenser slit S, while opposite gradients of the prism-like edges of the object, L and R, displace the direct image S' from its conjugate position to deviated positions S'_L and S'_R, where there are other interference colours than at S'. Hence, the images C', R', and L' of the object regions C, R, and L are in the colours that are cut out by the slit images S', S'_R, and S'_L from the fringe interference pattern I of the Wollaston prism W.

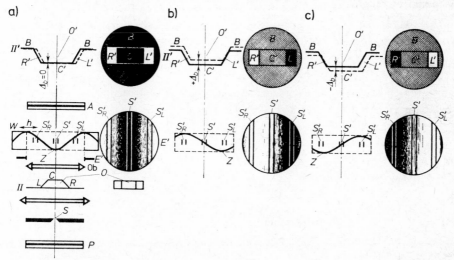

Fig. 7.10. Illustrating the discussion about the similarity between DIC microscopy and modulation contrast technique (see Section 6.4).

If the apex angle α of the prism W (Figs. 7.7 and 7.10a) is very small, then only a single fringe of the interference pattern I, e.g., the zero order fringe (Z), covers the exit pupil E' of the objective Ob. Suppose that the centre of this fringe is initially exactly coincident with the direct image S' of the condenser slit S.

Then the bias retardation $\Delta_b = 0$, and the background as well as the gradientless (central) region C of the object O appear as completely dark areas B and C', whereas the images L' and R' of the opposite object slopes, L and R, are bright. Next, let the zero order fringe Z be slightly decentred with respect to the objective exit pupil, as shown in Fig. 7.10b. In this case, the slit images S'_R and S'_L, respectively, transfer more and less light than S' to the image plane Π'. Consequently, the images R' and L' of the object slopes R and L will appear as brighter and darker regions than the image C' of the gradientless part C of the object under consideration. The relation between intensities of the image regions R' and L' can be reversed if the zero order fringe Z of the interference pattern is displaced to the opposite side of the objective exit pupil, as shown in Fig. 7.10c. This displacement is performed by transverse shifting (h) of the Wollaston prism W, and causes a change of sign of the bias retardation Δ_b.

The situation presented in Fig. 7.10b or 7.10c is quite similar to that which occurred in the modulation contrast system described previously (see Fig. 6.24a). The only difference is that now the Wollaston prism acts as a continuous amplitude filter, which modulates the transmittance (τ) across the objective exit pupil according to the square sine function, while the modulator in the Hoffman–Gross system modifies the transmittance by stages (100%, 75%, and 1% or 0% for bright, grey, and dark regions, respectively).

The feature described above is responsible for the optical shadow-casting effect, which produces a convincing illusion of tridimensionality in DIC images (compare Figs. XXXIV, XXXV, XXXVI).

A very important parameter of any DIC system is the instrumental extinction factor E. This quantity, for a double-refracting interference system such as shown in Fig. 7.7, is expressed by the ratio of intensities measured with the polars[4] parallel ($I_{||}$) and crossed ($I_\perp$), i.e.,

$$E = \frac{I_{||}}{I_\perp}. \tag{7.28}$$

If $I_\perp = 0$ ($E = \infty$), the background of the field of view of the polarizing DIC microscope is completely black for the bias retardation $\Delta_b = 0$. This ideal situation is never achieved in practice. The greatest practical limit for E is 10^6, when the optical system is specially selected, but usually this factor is not greater than 10^3. As E increases, so does the contrast of DIC images and the sensitivity of the DIC method [7.3].

---

[4] For the sake of brevity, the polarizer and analyser are frequently referred to as *polars*.

It is worth noting that Eqs. (7.17), (7.26) and (7.27) cannot be treated as a basis for accurate photometry of DIC images because these equations are only valid for an ideal optical system when $E = \infty$ and *optical leakage* $Q = 0$. The latter parameter is defined as $Q = I_{min}/I_{max}$, where I_{min} and I_{max} are the minimum and maximum observable (or recorded) intensities. For practical DIC systems, Q is always a non-zero constant for a fixed optical alignment (say, $Q = 0.015$). The problem of photometry of DIC images is discussed in more detail by Hartman *et al.* [7.4–7.6].

7.2. Nomarski DIC microscopy

Since the late 1960s, the Nomarski differential interference contrast microscopy has become increasingly popular. Today, the Nomarski DIC system is manufactured by many firms in Europe (C. Zeiss Oberkochen, VEB Carl Zeiss Jena, Nachet–Sopelem, Vickers, C. Reichert), Japan (Nikon, Olympus), and USA (American Optical Co., Bausch and Lomb). There are two basic versions of this system: for transmitted-light and for reflected-light microscopy.

7.2.1. Nomarski DIC system for transmitted light

The system shown in Fig. 7.7a incorporates a slit condenser diaphragm which reduces the condenser aperture, and therefore the resolving power of the microscope, and the effectiveness of the DIC method for optical sectioning. This defect is removed by applying two Wollaston prisms, one of which (W) is located behind the microscope objective Ob (Fig. 7.11a), while the other (Q) is placed below the condenser C. The latter is used as a compensator; it is similar to the prism W and orientated in such a manner as to cancel out the optical path differences between light wave components sheared by the birefringent prism W. The axial positions of these two prisms are adjusted so as to bring their planes of localization of interference fringes into coincidence with the front focal point F_c of the condenser C and the back focal point F'_{ob} of the objective Ob, respectively. The prism Q is preceded by a polarizer P, and W is followed by an analyser A. All these components are orientated as shown in Fig. 7.11b, where AA and PP denote the directions of light vibrations in the analyser and polarizer, E_W and E_Q are the edges of wedge angles α and α_Q of the birefringent prisms W and Q, respectively. As can be seen, the polarizer and analyser are crossed and their directions of light vibrations form an angle of 45° with the wedge edges E_W and E_Q of the birefringent prisms.

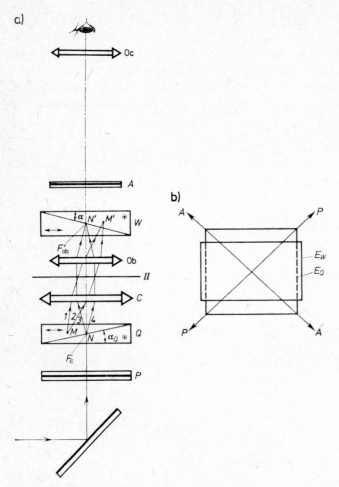

Fig. 7.11. The Nomarski DIC system using two typical Wollaston prisms.

Under these conditions, an effective interference with flat-tint field is achieved in the image plane of the objective Ob when the wedge angle α_Q of the compensator Q is fixed so to obtain infinitely enlarged interference fringes in the exit pupil of the objective Ob. To achieve this effect, referred to as *pupilar compensation*, the following condition should be fulfielled:

$$f_c \tan \alpha_Q = f'_{ob} \tan \alpha, \tag{7.29a}$$

or more generally

$$f_c \varepsilon_Q = f'_{ob} \varepsilon, \tag{7.29b}$$

where ε_Q and ε are the angular wavefront shears (expressed in radians) due to the birefringent prisms Q and W, respectively, f_c is the focal length of the condenser C, and f'_{ob} is that of the objective Ob. A derivation of Eqs. (7.29) may be found in Ref. [7.7]. The quantities ε_Q and ε are defined by Eq. (1.155).

As can be seen, the right-hand part of the relation (7.29b) expresses the transverse linear wavefront shear s related to the object plane Π (compare Eq. (7.20b)). By analogy, the left-hand part of the relation (7.29b) may be considered as a linear wavefront shear (s_Q) due to the compensator Q and related to the same object plane Π. Thus, the condition for the pupilar compensation may also be expressed as $s_Q = s$.

The compensator Q permits the use of the full condenser aperture. The result is that for each off-axis point M of the condenser aperture and its conjugate M' in the exit pupil of the objective Ob (Fig. 7.11a), the split rays 1 and 2, whose origin is at M, meet at M' and leave the prism W with the same optical path difference as the rays 3 and 4 whose origin is at the axial point N. This is why such a compensation of the optical path difference is referred to as "pupilar compensation".

The arrangement shown in Fig. 7.11 can only be used with low-power microscope objectives whose rear focal plane is localized outside their lens system. By contrast, the rear focal plane of middle-power and high-power objectives is inside their optical system and cannot be brought into coincidence with the plane

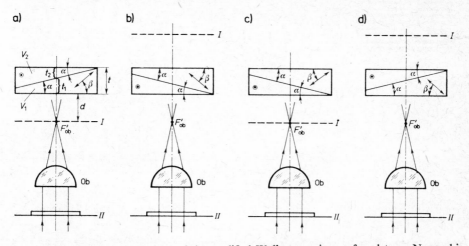

Fig. 7.12. Different configurations of the modified Wollaston prisms referred to as Nomarski prisms, whose interference fringes (I) are localized ahead of (a and c) or behind the prism (b and d); Ob—microscope objective, Π—its object plane, F'_{ob}—rear focal point of the objective.

of localization of interference fringes of a typical Wollaston prism. To overcome this difficulty, Nomarski used a birefringent prism whose interference fringes are localized outside it [1.36, 7.1, 7.2]. This modified Wollaston prism has been described in Subsection 1.6.5 (see Fig. 1.81), and its different combinations and possible uses with a microscope objective are illustrated in Fig. 7.12. Only the configurations shown in Figs. 7.12a and c are suitable for DIC microscopy.

The Nomarski prism is constructed so that its wedge components V_1 and V_2 satisfy the following condition:

$$(n'_e - n_o) t_1 = (n_e - n_o) t_2, \tag{7.30}$$

where t_1 and t_2 (Fig. 7.12a) are the axial thicknesses of the wedges V_1 and V_2, respectively; n_o and n_e are the main refractive indices (ordinary and extraordinary) of the birefringent crystal of which the prism is made, and n'_e is the extraordinary refractive index for light rays which strike normally the external surface of the wedge V_1 cut at an angle β to the optic axis of the crystal. The index n'_e is defined by Eq. (1.31), where $\xi = 90° - \beta$. The difference $n_e - n_o$ expresses the birefringence of the wedge V_2, which is cut parallel to the optic axis of the crystal, while $n'_e - n_o$ expresses the birefringence of the wedge V_1. The quantity $n'_e - n_o$ is smaller than $n_e - n_o$, hence the wedge V_1 must be thicker than V_2. Consequently, Eqs. (1.155), (1.156), (7.20b), and (7.22) now take the following forms:

$$\varepsilon = [(n'_e - n_o) + (n_e - n_o)] \tan \alpha, \tag{7.31}$$

$$b = \frac{\lambda}{\varepsilon} = \frac{\lambda}{[(n'_e - n_o) + (n_e - n_o)] \tan \alpha}, \tag{7.32}$$

$$s = \varepsilon f'_{ob} = f'_{ob}[(n'_e - n_o) + (n_e - n_o)] \tan \alpha, \tag{7.33}$$

$$\tan \alpha = \frac{0.82\lambda}{[(n'_e - n_o) + (n_e - n_o)] A f'_{ob}}. \tag{7.34}$$

For a given apex angle α and a given thickness t of the Nomarski prism, the distance d at which the interference fringes of this prism are localized depends primarily on the angle β. The distance d may be calculated from a number of equations derived from the Snell law of refraction in birefringent media. However, the calculation procedure is rather complicated (see Ref. [7.8], for example). Figure 7.13a illustrates the variations of d as a function of α for $t = 3.5$ mm and for three different values of β. On the other hand, Fig. 7.13b shows the variation of d as a function of β for three different values of α and $t = 3.5$ mm.

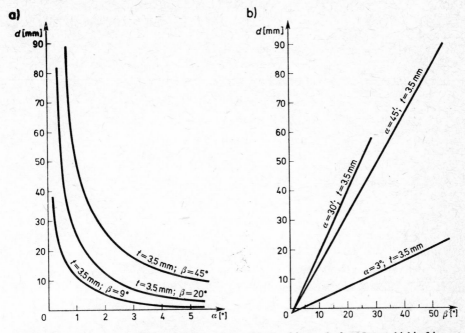

Fig. 7.13. Distance (d) of the localization of interference fringes I of a Nomarski birefringent prism (made of quartz) as a function of its apex angle α (a) and section angle β (b).

A typical Nomarski system for transmitted light DIC microscopy is shown in Figs. 7.14 and 7.15. From Eq. (7.34) it follows that the apex angle α of the Nomarski prism W should be equal to 30 minutes of arc for low-power microscope objectives, and to 60' for high-power objectives. Typically one takes $\alpha = 30'$ to 45', and a single Nomarski prism is used with all microscope objectives of magnifying power from $10\times$ to $100\times$. On the other hand, no single compensator Q can be used because each objective of a given focal length f'_{ob} needs an individual compensator with a specific apex angle α_Q resulting from Eq. (7.29a).

The system shown in Fig. 7.14 was refined by C. Zeiss Oberkochen [7.9] and introduced commercially as equipment (Fig. 7.15) for standard research microscopes. This device comprises a slide with a Nomarski prism and analyser, a DIC condenser with a number of compensators arranged in a turret, and a polarizer. The condenser usually serves for bright-field and phase-contrast microscopy as well; therefore, its turret comprises additionally a free opening and a number of annular diaphragms. As mentioned earlier, devices of this kind are now manufactured by many optical firms.

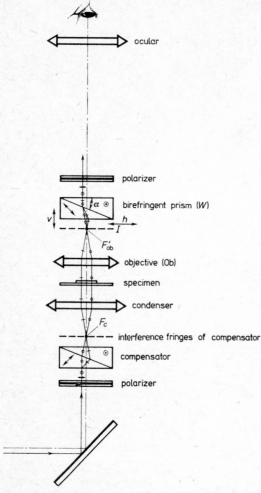

Fig. 7.14. Standard Nomarski system for transmitted-light DIC microscopy. The horizontal (*h*) and vertical (*v*) arrows show, respectively, the transverse and axial movement of the birefringent prism *W*. The first movement is for varying bias retardation and the second for bringing interference fringes *I* of the prism *W* into coincidence with the rear focal point F'_{ob} of different (low-, middle-, and high-power) objectives Ob.

The properties, advantages, and applications of the Nomarski system for transmitted-light DIC microscopy have been discussed by many authors [7.3, 7.9–7.46]. It is impossible to summarize this discussion in detail and only some general conclusions will be given here.

It was stated earlier that the transverse gradient of optical path difference across the object plane is responsible for the DIC image; phase objects are therefore perceived by their geometrical slopes and refractive index variations. As the Nomarski DIC system uses polarized light, optical anisotropy or birefringence is also a property which generates interference contrast depending on the angle made by the azimuth of the birefringent object with the direction of light vibrations in the polarizer and analyser.

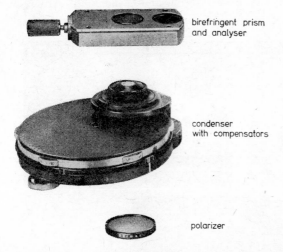

birefringent prism
and analyser

condenser
with compensators

polarizer

Fig. 7.15. Nomarski DIC device manufactured as an attachment to microscopes (e.g., Standard Universal, Standard RFL, Standard WL, Photomikroskop, Ultraphot II) available from C. Zeiss Oberkochen.

A predominant feature of DIC images is their relief appearance or shadow-cast effect. For transparent biological specimens, this effect results from both the gradient of optical path difference in the interior of cells and from surface irregularities. To avoid misinterpretation, it should be established whether the relief appearance represents the object surface topography or not. For this purpose, a combination of DIC and phase contrast is sometimes useful. Unlike phase contrast images, DIC images do not suffer from halo. For this and other reasons, the Nomarski DIC system is frequently preferable to phase contrast on some biological specimens since it reveals structures and minute details which are invisible when a conventional phase contrast microscope is used.

The contrast of DIC images is, however, not symmetrical and varies proportionally to the cosine of the angle made by the azimuth of the object and the direction of wavefront shear [7.3, 7.4, 7.6]. It is therefore necessary to examine unknown objects at several azimuths. Sometimes this defect even proves an advantage since some line structures (filaments of living cells, myofibrils in muscle fibres, fine striations of diatoms, etc.) are emphasized.

The relief appearance and shadow-casting effect may be optimized by varying bias retardation Δ_b. This instrumental operation is usually performed by sliding the birefringent prism W (Fig. 7.14) in a transverse direction as indicated by the arrow h. The shadow-casting effect is emphasized when one of the slopes of the transparent object in the direction of wavefront shear is brought to extinction (Figs. 7.16a and c). It will be readily seen that at the zero position of the prism W

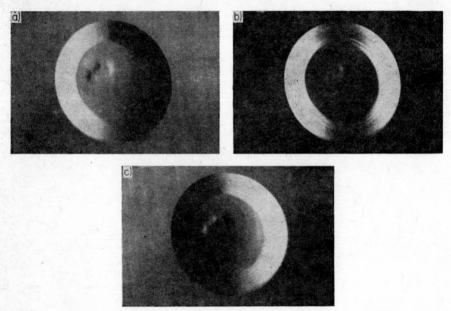

Fig. 7.16. DIC images of a truncated semispherical object illustrating the discussion of the shadow-casting effect.

(Fig. 7.14), bias retardation $\Delta_b = 0$ and the background is brought to extinction (polarizer P and analyser A are crossed), so that the object image (Fig. 7.16b; compare also Fig. IX) appears bisected by a dark fringe (at this setting a very small object could appear doubled). As bias retardation is added (or subtracted), the dark fringe will move toward the left-hand or right-hand slope of the object (Figs. 7.16a and c). When Δ_b is greater than that necessary to produce an optimum shadow-casting effect, the overall image contrast is reduced. No single bias retardation is ideal for all details in any objects, but for the majority of biological specimens the optimum values of Δ_b are within the range $\lambda/30$ to $\lambda/4$. In white light, we then have a background in specific grey interference colours ranging from nearly black, through iron-grey to lavender-grey or greyish blue tints (compare Fig. IXb). If phase objects with very large optical gradients are under study, bias retardation is not a critical quantity and may be selected at will within the range from $\lambda/10$ to 1.25λ (compare Fig. IXb). However, the maximum image contrast is always for grey or purple (I order) interference colours (Figs. IXb and d).

It is self-evident that optical gradients of opposite sign manifest themselves in DIC images as shadows in opposite directions. This feature is useful for evaluating the refractive properties of specimens under study; phase objects of lower refractive index, and hence phase-advancing relative to their surround, have shadow-casting in the opposite direction to that of optically denser (phase-retarding) objects.

It is always very important to adjust the DIC microscope so to obtain the highest possible extinction factor defined as[5] $E = I_{0.5\lambda}/I_0$, where I_0 and $I_{0.5\lambda}$ denote the background intensities at bias retardation $\Delta_b = 0$ and $\Delta_b = \lambda/2$ (or $-\lambda/2$), respectively. Extinction factors $E < 100$ are considered to be unacceptable for the study of most biological specimens, and for careful work $E > 200$ must be achieved [7.3].

Some further practical suggestions on the correct use of the Nomarski DIC system for transmitted-light microscopy may be found in Refs. [7.3] and [7.9].

The widest application of the Nomarski DIC method for transmitted-light microscopy is in human biology, zoology, and botany. In general, this method—like phase contrast—is primarily used for the study of unstained cytological and histological preparations, but it is also suitable for more complete identification of stained tissues, cells, and chromosomes (see Subsection 7.2.2).

In particular, cell biology [5.26, 7.25–7.28], protozoology [7.21–7.23], bacteriology [7.35], and plant morphology [7.29–7.33] are areas especially suitable for Nomarski DIC microscopy. The method is also useful for the definition of fluorescent details in immunofluorescent sections or fluorochromed cell culture monolayers [7.18], as well as for the visualization of some biological structures and processes which are invisible or barely observable when phase contrast or other contrasting techniques are used. Spindle fibres in living dividing cells [7.25, 7.27], microtuble-related motility in the reticulopodial network

---

[5] This definition is adequate to that expressed earlier by Eq. (7.28). However, some authors define this factor as $1/E$ (see Ref. [7.6], for instance).

[7.40], and exoplasmatic fibrils [7.41] are examples of such structures. Moreover, Nomarski DIC microscopy is almost indispensable for the testing of spermatozoa of animal sperm samples used for artificial insemination; some specific defects of the spermatozoa are clearly visible in their DIC images only [7.38, 7.39].

The Nomarski DIC system for transmitted light is also an useful tool in materials sciences microscopy; a paper [7.42], for instance, reports some advantages of this system to the study of thin sections of soil samples; a research group from Technical University of Mining and Metallurgy in Cracow (Poland) used DIC microscopy in their study of the texture and sinterability of MgO powders and of the effect of glaze on strength of high-tension porcelain [7.45, 7.46]; investigations of nuclear tracks in solids were reported as well [7.43].

7.2.2. Amplitude DIC microscopy

It was previously stated that the polarizer and analyser of a Nomarski DIC device were crossed and that their directions of light vibrations formed an angle of 45° with the wedge edges of two birefringent prisms (compare Figs. 7.11 and 7.14). The crossed polars constitute an optimum configuration for the observation of phase objects. At the same time, the system was also said to be suitable for the examination of phase-amplitude objects which absorb light in some degree. Here, however, the polars should not be crossed. In order to achieve a highly contrasting image in any form of interference microscopy, the interfering light waves must have the same amplitude. An absorbing object reduces the amplitude of transmitted light and this reduction should be compensated for if we wish to obtain high interference contrast. In the Nomarski DIC system, the compensation is simply achieved by rotating the polarizer, starting from its crossed position. The higher the light absorption by the object under study, the greater should be the rotation of the polarizer.

The adjustments required when observing light absorbing objects are as follows: (1) polars are crossed and bias retardation is adjusted (by moving the birefringent prism laterally) to produce an optimum shadow-cast effect; (2) the polarizer is rotated from its crossed position until light absorbing details of the object under study appear maximally dark. This is illustrated in Fig. XXXVII. The degree of polarizer rotation required to achieve the optimum image contrast depends primarily on local light absorption and the gradient of absorption across the object. When white light is used for observing stained preparations, this procedure does not change the colouring of the object details significantly, but only enhances colour contrast and improves resolution (some authors speak of the super-resolution of the Nomarski DIC system [7.3, 7.14]).

Amplitude DIC procedure is especially recommended for studying thin weakly-stained microtome sections of tissues or smears of cells, stained chromosome spreads, chloroplasts in living protists, pigment globules in living metazoan cells, and many other biological preparations qualified as "difficult" absorbing objects [7.3, 7.47].

7.2.3. The image of a light point in Nomarski DIC microscopy

It has been stated that the resolving power of the Nomarski DIC system is better than that of a common bright-field microscope. An useful aid to understanding this fact is given by Galbraith in his paper [7.14] on the computer simulation of the image of a point of light in Nomarski DIC microscopy. The essential results are shown in Figs. 7.17 and 7.18, which illustarate the three-dimensional

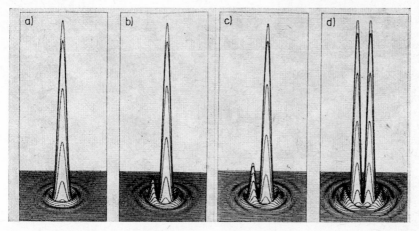

Fig. 7.17. Three-dimensional intensity distributions of the image of a point light object [7.14]: a) conventional bright-field microscope; b) DIC microscope, wavefront shear $s = 0.5r_{Airy}$, amplitude ratio $R = 0.333$; c) as b) but $s = 0.55r_{Airy}$ and $R = 0.366$; d) as b) but $s = 0.5r_{Airy}$ and $R = 0.5$ (by courtesy of W. Galbraith).

and two-dimensional intensity distributions of the image of a light point in a standard (S) microscope (Figs. 7.17a and 7.18a) and in the Nomarski DIC microscope with different *amplitude ratios* of two components of the sheared light wave and zero bias retardation when the amount of shear s is 0.5 to 0.6 times the radius (r_{Airy}) of the Airy disc (Figs. 7.17b–d and 7.18b–d).

The amplitude ratio (R) mentioned above is defined as the amplitude of one wave component divided by the sum of the amplitudes of two components. If one polar of the Nomarski DIC system is offset from the crossed position by an angle ϑ, then $R = (1 - \tan\vartheta)/2$. For $\vartheta = 0$, the polars are crossed and $R = 0.5$; the two components are of equal intensity, and the DIC image of the light point has two equal intensity maxima separated by the shear distance s (Figs. 7.17d and 7.18d). As the polars are gradually offset, one of these maxima takes over light from the other, and when the polars are offset 45°, the amplitude ratio $R = 0$, and one maximum is completely extinguished (Figs. 7.17a and 7.18a); if it is offset 18.45°, the amplitude ratio $R = 0.333$, and one wave component has half the amplitude of the other (Figs. 7.17b and 7.18b). This would be a suitable setting for a pure amplitude object such as a bright point on a dark background.

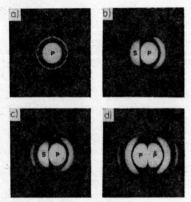

Fig. 7.18. The same four computer-calculated images of a point light object as in Fig. 7.17a–d, but displayed as two-dimensional intensity pictures [7.14]: a) conventional bright-field microscope; b) DIC microscope, $s = 0.5r_{Airy}$; $R = 0.333$; c) DIC microscope, $s = 0.55r_{Airy}$, $R = 0.366$; d) DIC microscope, $s = 0.5r_{Airy}$, $R = 0.5$ (by courtesy of W. Galbraith).

As can be seen from Figs. 7.17 and 7.18, the DIC maxima are narrower in the shear direction than the S maximum. This phenomenon allows the DIC system to achieve better resolution than an ordinary microscope. The gain in resolution is more marked when $R < 0.5$ and the wavefront shear $s = 0.5r_{Airy}$ to $0.6r_{Airy}$. However, in most DIC instruments $s = 0.6r_{Airy}$ to $1r_{Airy}$ and is not adjustable. Small shear gives superior image quality of small objects, but image brightness decreases.

As Galbraith and David have stated [7.13], a pure amplitude object requires amplitude compensation (see Subsection 7.2.2) and bias retardation $\Delta_b = 0$,

while a pure phase object requires phase compensation (a suitable bias retardation) and amplitude ratio $R = 0.5$. Complex (phase–amplitude) objects require
both phase and amplitude compensation. A bright point is, of course, a pure
amplitude object and the bias retardation should be equal to zero. Otherwise,
the DIC image approaches more closely to the S image (Figs. 7.17a and 7.18a)
and super-resolution is lost [7.14]. The background is totally opaque in the
simulations shown in Figs. 7.17 and 7.18, and in this case the best amplitude
ratio R is 0.333, so that one wave component of the sheared light wave has half
the intensity of the other.

The reverse situation—when a dark point object appears on a bright background—is more complicated in ordinary microscopy, but in DIC microscopy
the background may be maximally darkened as before, and the dark object
then appears as a bright double image similar to Figs. 7.17d and 7.18d, in reversed
contrast. The image of the dark point has an amplitude gradient, and is therefore
not as dark as the background at the settings $R = 0.5$ and $\Delta_b = 0$. The image
may also be made darker against a brighter background ($R \neq 0.5$ or $\Delta_b \neq 0$)
and emphasized by the characteristic shadow-cast effect.

7.2.4. Nomarski DIC system for reflected light

This version of Nomarski DIC microscopy is simpler than that for transmitted
light since it comprises only a single birefringent prism W (Fig. 7.19), which
is traversed twice by the light beam; first, when light emerging from the epi-
illuminator is incident on the object under study, and second, when light B_r
comes back after reflection from the object surface. For the incident beam B_i,
the objective Ob manifests itself as a condenser and the prism W acts as it was
a compensator. The system is therefore self-compensating and a large aperture
of the epi-illuminator may be used.

The polariser P is adjusted so as to yield two orthogonally polarized light
beams of equal amplitude at the object O. These beams are represented by two
rays 1 and 2 in Fig. 7.19. Their plane of apparent splitting, i.e., the plane I of localization of interference fringes of the birefringent prism W, is coincident with
the rear focal point F'_{ob} of the objective Ob. Normally, the polarizer P and analyser A are crossed and adjusted so that their directions of light vibration form
an angle of 45° with the incidence plane, i.e., the plane determined by the normal
to the semitransparent mirror BS and the axis of the objective Ob. The birefringent prism W is orientated so that one of its two principal sections is coincident
with the incidence plane defined above (some other adjustment variants of these

elements are also possible). If the object surface contains some details which absorb light, the polarizer may be rotated starting from its crossed position and adjusted so as to achieve the optimum amplitude DIC image (see Subsection 7.2.2).

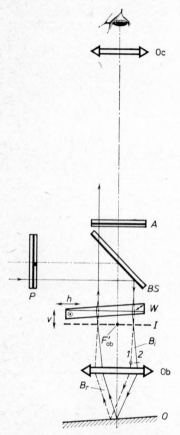

Fig. 7.19. Nomarski DIC system for reflected light.

A serious problem which reduces the performance of the Nomarski DIC system using an epi-illuminator is stray light generated by reflections of illuminating rays at the surface of optical elements. Unwanted reflections are produced, in particular, by the plane surfaces of the birefringent prism W. To overcome this defect, the prism is not placed at right angles to the objective axis, but is slightly slanted. The slant angle is adjusted so as to reject reflections outside the field of view of the ocular Oc. It is important, moreover, to orientate the plane of apparent splitting (I) so that it is as nearly perpendicular as possible to the optic axis of the microscope objective Ob (the plane I of a typical Wollaston or Nomarski prism is inclined with respect to the external surfaces of this prism; see Subsection 1.6.5 for details). The lateral translation (h) of the birefringent prism W along the wavefront shear direction allows bias retardation to be varied, thus the shadow-cast effect and overal image contrast can be optimized, while the vertical translation (v) adjust the prism to coincidence of its plane I with F'_{ob} when one objective Ob is replaced by another.

Fig. 7.20. Nomarski DIC device manufactured as an attachment to metallurgical microscopes (e.g., MeF2) available from C. Reichert Wien.

The Nomarski DIC device for reflected light was first manufactured by C. Reichert Wien as part of the equipment (Fig. 7.20) of the MeF metallurgical microscope. Today this device is commercially available from other firms as well, and provides a simple technique for the study of the surfaces of metallurgical specimens, semiconductor materials, and microelectronic circuits [7.2, 7.48, 7.49, 7.52, 7.56–7.59], and also of some opaque surfaces of hard biological tissues such as tooth and bone. However, it is primarily recognized as an extremely sensitive tool for the qualitative study of the surface roughness of reflective specimens [7.2, 7.48, 7.49, 7.52, 7.56, 7.57]; vertical microaspirities as small as 0.5 nm in root mean square height can be detected [7.52]; moreover, the method is non-contacting and non-destructive. At the same time, the Nomarski reflection microscope provides a useful technique for the measurement of slopes of microscopic surface features of metallographic and other specimens [7.49, 7.54, 7.55]. It is also useful for showing pyramidal hardness impressions in enhanced and vivid contrast [7.50, 7.51]. When bias retardation is set so to give a grey-green background, the system is capable of yielding high contrast in the image and would then appear to be better than phase contrast for the non-destructive demonstration of stacking faults in silicon epitaxial wafers [5.84–5.87].

7.3. Video-enhanced DIC microscopy

Video-enhanced contrast (VEC) microscopy is based on the use of a specialized TV camera by means of which the object image may be enhanced in contrast and recorded by using a video recorder of high quality. This technique has primarily been developed by Allen (Fig. 7.22) and is therefore known as *Allen video-enhanced contrast* (AVEC) *microscopy* [7.40, 7.60]. Its history was, however, initiated by Parpart [7.61] and Flory [7.62], who first used a videcon camera for displaying a microscopical image. Another significant contribution to microscopy of this kind was made by Inoué [7.63].

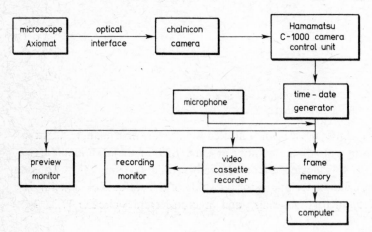

Fig. 7.21. A block-diagram of the AVEC-DIC system [7.60].

Fig. 7.22. Robert Day Allen (1927—1986).

The AVEC-DIC method takes advantage of a specially designed video camera (the Hamamatsu C-1000, for instance), which can reject stray light and thus improve both contrast and resolution in the final microscope image. Figure 7.21 illustrates an arrangement of the Allen video microscope system described in Ref. [7.60]. The video camera was a C-1000-01 binary, computer-compatible chalnicon camera driven by a polyprocessor frame memory and minicomputer (Hamamatsu Systems).

The steps necessary for obtaining the optical condition for AVEC-DIC images are as follows [7.60]: (1) the specimen under study is brought into focus and illuminated according to the Köhler principle (see Subsection 2.3.4); (2) bias retardation is adjusted to be equal to about $\lambda/10$; (3) the condenser iris diaphragm is opened to match the numerical aperture A_c of the condenser with that, A, of the objective (hence, the AVEC-DIC technique is not subject to the customary condition $A_c = 2A/3$ for a compromise between resolution and image contrast; see Subsection 3.8.4); (4) the microscopical image is then sent to the video camera, taking care to keep the image brightness below the saturation level of the camera; (5) the offset is adjusted to reduce monitor image brightness to optimize visual contrast; (6) electronic gain and additional offset are added, if desired, to further enhace image contrast. At this stage, the enhanced image suffers from a constant pattern of mottle (noise) due to inaccessible dust particles and other cosmetic defects present on even high quality lenses. This serious limitation is overcome by storing the mottle pattern in the frame memory (frame store) with the specimen out of focus and then continuously subtracting this pattern from each succeeding frame to clear the image of the specimen brought into focus. This is a technical innovation of basic importance. The frame memory contains an internal offset, which is separate from that of the camera control unit. Using that offset, the background may be adjusted to a neutral grey. When the specimen is brought into focus, the mottle entirely disappears, and a major improvement in quality of the final image is achieved. The analogue signal output of the frame memory can also be recorded on a video cassette type.

The image contrast definitions expressed by Eqs. (3.64) apply in a similar manner to conventional DIC microscopy. Here the contrast is, however, a function of the gradient of the optical path difference across the object in the direction of the wavefront shear. Let us consider Eq. (3.64b), for instance. The AVEC method enables this equation to be modified using video manipulations. One of these manipulations is the reduction of background intensity (or brightness) I'_b without diminishing the difference $I'_b - I'_o$ in intensity of the background (I'_b) and that (I'_o) of the object under study. The modified formula (3.64b) for video-enhanced contrast C_v' may therefore be expressed as

$$C_v' = \frac{I_b' - I_o'}{I_b' - I_v'}, \tag{7.35}$$

where I_v' is the electronic reduction in signal level introduced by turning an offset knob, which changes the clamp level of a DC restoration circuit [7.60]. This operation has the effect of reducing the denominator of Eq. (7.35), compared with Eq. (3.64b), thus increasing visual contrast. This electronic manipulation and subtracting of the mottle pattern make visible (i.e., detectable) many linear elements and particles that are of an order of magnitude smaller than the resolution limit and not visible in conventional DIC images, regardless of the brightness level at which they are presented.

For further details concerning this new technique, the reader is referred to the original papers by Allen *et al.* [7.40, 7.60].

7.4. Differential interference contrast microscopy with continuously variable wavefront shear

For a fixed objective magnification, the wavefront shear in the Nomarski DIC system is constant. In microscopy practice, there are usually preparations which contain small and extended objects with various optical path gradients and different spatial frequencies. A constant wavefront shear equivalent roughly to the resolving power of the microscope objective cannot, of course, be suitable for all objects. In particular, when very fine structures are investigated, the wavefront shear should be smaller than the limit of resolution of the microscope. Low spatial frequency objects, on the other hand, and extended structures with small optical-path gradients need larger wavefront shears; if not, relief appearance or the shadow-cast effect of the DIC images are unsatisfactory. In short, a DIC microscope with a constant wavefront shear for a given objective magnification cannot produce optimum conditions for objects of different sizes and various optical-path gradients.

Taking into considerations these facts, many microscopists are likely to be interested in a system of DIC microscopy with continuously variable wavefront shear. A system of this kind for both transmitted and reflected light microscopy has been developed by the author of this book [7.64, 7.65]. Here, it will be termed *VADIC (variable DIC) system.*

7.4.1. Transmitted-light VADIC system with pupilar compensation

This system is schematically shown in Fig. 7.23. Its basic elements are two bire-fringent prisms W_1 and W_2, separated by a half-wave plate H_1, and two compensators Q_1 and Q_2, separated by another half-wave plate H_2. Both the birefringent prisms and compensators are Wollaston prisms as modified by Nomarski. They are made of quartz crystal.

One of the birefringent prisms (W_1) is located close behind the optical system of the objective Ob at a constant distance d_1 and is rotatable round the objective axis; the rotation changes the wavefront shear. The other birefringent prism (W_2) is placed in the microscope tube at a variable distance d_2 and can be translated in two directions, transverse (h) and parallel (v) to the objective axis. The transverse translation shifts the phase difference between sheared light waves and, in particular, introduces variable bias retardations Δ_b, while the parallel translation adjusts the interference system to uniform field (flat-tint) interference when one objective is replaced by another. Each of these prisms has an outside localizing plane of interference fringes as is normally the case in a Nomarski prism. These fringes are marked by I_1 and I_2 in Fig. 7.23, and are brought into coincidence with the back focal point F'_{ob} of the objective.

The compensators Q_1 and Q_2 are similar to the birefringent prisms W_1 and W_2, respectively. One of them (Q_1) is rotatable round the condenser axis and cancels (across the exit pupil of the objective) the optical path difference between the light wave components (ordinary and extraordinary) sheared by the rotatable birefringent prism W_1. The other compensator (Q_2) is immovable and orientated in such a manner as to cancel out the optical path difference between light wave components sheared by the sliding birefringent prism W_2. The distances d_3 and d_4 of these compensators are adjusted so as to bring their own interference fringes I_3 and I_4 into coincidence with the front focal point F_c of the condenser C.

An initial ("zero") orientation of the compensators Q_1 and Q_2 with respect to prisms W_1 and W_2 is assumed as shown in Fig. 7.23. Next, let the initial ("zero") positions of the analyser A, half-wave plate H_1, polarizer P, and half-wave plate H_2 be defined as shown in Fig. 7.24, where AA and PP denote directions of the light vibrations in the analyser and polarizer; $F_1 F_1$ and $F_2 F_2$ are one of the principal axes of the half-wave plates H_1 and H_2 ($F_1 F_1$ and $F_2 F_2$ are the fast axes, for example); E_1, E_2 and E_3, E_4 denote the edges of wedge angles α_1, α_2 and α_3, α_4 of the upper wedges of the birefringent prisms and compensators, respectively. As can be seen, the polarizer and analyser are crossed and their directions of light vibrations form an angle of 45° with the wedge edges E_4

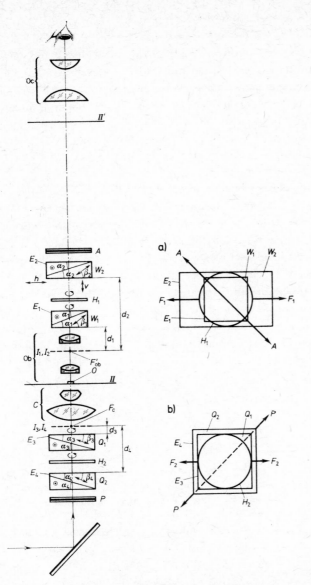

Fig. 7.23. Optical system of a differential interference contrast microscope with continuously variable wavefront shear and pupilar compensation—the VADIC system [7.64].

Fig. 7.24. Initial (zero) orientation of principal elements of the VADIC system shown in Fig. 7.23: a) next-after-the-objective part, b) before-the-condenser part.

and E_2 of the compensator Q_2 and birefringent prism W_2, respectively. Moreover, the directions of light vibrations of the half-wave plates are in their zero
position parallel to the principal sections of the birefringent prism W_2 and compensator Q_2. Effective interference with a flat-tint field is achieved in the image
plane Π' when the construction parameters (wedge angles α_3, α_4 and section
angles β_3, β_4) of the compensators Q_1 and Q_2 are fixed so as to obtain infinitely
enlarged interference fringes in the exit pupil of the objective Ob. To obtain
this effect (i.e., pupilar compensation), the following condition should be fulfilled:

$$\varepsilon f'_{ob} = \varepsilon_c f_c, \tag{7.36}$$

where ε is the resultant angular wavefront shear produced by the combination
of two birefringent prisms W_1 and W_2, f'_{ob} is the focal length of the objective Ob,
ε_c is the resultant angular wavefront shear due to the combination of two compensators Q_1 and Q_2, and f_c is the focal length of the condenser C.

As in Eq. (7.20b), the left-hand part of the relation (7.36) expresses the transverse linear wavefront shear (s) related to the object plane Π (Fig. 7.23), while
the right-hand part of this relation can be treated as a resultant linear wavefront
shear (q) occurring in the same object plane Π due to the compensators Q_1
and Q_2. Thus, a general rule for pupilar compensation can be simply expressed
with the aid of a vectorial diagram as shown in Figs. 7.25 a and b, in which

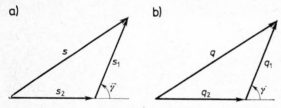

Fig. 7.25. Vectorial diagram showing the principle of wavefront shear variation and pupilar
optical path difference compensation in the VADIC system shown in Fig. 7.23: a) orientation
of the birefringent prisms W_1 and W_2, b) orientation of the compensators Q_1 and Q_2.

vectors s_1 and s_2 represent the directions and values of wavefront shear produced
in the object plane Π by the birefringent prisms W_1 and W_2, whereas vectors
q_1 and q_2 represent directions and values of wavefront shear produced in the
same plane Π by the compensators Q_1 and Q_2, respectively. In order to obtain
complete compensation in the exit pupil of the objective, the resultant vectors s
and q should be parallel and equal in length. For the double refracting system
considered here, this condition is achieved when $s_1 = q_1$ and $s_2 = q_2$. Thus the
following conditions can be formulated for the compensators Q_1 and Q_2:

$$\varepsilon_{c1} f_c = \varepsilon_1 f'_{ob},$$
$$\varepsilon_{c2} f_c = \varepsilon_2 f'_{ob}, \tag{7.37}$$

where ε_{c1} and ε_{c2} are the angular wavefront shears produced by the compensators Q_1 and Q_2, respectively, and ε_1 and ε_2 are those due to the prisms W_1 and W_2. The directions of the vectors $\mathbf{s}_1$, $\mathbf{s}_2$, $\mathbf{q}_1$, and $\mathbf{q}_2$ are selected so that their origins and ends match the ordinary and extraordinary light rays, respectively (or vice versa).

Under these conditions, the general working principle of the VADIC microscope can be summarized as follows. A light wave, polarized linearly by the polarizer P (Fig. 7.23), is split by the compensators Q_1 and Q_2 into two wavefronts polarized at right angles. When passing through an object O under examination, the sheared wavefronts are subjected to a phase shift corresponding to the optical path difference between the object O and its surrounding medium. The deformed wavefronts enter the objective Ob and are recombined by the birefringent prisms W_1 and W_2. When passing through the analyser A, both wavefronts interfere with each other and make visible the transparent object O in the form of two images laterally duplicated to a greater or lesser degree. If the image duplication is much smaller than the lateral size of the object, differential interference contrast is obtained. Then any change of the interference colour (or brightness) in the object image shows up a gradient of the optical path difference in the direction of the resultant image duplication $\mathbf{s}$ (Fig. 7.25a).

This image duplication can be arbitrarily changed by rotating the birefringent prism W_1 through an angle γ, starting from its initial position (Figs. 7.24 and 7.25). The value of the resultant image duplication is generally equal to

$$s = \sqrt{s_1^2 + s_2^2 + 2s_1 s_2 \cos \gamma}, \tag{7.38}$$

where s, s_1, and s_2 are the lengths of the vectors shown in Fig. 7.25a. However, in order to maintain pupilar compensation and effective interference in the image plane Π' (Fig. 7.23), the compensator Q_1 should be simultaneously turned through the same angle γ. Moreover, the half-wave plates H_1 and H_2 should be rotated through an angle $\gamma/2$, starting from their initial orientations (Fig. 7.24), or through $(\gamma - 90°)/2$, $(\gamma - 180°)/2$, and $(\gamma - 270°)/2$ if $90° < \gamma < 180°$, $180° < \gamma < 270°$, and $270° < \gamma < 360°$, respectively. The maximum range of rotation of the half-wave plates needed to cover all the angles of rotation of the prism W_1 and compensator Q_1 is then no larger than $45°$.

The half-wave plate H_2 is rotated in order to align the vibration directions of sheared light waves leaving the compensator Q_2, parallel to the principal sections of the compensator Q_1; similarly, rotating the half-wave plate H_1 enables the vibrations of light waves issuing from the birefringent prism W_1 to be aligned parallel to the principal sections of the prism W_2.

When the prism W_1, together with compensator Q_1, is set at angles $\gamma = 0°$, 90°, 180°, and 270°, both half-wave plates remain at their initial positions or can even be removed from the double refracting interference system. These four particular settings will consequently be called *additive, crossed left-handed, subtractive*, and *crossed right-handed*, and they give the resultant image duplications $s = s_1 + s_2$, $s = \sqrt{s_1^2 + s_2^2}$, $s = s_1 - s_2$, and $s = \sqrt{s_1^2 + s_2^2}$.

A microscope incorporating the above principle has been constructed on the basis of a biological microscope manufactured by the Polish Optical Works (PZO), Warsaw. It incorporates a carriage with the birefringent prism W_2 mounted in a unit between the nose-piece and binocular tube. The basic parameters of this prism are as follows: wedge angle $\alpha_2 = 45'$, section angle $\beta_2 = 35°$, thickness $t = 3.5$ mm. It gives an image duplication, relative to the object plane, $s_2 \approx 35/M_{ob}$ μm, where M_{ob} is the objective magnifying power. The carriage, together with the prism W_2, can be slid in a direction perpendicular (h, Fig. 7.23) and parallel (v) to the objective axis. This microscope is fitted with four objectives of magnifying power $M_{ob} = 10, 20, 40$, and $100 \times$ (oil immersion). Each of these objectives are provided with a rotatable birefringent prism W_1 (Fig. 7.23), whose basic parameters are as follows: $\alpha_1 = 30'$, $\beta_1 = 10°$, $t = 3.5$ mm. This prism gives an image duplication, relative to the object plane, $s_1 = 27/M_{ob}$ μm, and when rotated round the objective axis, produces, together with the prism W_2, the following image duplications: 0.8 to 6.2 μm, 0.4 to 3.1 μm, 0.2 to 1.7 μm, and 0.08 to 0.62 μm for objectives of magnifying power 10, 20, 40, and $100 \times$, respectively. For each of these objectives, two compensators Q_1 and Q_2 (Fig. 7.23) were calculated in accordance with the conditions (7.37). The compensators are arranged in two switch-driven carriages mounted below a condenser of aperture 0.9 (in air). Unfortunately, this microscope is not available commercially to date and only exists as a prototype.

The most important feature of the VADIC system described above is that the differential wavefront shear is arbitrarily varied, thus image contrast, sharpness, and shadow-cast effect can be suitably adjusted to differences in size and structure frequencies occurring in the specimen.

Some advantages of the system are illustrated by the photomicrographs. Figures XXXVIIIa to d illustrate a series of VADIC images of the same fragment of an unstained and relatively thick tissue section mounted in Canada balsam between a slide and cover slip. This is a testicle labule of a rat. The testicle labule as a whole shows up better if the wavefront shear is increased. However, the definition and resolution of fine structures of the labule then deteriorate. The best visibility of spermatogenic cells and spermatozoa is obtained in photomicrographs c and d, where the wavefront shear is small (1 and 0.4 μm). Figure XXXIX shows three images of the same epithelial cell of buccal mucosa, which illustrate the effect of optical sectioning. The photomicrographs were taken with an objective $100 \times /1.25$ (oil immersion) focused at: a) nucleus, b) upper surface and c) lower surface of the cell. A characteristic feature of this cell is an abundance of proto-plasm granules, as shown in Fig. XXXIXa, and the ridged and folded structure of its membrane, as shown in Figs. XXXIXb and c. This structure can be parti-cularly well observed (using the VADIC microscope) when the wavefront shear in the object plane is equal to about 0.2 μm. As can be seen from photomicro-graphs of Fig. XXXIX, the optical sectioning is very effective. The fine ridged structure of the cell membrane is well outlined without appreciable disturbance from out-of-focus granules in the interior of the cell.

7.4.2. Transillumination VADIC system with condenser slit diaphragm

The system with variable wavefront shear presented above appears to have many advantages and is likely to be very useful for qualitative and descriptive investiga-tions of different isotropic specimens in transmitted light. However, its usefulness is limited where the study of anisotropic objects is concerned because orthogonally plane-polarized light wave components are, in fact, only slightly laterally sheared in the object plane and the specimen is transilluminated by a light beam with variable polarization. Thus, for a complete examination of anisotropic objects (birefringent fibres and microcrystals, for instance), the VADIC system shown in Fig. 7.26 is preferable. The latter differs from the former (Fig. 7.23) in that a slit condenser diaphragm D_c is used instead of the compensators Q_1 and Q_2 separated by the half-wave plate H_2. Hence, only one plane-polarized beam issuing from the slit S illuminates the object O under study. The diaphragm D_c is located in the front focal plane of the condenser C and its role is the same as in the DIC system described earlier in Subsection 7.1.4 (see Fig. 7.7a), except that the slit is rotatable round the condenser axis.

Let the initial orientation of the slit S be defined as shown in Fig. 7.27a. If the birefringent prism W_1 is rotated through an angle γ and the half-wave plate H_1 through $\gamma/2$ (Fig. 7.27b), then the slit S should always be at right angles to the direction of the resultant wavefront shear s. The angle ζ between the actual and initial orientations of the slit equals that between the vectors s and s_2, which

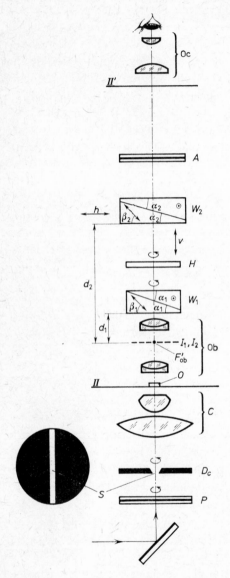

Fig. 7.26. The VADIC system with a condenser slit diaphragm (see text and Fig. 7.23 for explanation).

represent, respectively, the resultant wavefront shear and the shear produced only by the birefringent prism W_2. This angle ζ can be calculated from the geometry of Fig. 7.27b and is expressed by

$$\tan \zeta = \frac{s_1 \sin \gamma}{s_2 + s_1 \cos \gamma}.$$ (7.39)

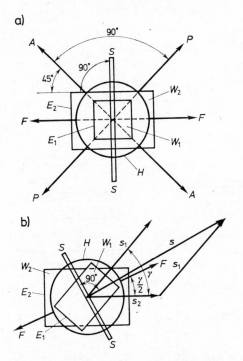

Fig. 7.27. Diagrams illustrating a) the initial orientation of principal elements of the VADIC system shown in Fig. 7.27, b) the principle of wavefront shear variation.

In order to obtain effective interference in the image plane Π' (Fig. 7.26), the vibrations of light leaving the polarizer P should always be at 45° to the principal directions of light vibration in the birefringent prism W_1. Hence, if this prism is rotated through an angle γ (Fig. 7.27b), the polarizer P must also be rotated through an equivalent angle starting from its initial position shown in Fig. 7.27a; in other words, the direction PP of light vibrations in the polarizer should form an angle of 45° with the vector $\mathbf{s}_1$, which represents the wavefront shear produced only by the birefringent prism W_1. It is worth noting that in the

VADIC system with compensators the polarizer need not be rotated (Fig. 7.23). This also holds good for the system with a condenser slit diaphragm (Fig. 7.26) when the birefringent prism W_1 is orientated subtractively ($\gamma = 180°$) and in crossed position ($\gamma = 90°$ or $270°$) with respect to the prism W_2. In these particular instances, the half-wave plate H_1 is not required.

A slit condenser diaphragm makes it possible to obtain a high instrumental extinction factor (see Eq. (7.28)) and good image contrast. However, the slit reduces the condenser aperture and consequently the resolving power of the microscope in the direction perpendicular to the slit, although for certain types of objects this defect can be greatly reduced or even quite removed by suitably decentring the slit diaphragm. For optical sectioning, this system is, however, considerably less effective than the VADIC system with compensators.

7.4.3. VADIC system for reflected light

This version of VADIC microscopy does not differ greatly from that described earlier for transmitted light except that a vertical illuminator with a polarizer is installed between the binocular tube and unit which follows the objective nose-piece and comprises a carriage with the birefringent prism W_2 (Fig. 7.28). The objectives with the rotatable birefringent prism W_1 are the same as for transmitted light, except for being corrected for the zero thickness of the cover-slip.

The initial ("zero") position of the double-refracting and polarizing elements is defined as follows: birefringent prisms W_1 and W_2 are orientated as shown in Fig. 7.28 (additive position); the privileged directions of light vibrations of the half-wave plate H are parallel to the principal sections of the prism W_2; polarizer P and analyser A are crossed, and their directions of light vibrations form an angle of 45° with the principal sections of the birefringent prism W_2; the semi-reflective plate (beam-splitter) BS is so arranged that its plane of light incidence is perpendicular or oriented at an angle of 45° to the direction of light vibrations in the polarizer P. A light beam issuing from the source LS and polarized linearly by the polarizer P is reflected by the plate BS towards the objective Ob. Passing through the birefringent prisms W_2 and W_1, the beam is sheared into two components polarized perpendicularly to each other. After being reflected onto the object under examination O, these two components pass back through the objective, birefringent prisms, semireflective plate, analyser A, and reach the ocular Oc. When interfering with each other, they make visible any surface detail of the object O in the form of two overlapped images differentially sheared to a greater

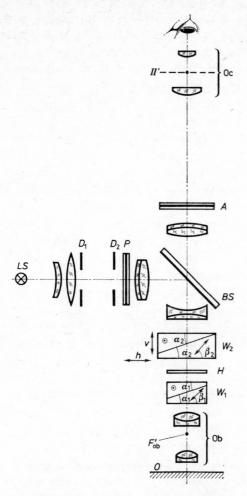

Fig. 7.28. Optical system of the reflected-light VADIC microscope [7.65].

or lesser degree. Variable wavefront shearing is carried out by rotation the birefringent prism W_1. This is the same operation as in the transillumination
VADIC system and does not require detailed explanation here. If the prism W_1
is orientated additively or subtractively, or is crossed with the prism W_2, the
half-wave plate H is not required.

Each of the birefringent prisms produces across its length a monotonically
varying phase difference between ordinary and extraordinary wave components.
Thus, the single passage of a light wave through the prisms W_2 and W_1 gives

rise to parallel and equidistant straight interference fringes in the exit pupil of the objective. But for the forward and backward rays this phase difference is of opposite sign and cancels out. Thus, the twofold passage of light waves through birefringent prisms allows these interference fringes to be completely overspread. Such a pupilar phase-difference autocompensation allows an extended light source to be used without any slit or pinhole diaphragm, as in the case of the Nomarski DIC system for reflected light (see Subsection 7.2.4).

The VADIC system as shown in Fig. 7.28 functions satisfactorily with an ordinary vertical illuminator, which incorporates an iris aperture diaphragm D_1 and a field diaphragm D_2, enabling one to obtain the correct Köhler principle of illumination. In order to reduce stray light generated by reflections of illuminating rays and accepted by the ocular Oc, the prism W_2 is not positioned strictly at right angles to the objective axis, but is slightly slanted.

Some interesting results of observations using this microscope are illustrated in Figs. XL–XLIII. The photomicrographs of Fig. XL present triangle stacking faults in an epitaxial Si film. Although this film was unetched, the triangles are very clearly visible. However, this visibility depends greatly upon both the degree and direction of wavefront shear. In all these photomicrographs, the value of shear was the same and equal to about 1 μm, which appeared to be the optimal value for these objects. The direction of shear, however, differed for the various images and was successively vertical, at 45°, and horizontal. When we compare the images, it can be seen that one of the triangle edges stands out more clearly than the others. This edge is clearest when the direction of wavefront shear is perpendicular to it (photomicrograph to the right). This means that in this direction the edge produces a steep optical-path gradient.

The photomicrographs of Fig. XLI present variously sheared VADIC images of a silicon single-crystal substrate (wafer) mechanically polished and smoothly etched. Comparison of these images reveals a correspondence between the photomicrographs b, c, and d, in which there are nearly equally clearly visible scratches, microetched pits, and some wide surface deformations (elevations and depressions). These last defects appear more clearly when the wavefront shear increases and, with the exception of one steep hill, are hardly visible in the first photomicrograph (Fig. XLIa), where the wavefront shear is small. However, in this picture the scratches and boundaries of their etched cells stand out most sharply.

The photomicrographs of Fig. XLII present another region of the same specimen as shown previously (Fig. XL). Now, the patterns consist of some point contaminations, two triangle stacking faults, and several wide surface irregularities (shallow depressions and easy hillocks). These last defects are, however, visible when the wavefront shear is equal to about 2 μm or more. By using a large (non-differential) wavefront shear, as shown in Fig. XLIIc, it is possible to determine the optical path difference δ between the triangle stacking faults and their surrounding field (see Chapter 16 for measuring procedure). In this case, δ = 0.04λ.

In the photomicrographs of Fig. XLIII, the surface topography of the epitaxial Si film hardly differs from that shown in Fig. XLII, except that now surface depressions and elevations have slightly greater slopes.

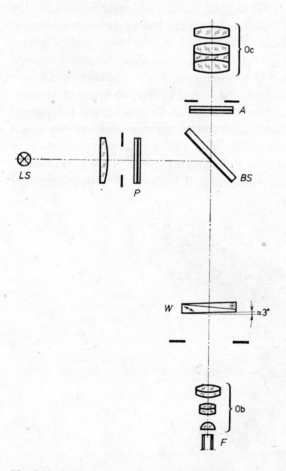

Fig. 7.29. Optical system of a specialized DIC microscope for testing cross-sections of optical fibres [7.66].

In conclusion, the incident-light VADIC system presented above appears to have many advantages compared to systems using a fixed wavefront shear. It is capable of giving maximum image sharpness, relief appearance, and optimum detectability of surface structure and irregularities of different size and various slopes.

7.5. A DIC microscope for testing cross-sections of optical fibres

The principle of the Nomarski DIC method for reflected light microscopy has recently been applied in optical fibre technology for testing the quality of cross-sections of optical fibres (or optical cables), which should be coupled or welded with each other effectively. This quality assessment is necessary to obtain low-loss connections of optical cables used in telecommunication by guided light waves.

Figure 7.29 illustrates schematically the optical system, while Fig. 7.30 shows the appearance of this new instrument developed in Central Optical Laboratory, Warsaw [7.66]. It comprises a single objective Ob whose magnifying power and numerical aperture are, respectively, $40\times$ and 0.65, and a single ocular Oc of magnifying power $15\times$. Thus, the total magnifying power of the microscope is equal to $600\times$. A Nomarski birefringent prism W is placed between the objective

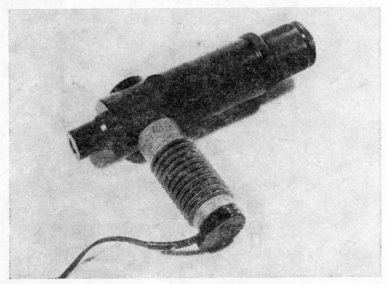

Fig. 7.30. A specialized DIC microscope whose optical system is shown in Fig. 7.29.

and 45°-slanted semitransparent mirror *BS* of a simple vertical illuminator. The splitter *BS* is preceded by a polarizer *P* and followed by an analyser *A*. The prism *W* is slightly inclined with respect to the objective axis and arranged so to be slid transversely. This movement permits the bias retardation to be controlled within the range equal to about $\pm 1.5\lambda$. A low-voltage light source *LS* (6 V/15 W) and a collective lens supplement this highly specialized instrument. The optical fibre or optical cable to be tested is attached to the objective by a special holder. Focusing is performed by moving the objective axially.

Because of the high sensitivity of the DIC method, this microscope permits one to observe very fine irregularities and surface defects of optical fibre cross-sections and therefore to prepare and accept for connection only two fibres whose face surfaces are perfectly flat, smooth, without cracks and other defects, which attenuate the energy of guided light waves.

This microscope is commercially available from the Central Optical Laboratory, Warsaw.

*

*   *

In the literature, some other DIC systems are described for microscopy in both transmitted and reflected light. The Françon interference ocular with a Savart polariscope [7.67, 7.68] and Françon–Yamamoto interference microscope with compensation in the object space [7.69] were at one time manufactured by some firms in France and Japan. Today, however, these instruments have been almost completely replaced in practice by the Nomarski DIC system, although their performance is in many cases comparable with that of the Nomarski devices [7.70].

For the interested reader, more details regarding DIC microscopy suggested by Françon and other researchers may be found in Refs. [1.31] and [1.37].

8. Reflection Contrast Microscopy

This highly specialized technique of reflected-light microscopy is known under various names, such as *interference reflection microscopy* [8.1–8.4], *reflection interference contrast* [8.5, 8.6], *surface contrast microscopy* [8.7, 8.8], *surface reflection interference microscopy* [8.9], although the most popular is *reflection contrast microscopy* [8.10–8.16]. It was initially introduced for studying thin cells and some cell processes (adhesive behaviour of biological material, cell–cell or cell–substrate contacts) in living preparations [8.1–8.3, 8.7, 8.8, 8.10], and later some new applications were developed for the evaluation of stained cyto- and histological objects [8.9, 8.14].

8.1. Optical principles of RC microscopy

Reflection contrast (RC) microscopy is based on both light reflection phenomena and the effects of thin film interference, which were discussed in Volume 1 (see Subsections 1.5.4 and 1.6.3). In general, equipment for the RC method consists of an inverted reflected-light microscope containing some additional optical elements which completely reduce stray light reflected by the lens and glass surfaces of the preparation. Since RC microscopy is mainly used in the study of cells and tissues cultured in vitro, it will therefore be useful to discuss its principle within the scope of such use.

8.1.1. Light reflection and interference in RC image formation

Figure 8.1 illustrates a situation where a cell C is in contact with a substrate S (a glass plate, for example) and surrounded by an immersion medium M. Let n_c, n_s, and n_m denote the refractive indices of the cell, substrate, and immersion medium, respectively. All these indices are not equal to each other, say $n_m < n_c < n_s$. In this case, only three interfaces are of major importance as a source of light reflection responsible for the visualization of the cell. These are: substrate–cell (s–c) interface, cell–immersion medium (c–m) interface, and substrate–immersion medium (s–m) interface.

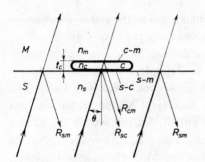

Fig. 8.1. Schematic cross section of a cell spread on a glass substrate.

The reflectance ϱ due to an interface is defined by the refractive indices of the two media that are in contact. This quantity is, in general, expressed by Fresnel equations (1.79). If, however, the angle of incidence θ is equal or close to zero, Eq. (1.84) can be used for the situation shown in Fig. 8.1, and rewritten as

$$\varrho_{sc} = \left(\frac{n_c - n_s}{n_c + n_s}\right)^2, \tag{8.1a}$$

$$\varrho_{cm} = \left(\frac{n_m - n_c}{n_m + n_c}\right)^2, \tag{8.1b}$$

$$\varrho_{sm} = \left(\frac{n_m - n_s}{n_m + n_s}\right)^2, \tag{8.1c}$$

where ϱ_{sc}, ϱ_{cm}, and ϱ_{sm} are the reflectances for the interfaces s–c, c–m, and s–m, respectively. If $n_s > n_c > n_m$, then $\varrho_{sm} > \varrho_{sc}$ and $\varrho_{sm} > \varrho_{cm}$; thus cell-free areas ($s$–$m$ interface) represent the brightest zones, and the cell may be perceived as an area less bright than the background. This is, however, true only when the cell is sufficiently thick (much thicker than the wavelength λ of light used). Otherwise, the cell manifests itself as a thin film, and rays R_{cm} reflected at the c–m interface can interfere with those (R_{sc}) reflected at the s–c interface. Consequently, the cell area becomes modified in brightness by thin film interference effects. This modification tends toward greater or lesser brightness than would follow from Eqs. (8.1). The local brightness of the cell area depends on the optical path difference Δ between rays R_{sc} and R_{cm}. In general, this quantity is expressed by Eq. (1.130), in which n should now be replaced by n_c. After a simple trigonometrical transformation, Eq. (1.130) may be rewritten as

$$\Delta = 2t \sqrt{n_c^2 - n_s^2 \sin^2 \theta}. \tag{8.2}$$

By estimating Δ, it is possible to evaluate the thickness t of the cell if its refractive index n_c is known or, vice versa, n_c may be evaluated if t is known (n_s and θ are the constant parameters, which can be exactly defined for a given experiment).

Estimating the optical path difference Δ is a simple operation if the interference fringes of equal thickness occur (see Subsection 1.6.3 and Fig. 1.64). Otherwise, some photometric procedures are necessary. Moreover, an analysis of phase jumps at interfaces must be carried out. If, for instance, $n_c > n_m$ and $n_s > n_c$, no phase jump occurs between rays R_{sc} and R_{cm}. If, however, $n_c < n_m$ and $n_s > n_c$, rays R_{cm} suffer from a jump of phase by π at the c–m interface, and a term equal to $\lambda/2$ must be added to the optical path difference Δ defined by Eq. (8.2).

Let us assume that the cell is irregular in thickness t, and that its refractive index n_c is greater than that (n_m) of the immersion medium, and $n_s > n_m$. Given these assumptions, bright interference fringes cover the cell where its thickness t is defined by

$$t = \frac{m\lambda}{2\sqrt{n_c^2 - n_s^2 \sin^2 \theta}}. \tag{8.3}$$

Here m is the interference order ($m = 0, 1, 2, \ldots$). Similarly, the centres of dark interference fringes run through those points where

$$t = \frac{(2m+1)\lambda}{4\sqrt{n_c^2 - n_s^2 \sin^2 \theta}}. \tag{8.4}$$

By identifying m, the local cell thickness may be determined if other parameters (λ, n_c, n_s, and θ) are known.

The above relations are valid for a system for which there is no phase jump at the s–c or c–m interface. This is the case when $n_c > n_m$ and $n_s > n_c$. If, however, $n_s > n_m$ but $n_c < n_m$, then a phase jump equal to π occurs at the c–m interface, and bright interference fringes run across these points of the cell where its thickness is defined by Eq. (8.4), while dark fringes run across the cell points which satisfy Eq. (8.3). This is the reverse of the previous situation ($n_c > n_m$).

Figure 8.2 illustrates a situation which differs slightly from Fig. 8.1 in that the cell C is no longer in contact with the substrate S, but is separated by a thin layer of the immersion medium M. Here too are three interfaces of major importance as a source of reflected light responsible for the formation of the RC image. These are: substrate–immersion medium (*s–m*), immersion medium–cell (*m–c*), and cell–immersion medium (*c–m*). Two of them (*s–m* and *c–m*) remain unchanged; the *s–c* interface, however, is now replaced by the *m–c* interface, which introduces some additional factors to be considered.

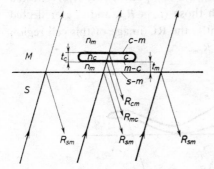

Fig. 8.2. As in Fig. 8.1, except that there is no direct contact between the cell and glass substrate.

If the distance t_m between the substrate and the cell is of the order of light wavelength, the subcell layer of the immersion medium functions as a thin film and produces its own interference pattern. This pattern enables the distance t_m to be estimated from Eqs. (8.3) and (8.4), in which n_c should only be replaced by n_m. This is possible when the cell is thick enough to avoid interference of the ray R_{cm} with R_{mc} and R_{sm}. Otherwise, the cell can produce its own thin film interference pattern which influences that produced by the subcell film of the immersion medium; a resultant RC image follows from interference of light wave components reflected from three interfaces: *s–m*, *m–c*, and *c–m*. Such an interference image is complicated and unsuitable for an accurate estimation of the cell–substrate distance t_m or the cell thickness t_c.

Preparations such as shown in Figs. 8.1 and 8.2 are somewhat theoretical. A more realistic situation is represented by Fig. 8.3, where several characteristic regions may be distinguished: (1) a thin portion of the cell *C* is in contact with the substrate *S*, the RC image of this portion results from interference of rays R_{sc} and R_{cm}; (2) another thin cell fragment is slightly separated from the substratum, the RC image is a result of interference of rays R_{sm}, R_{mc}, and R_{cm} reflected from three interfaces; (3) a thick portion of the cell is in contact with the substratum, rays R_{sc} and R_{cm} cannot interfere effectively as the optical path difference produced by them is too large for moderately coherent light, thus the RC image results only from the difference in light reflection at the *s–c* interface (rays R_{sc}) and the *s–m* interface (rays R_{sm}); (4) another thick cell fragment is slightly separated from the substratum, and the subcell layer of the immersion medium acts as a thin film whose interference pattern may be observed; here ray R_{cm} does not interfere effectively with R_{sm} and R_{mc} as well; (5) a thin cell portion is significantly removed from the substratum, and in this case the light wave component (ray R_{sm}) reflected from the *s–m* interface cannot interfere with those (rays R_{mc} and R_{cm}) reflected from the *m–c* and *c–m* interfaces; consequently, the RC image of this cell region results from interference of rays R_{mc} and R_{cm}.

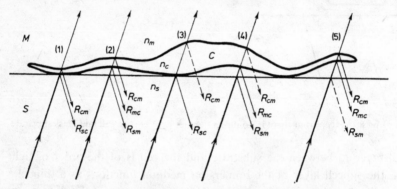

Fig. 8.3. A more typical situation than those shown in Figs. 8.1 and 8.2.

The above discussion shows that the RC image of the whole cell may be highly complicated and its accurate quantitative evaluation is rather difficult.

8.1.2. Discrimination between pure reflection and interference

In a RC image, pure reflection phenomena and interference effects may be distinguished by varying the optical conditions of the experiment [8.16]. One way depends on two-wavelength illumination. If thin film interference takes place and its fringes are infinitely enlarged or not perceived, then the intensity of the RC image is altered when one light wavelength (λ_1) is replaced by the other (λ_2). This intensity alternation results from the fact that the optical path difference is highly dependent of the wavelength. Therefore, all regions of the cell which are not in contact with the substrate (regions 2, 4, and 5 in Fig. 8.3, for instance) manifest themselves as areas of variable intensity, while some contact zones (3 in Fig. 8.3, for example) remain unchanged when the wavelength of used light is varied.

When not influenced by interference, the intensity of reflected light is approximately independent of the wavelength of incident light. In reality, however, spectral dispersion of the refractive index causes a change of reflected intensity, but this alteration is very small and can be neglected. For quantitative evaluation of intensity changes, the cell–free area of the preparation can be used as a reference for measuring intensities in different cell regions, provided unpolarized or circulary polarized light is used; otherwise, the problem of polatization-dependent reflection arises.

The second way of distinguishing cell–substrate contacts depends on the use of variable illuminating apertures. As with the two-wavelength method, thick cell portions that are in contact with the substrate are clearly delineated, while regions perturbed by interference change their intensity with altered illuminating apertures.

8.1.3. Evaluation of the RC image by measurement of its reflectivity

To use reflectivity for the quantitative evaluation of RC images, all reflected light must be accepted by the microscope objective. This condition is fulfilled for the reflected-light microscope with a bright-field illumination system if rays emerging from the objective reflect from a plane interface which is at right angles to the optic axis of the objective. However, when the interface is not perpendicular, the objective gathers only these reflected rays whose direction of propagation does not leave its aperture angle.

One of the methods used to evaluate the reflectivity of a preparation is based on the densitometry of a photographic negative of the RC image. Photographic density over background in cell–free areas of the preparation is used as a reference reflectance (ϱ_{sm}). If, for instance, $n_s = 1.515$ and $n_m = 1.335$, then Eq. (8.1c) gives us $\varrho_{sm} = 0.381\%$. This procedure needs a photographic material with a large dynamic range, since the negative must be recorded within the straight segment of the Hurter–Driffield curve (see Fig. 11.3b in Chapter 11).

When the reflectance ϱ_{sc} is measured, the refractive index n_c of the cortical cytoplasm of a cell may be determined from Eq. (8.1a), which can be rewritten as

$$n_c = n_s \frac{1 + \sqrt{\varrho_{sc}}}{1 - \sqrt{\varrho_{sc}}}, \tag{8.5}$$

where $\sqrt{\varrho_{sc}}$ is a negative or positive quantity depending on the relation between two refractive indices n_s and n_c. If $n_s > n_c$, as is usually the case, $\sqrt{\varrho_{sc}}$ is a negative value. Let us assume, for instance, $n_s = 1.525$ and $\varrho_{sc} = 0.00287$ $(= 0.287\%)$, then $\sqrt{\varrho_{sc}} = -0.054$, and $n_c = 1.37$.

The above consideration are valid for a RC image which is not disturbed by interference. As was stated previously, rays reflected from the m–c or c–m interface can interfere with those reflected from the s–m interface. Cells cultured in vitro are usually attached to glass substrate whose refractive index is equal to about 1.515. An aqueous culture of refractive index lower rather than 1.4 is used as an immersion medium. In this case, the highest reflection intensity occurs at the s–m interface, where the cell-free areas of the preparation are to be found. Consequently, cell-occupied areas brighter than cell-free ones can only be produced by constructive interference, while darker areas may result from either destructive interference or a reduced reflection caused by cell fragments being in contact with the glass substrate. Therefore, in order to measure reflectivity, it is first necessary to analyse the character of the reflection.

8.2. Instrumentation of RC microscopy

The beginnings of RC microscopy go back to Ambrose [8.7], who used a special light microscope for studying the locomotion and adhesion of cells in tissue cultures, and especially to Curtis [8.1], who used a type of interference reflection microscopy with a high-intensity vertical illuminator (mercury arc lamp) for the study of the attachment of living cells to glass. The latter system was improved by Izzard and Lochner [8.3], mainly by increasing the numerical aperture of illumination. Next, Ploem introduced oblique illumination by using an annular aperture diaphragm in combination with phase contrast microscopy [8.17]. Further improvements, largely the work of Ploem [8.10], proved the basis for the design of the modern RC microscope, which has been available from E. Leitz Wetzlar for several years [8.11, 8.12].

8.2.1. Leitz–Ploem RC microscopy

The Leitz–Ploem RC equipment consists of the following units and elements (Fig. 8.4): inverted biological microscope Diavert, reflected-light illuminator Ploemopak 2.2 (or 2.1) with HBO50- or HBO100-lamp and annular aperture diaphragm, transmitted-light illuminator with a 100 W halogen lamp, and phaco-condenser (not shown in Fig. 8.4), special *immersion-contrast objectives* NPL Fluotar 50/1.00 Oil Phaco 2RK and NPL Fluotar 100/1.32 Oil Phaco 2RK with rotatable quarter-wave plate, polarizer, and analyser.

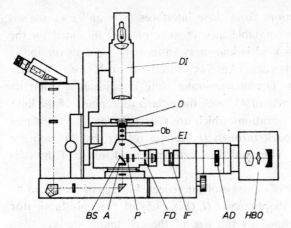

Fig. 8.4. Leitz Diavert microscope equipped with reflection contrast device according to Ploem; *HBO*—high-pressure mercury lamp (50 or 100 W), *AD*—annular diaphragm, *IF*—selective interference filter (peak wavelength $\lambda = 546$ nm), *EI*—Ploemopak epi-illuminator, *FD*—field diaphragm, *P*—polarizer, *BS*—interference beam-splitter (dichroic mirror), Ob—immersion contrast objective, *O*—specimen under study, *DI*—dia-illuminator, *A*—analyser.

Figure 8.5 shows the optical pathway in the Leitz–Ploem reflection contrast microscope. The intense light beam emitted by the Hg-lamp is deflected by the special *interference beam-splitter BS (dichroic mirror)*[1] of the epi-illuminator onto the oil immersion objective Ob below the microscope stage *T*. Part of this light is reflected by the glass surfaces of the objective lenses and becomes a source of stray light (*SL*). To eliminate this from the image of the object under study, a polarizer *P* is used before the beam-splitter *BS* to polarize linearly the light beam emitted by the Hg-lamp. Another polarizer (analyser) *A* is placed below *BS* and crossed with *P* so that the unwanted stray light *SL* is extinguished. The space between the front lens of the objective Ob and the substrate *S* of a preparation is filled with immersion oil *I* to give homogeneous immersion (see Subsection 2.3.5 in Volume 1 for definition). A glass cover slip is commonly used as the substrate.

Thanks to homogeneous immersion, the first distinct interface encountered is between the upper surface of the glass slip *S* and the mounting (culture) medium *M* (the *s–m* interface) or the cells *C* attached to the slip *S* (the *s–c* interface). The reflection of light at these two interfaces is defined by Eqs. (8.1c) and (8.1a), respectively, and is of low intensity. A very intense light source must therefore be used.

[1] A detailed description of a beam-splitter of this kind is given in Chapter 9 (see Subsection 9.3.3).

In order to visualize reflections from these interfaces (s–m and s–c), the oil immersion objective Ob has a rotatable quarter-wave plate Q mounted on the front lens. An objective of this kind is known as an *antireflection system* and is available from C. Zeiss Oberkochen (Antiflex objectives) or E. Leitz Wetzlar (*immersion-contrast objectives*). The quarter-wave plate is adjusted so that its axes (fast and slow) form an angle of 45° with the polarization plane of the light incident on it; thus, the light vibrations which are reflected from the interfaces of the preparation and pass back through Q become rotated by 90° and are transmitted by the analyser A. The result is a well contrasted image of the cells under study (Fig. XLIV).

An additional improvement of contrast of the reflected image of a cell can be obtained by closing the field diaphragm FD (Fig. 8.5) of the epi-illuminator to the necessary minimum defined by the size of the cell under study.

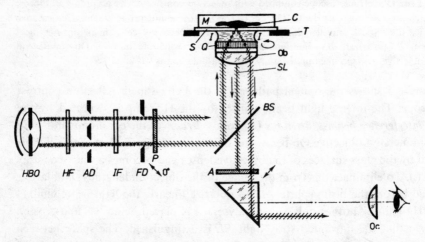

Fig. 8.5. Pathway of light in the Leitz-Ploem reflection contrast microscope; *HBO*—high-power light source, *HF*—heat filter, Oc—ocular (see text and Fig. 8.4 for further explanation).

The performance of reflection contrast microscopy can be further bettered by using axially symmetric oblique illumination. This is achieved by inserting an annular aperture diaphragm AD into the epi-illuminator. The opaque central circle of this diaphragm is adjusted so that its image in the exit pupil of the objective Ob is larger than the outer diameter of the phase ring. Therefore, this ring does not significantly disturb the RC image. Antireflection phase-contrast objectives are especially recommended when a comparison between the reflection-contrast and transmitted-light phase-contrast images may be useful. This combination is obtained simply by using additionally a *LWD* phase condenser mounded above the microscope stage, and by changing illuminating systems (Fig. 8.4).

With a proper set of filters, the Ploemopak epi-illuminator can also serve for fluorescence microscopy [8.10]. In fact, this illuminator was originally designed for use in highly advanced fluorescence microscopy with epi-illumination (see Chapter 9). Thus, the Leitz–Ploem system makes it possible to perform reflection-contrast microscopy, phase contrast technique with dia-illumination, and fluorescence microscopy with epi-illumination consecutively on the same specimen.

8.2.2. Other systems for RC microscopy

Equipment for RC microscopy is frequently the result of "do-it-yourself" ingenuity. With the exception of E. Leitz Wetzlar, there are no firms making specialized instruments for the technique, although there are a few manufactures whose products may be suitably combined for RC microscopy. Opas and Kalnins, for instance, performed RC microscopy by using either a Photomicroscope II or an IM35 inverted microscope available from C. Zeiss Oberkochen. The Photomicroscope II was equipped with an insert carrying the aperture diaphragm by which the illuminating numerical aperture could be controlled within the range 0.37–1.17. The Hg200-lamp in combination with appropriate filters was employed as a source of monochromatic light, while a 100 W halogen lamp provided white light illumination. For both the RC technique and fluorescence microscopy, Antiflex Neofluar objectives (especially 63/1.25 oil immersion objective) were used in combination with the III RS epi-illuminator (epi-condenser). Stray light in the above system was suppressed according to the procedure described earlier in Subsection 8.2.1.

Another system made use of the IM35 inverted microscope fitted with a 100 W halogen lamp, 2FL filter housing with a semitransparent mirror, an intermediate tube with iris diaphragm for varying the illuminating numerical aperture, and the Antiflex Neofluar oil immersion objectives.

According to Opas and Kalnins, the 63/1.25 Antiflex Neofluar (C. Zeiss Oberkochen) lent itself very well to the combined RC technique and fluorescence microscopy, but produced a disturbing field curvature, while the NPL Fluotar 50/1.00 Oil Phaco 2RK objective (E. Leitz Wetzlar) provided a flat image field, but was restricted to lower illuminating numerical apertures.

On another occasion [8.18], Opas successfully used He–Ne laser illumination for interference reflection-contrast microscopy of the adhesion of Amoeba proteus.

8.3. Application of RC microscopy

As was mentioned earlier, RC microscopy is mainly used in cell biology for the study of the adhesive behaviour of cellular material, especially cell–substrate or cell–cell contact phenomena. Ambrose was one of the first researchers who suggested these studies [8.7, 8.8]. Next, Curtis successfully applied RC microscopy in his studies of chicken heart fibroblasts [8.1]. He used a simplified form of the theory (normal incidence of light) for observed RC image to obtain some quantitative parameters relating to cell–substrate contacts. Izzard and Lochner [8.3], however, showed that in this study the application of the simplified theory and the disregard of angle-of-incidence effects could lead to significant discrepancies and a serious underestimation of the distances involved [8.19–8.22].

The attachment of mouse peritoneal macrophages and cultured rat fibroblasts to glass slips was studied by Ploem [8.10]. The RC images showed well-defined boundaries of the cell–surface complex. The rat fibroblasts presented numerous small marginal extensions and very long thin extensions with lengths of up to several times the cell diameter. The mouse peritoneal macrophages showed areas of attachment, and interference fringes indicated a slope of the cell–surface complex facing the glass slip. Ploem concluded that RC microscopy provided a sensitive instrument for studying the cell–surface complex of living cells and its dynamic responses to various stimuli.

Thick fibroblasts, their contacts with the substrate, and relation to the microfilament system were also examined by Heath and Dunn [8.23], while Cottler-Fox et al. studied the process of epithelial cell attachment to the glass substrates [8.24].

Next, adhesion patterns of cell interaction [8.25], cell-to-substrate adhesions during spreading and locomotion of carcinoma cells [8.26], and in vitro motility of cells from human epidermoid carcinomas [8.27, 8.28] were studied by Haemmerli, Ploem, and other researchers.

Mammalian cells in culture (BHK-21, PtK2, Friend, human glia, and glioma cells) were also studied by Breiter-Hahn et al. [8.16]. Reflection contrast images photographed at two different wavelengths (546 and 436 nm) or at two different angles of light incidence allowed discrimination between pure reflected rays and light that was both reflected and modulated by thin film interference. The authors stated that the reflected-light intensity from the preparation interfaces was influenced by the cell–substrate distance t_m (see Fig. 8.2), as well as by the angle of incidence and wavelength of light used if t_m was at least equal to 50 nm. The reflectance at the s–m interface was used as a reference standard for calculating the refractive index of the cortical cytoplasm. Refractive indices were found to be higher (1.38–1.40) at the so called "focal contacts" (points of insertion or centres of stress fibre organization), than in areas of close contact (1.354–1.368), while between focal contacts in areas of the cortical cytoplasm not adherent to the glass substrate, refractive indices appeared to be within the range 1.353–1.368. These results were thought to derive from a microfilamentous network within the cortical cytoplasm.

These estimates were made by Breiter-Hahn et al. assuming a normal incidence of light and ignoring its actual conical shape. Next, Beck and Breiter-Hahn [8.22] showed that this assumption was only valid under certain conditions, depending on the accuracy of measurement required; an angle of incidence θ equal to about 30° results in an error of about 10%, and $\theta \approx 50°$ in one greater than 50%.

A further interesting example of the application of RC microscopy was given by Piper and Pera [8.15], who suggested a procedure for the reconstruction of the surface profile of normal an pathological erythrocytes from their RC images. Independently, Breiter-Hahn et al. [8.4] used a similar procedure for reconstructing the thickness profile of cultured cells.

Current interest in the organization and function of the cytoskeleton has led to the application of RC microscopy to the observation of high resolution images of detergent resistant residues of cultured cells stained with protein dyes [8.9]. Somewhat earlier interesting advances were made in histology, hematology, and cytogenetics by using microscopy of this kind to the study of fixed and stained cells and tissues [8.13, 8.14]. An improvement of image contrast against both bright field and phase contrast was observed particularly in red-stained structures (compare Fig. XLV). In HE-stained histological preparations, the elastic fibres become clearly visible, while other weakly red-stained structures (collagen fibres, muscle fibres, erythrocytes, and bone) are also distinctly accentuated in RC images. In blood smear specimens stained after May-Grunwald and Giemsa, the platelets and the granules of leukocytes show a strong light reflection. Moreover, the structure of chromosomes stained according to the Feulgen procedure is particularly clearly demonstrable in RC images. Although reflection contrast is not suitable for the specific demonstration of some structures, it can emphasize the contrast between slight colour variations that would otherwise not be optically discernible so as to enhance the amount of information obtainable from ordinary stained specimens.

For the interested reader, more details regarding the applications of reflection contrast microscopy may be found in an extensive review article "Interference reflection microscopy in cell biology: methodology and applications" by H. Ver-schueren, published in *J. Cell Sci.*, **75** (1985), 279–301. This article shows that RC microscopy is a useful innovation within major trends in modern light microscopy, and its importance lies primarily in its unique suitability for studying cell-substrate adhesion of living, moving cells. However, much confusion remains about the principles underlying formation and interpretation of images in RC microscopy. Such a situation prevents this technique from becoming more popular among cell biologists.

*

*   *

While reflection contrast microscopy can certainly be regarded as a useful technique for investigating the mechanism of cell adhesion to the glass substrate and of cell movements and related phenomena, any quantitative results obtained by this procedure must approached with a critical eye. Acceptance of the idealized procedures of quantitative processing of RC images must depend on preinter-pretations that are only possible when the object under study is sufficiently known from observations made by other methods [8.22]. After all, RC microscopy is still a relatively young technique which offers scope for much further research and detailed applications.

In particular, recent examples of greatly effective observations of fixed, stained or unstained preparation, such as histological sections, blood smears, chromo-somes, cytoskeletal structures, suggest that RC microcsopy will be further devel-oped as a useful research technique for an increasing number of microscopists.

9. Fluorescence Microscopy

Fluorescence is a kind of luminescence that occurs when a substance is excited by radiation. The wavelength of the emitted fluorescent light is normally longer than that of the exciting radiation (see Section 1.3 in Volume 1). Hence, *fluorescence microscopy* is a specialized technique for the study of substances which can be made to fluoresce.

The beginnings of fluorescence microscopy go back to before 1904, when A. Köhler first made some study of it. The earliest notable advance, however, came in 1911, when K. Reichert (the founder of the famous firm manufacturing microscopes in Vienna) and O. Heimstädt demonstrated the first practical fluorescence microscope. This type of microscopy was then shown to be a very useful tool for the study of plant cells and tissues which, as a rule, show relatively intense *primary fluorescence* when illuminated with ultraviolet and sometimes even violet light. The term "primary fluorescence" is used for the *inherent fluorescence* of the molecules of a substance. In general, all materials are inherently fluorescent but in some instances their spectrum of fluorescence is too weak for visual observation or invisible since it lies in the infrared or even ultraviolet region. Primary fluorescence is also called *autofluorescence*. The latter term is used especially to describe unwanted fluorescence (autofluorescence of microscope lenses, for instance). Unlike botany, zoology did not initially offer much scope for fluorescence microscopy, since animal cells and tissues are characterized by weak primary fluorescence. In 1933, this limitation was overcome by M. Haitinger, who was the first to use *secondary fluorescence* for the study of biological specimens [9.1]. Fluorescence of this kind is produced by staining the specimens with fluorescent dyes known as *fluorochromes*. These are dyes which are used in a similar fashion to histochemical stains. Depending on their nature, they attach to specific substances of some specimen areas and leave others unstained. This process is referred to as *fluorochromization*. It is worth noting that even extremely low concentration of fluorochromes (10^{-18} g, for example) can be detected with a fluorescence microscope.

In 1941, there came a milestone in the progress of fluorescence microscopy

when A. H. Coons proposed the specific fluorochromization or "labelling"
of antibodies [9.2–9.4], and next developed, together with N. H. Kaplan, the
famous FITC (*fluorescein isothiocyanate*) procedure for identifying specific
proteins in biological samples. The Coons method, known as *immunofluorescence
microscopy*, began to be widely used in the late 1950s and early 1960s, and is now
standard practice.

An important contribution to fluorescence microscopy was also made by
E. M. Brumberg [9.5, 9.6], who introduced interference filters (*dichroic mirrors*)
as part of the fluorescence equipment, and J. S. Ploem [9.7–9.10], who developed
his epi-illuminator (*Ploemopak 1* and *Ploemopak 2*) with interchangeable dielectric
mirrors and interference filters for both single- and two-wavelength fluorescence
microscopy.

9.1. General principles and types of fluorescence microscopy

In fluorescence microscopy, excitation light (*EL*) incident on a specimen under
examination is removed, while fluorescence light (*FL*) is led to the observer's
eye (Fig. 9.1). This requires a light source (*LS*) providing ample short-wavelength
light (UV and/or blue-violet), such as a high-pressure mercury lamp (HBO50

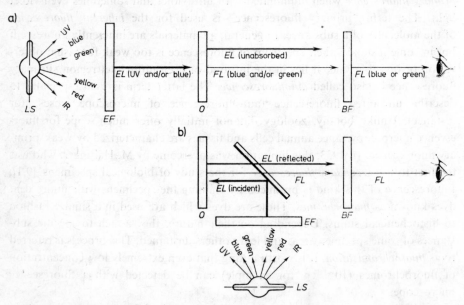

Fig. 9.1. Principle of fluorescence microscopy: a) dia-excitation. b) epi-excitation.

or HBO200). Frequently, a high-power (12 V/100 W) halogen lamp is also used, e.g., for blue excitation. The wavelengths required for fluorescence excitation are selected by an *exciter (primary) filter (EF)*, which transmits only exciting radiation and suppresses all wavelengths unsuited to excite fluorescence. Part of the exciting radiation is absorbed by the specimen *O* and simultaneously reemitted at altered (longer) wavelengths as fluorescence light *(FL)*. The latter is transmitted by a *secondary* or *barrier filter (BF)*. The rest of the exciting light which passes through (Fig. 9.1a) or reflects from the specimen (Fig. 9.1b) must, of course, be absorbed by this filter. Thus a coloured image is observed against a dark background. The darker the background of the field of view, the higher the performance of the fluorescence microscope. This, in turn, depends primarily on the selection of exciter and barrier filters. The problem is illustrated by Fig. 9.2, on which curve *EL* and *BL*, respectively, represent the transmission spectra of the exciter and barrier filters, while dashed curve represents the emission spectrum of fluorescence light. The latter should lie within the spectrum *BL* of the barrier filter, while, at the same time, the spectra *BL* and *EL* should not overlap. In practice, the latter condition cannot be ideally fulfilled, and there is always a small overlapping region *OR*; the smaller this is, the better the performance of the fluorescence microscope.

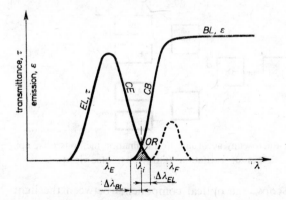

Fig. 9.2. Illustrating the problem of the suppression of exciting light in the image plane of a fluorescence microscope.

Let us suppose that τ_{i1} and τ_{i2} are transmittances of the exciter and barrier filters for the wavelength λ_i for which curves *EL* and *BL* intersect. The resultant transmittance τ_i for λ_i is defined by $\tau_i = \tau_{i1} \tau_{i2}$ (compare Eq. (9.3)). The overall brightness of the background is, however, defined by the area of the overlapping region OR, i.e., by the total (integral) transmittance τ_t in the spectral range from $\lambda_i + \Delta\lambda_{EL}$ to $\lambda_i - \Delta\lambda_{BL}$, where $\Delta\lambda_{EL}$ and $\Delta\lambda_{BL}$ are the distances from λ_i to wavelength values for which $\tau_1 = 0$ and $\tau_2 = 0$. It is clear that

$$\tau_t = \int_{\lambda_i - \Delta\lambda_{BL}}^{\lambda_i} \tau_1(\lambda)\tau_2(\lambda)\,d\lambda + \int_{\lambda_i}^{\lambda_i + \Delta\lambda_{EL}} \tau_1(\lambda)\tau_2(\lambda)\,d\lambda. \tag{9.1}$$

For a fluorescence microscope of high quality, the integral transmittance τ_t should be lower than 0.00005. This value can only be achieved by using special (interference) filters whose cut-off edges (CE and CB, Fig. 9.2) are very abrupt.

The specimen under examination may be excited either by transillumination (Fig. 9.1a) or epi-illumination (Fig. 9.1b), so that we can speak of dia- and epi-fluorescence microscopy. Each of these types of illumination can be achieved by using either bright-field (BF) or dark-field (DF) illuminating systems. Dark-field epi-illumination is, however, less useful than the other types and will not be considered here. Depending on its nature, the specimen may be excited either by UV radiation or blue-violet light. In general, primary fluorescence is excited by ultraviolet and secondary (fluorochrome) fluorescence by blue-violet (B-V) or even green light. The above types of fluorescence microscopy are summarized in a block diagram in Fig. 9.3.

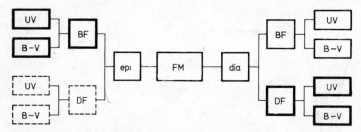

Fig. 9.3. Classification of fluorescence microscopy according to illumination methods (dia, epi, bright-field, dark-field) and spectral ranges of excitation (ultraviolet, blue-violet).

For UV fluorescence microscopy, the optical components between the light source and specimen (lamp collector, substage condenser) are usually made of quartz glass, since ordinary optical glasses absorb UV radiation of wavelengths smaller than 360 nm. For this reason, UV fluorescence microscopy with a bright-field type of epi-illumination is inconvenient; objectives should be made of quartz glass and/or other materials transparent for UV radiation and free from auto-fluorescence. The latter condition must also be fulfilled for UV fluorescence microscopy with bright-field dia-illumination because, in addition to fluorescent light, the exciting UV radiation is also present between specimen and objective (Fig. 9.1a). If, as is usually the case, UV light is intense, special objectives free from autofluorescence must also be employed. This inconvenience is reduced when dark-field illumination is applied. A dark-field condenser has the advantage that a darker background of the field of view may be obtained, and only a small portion of UV radiation scattered forwards by the specimen enters the objective.

In UV fluorescence microscopy, the mountant media must also be free from autofluorescence. This requirement, therefore, precludes the use of Canada balsam; instead, water or glycerol and pure liquid paraffin are suitable for temporary mounts of preparations. Cederwood oil also suffers from autofluorescence and is unsatisfactory as an immersion liquid for oil immersion objective and condensers. Instead, glycerol or sandalwood oil, saturated liquid paraffin or special synthetic immersion oil (Cargille immersion oil A) should be used.

All restrictions mentioned above are less important if fluorescence is excited with blue light, as is usually the case in fluorochrome technique (secondary fluorescence); to begin with, normal glass optics can be used before and behind the specimen. In this case, transmitted-light dark-field illumination is also more advantageous than bright-field illumination. From a dark-field condenser (see Fig. 6.6), the exciting rays fall onto the specimen under such an acute angle that they do not enter the microscope objective when they pass directly by the specimen. Exciting radiation and fluorescent light are therefore spatially separated from each other, and the small unwanted portion of scattered exciting rays can very easily be removed from the fluorescent light with a barrier filter. Dark-field fluorescence is the technique in general use today, especially for FITC fluorescence, because of the small distance $\lambda_F - \lambda_E$ (Fig. 9.2) between the peak wavelengths of the exciting and emission spectra of this fluorochrome.

One may ask, which is preferable, dia- or epi-fluorescence microscopy? Up to two decades ago, there was no such alternative—fluorescence microscopy with dia-illumination was in a privileged position. The situation has, however, changed with improving fluorochrome techniques and the development of interference filters more suitable for FITC immunofluorescence microscopy. Initially the FITC method and its improvements were introduced in the transmitted-light dark-field technique, but certain difficulties in adjusting the dark-field condensers were found unacceptable by many users. The front surface of these condensers must be connected by immersion oil with the bottom surface of the specimen slide. During this operation small air bubbles frequently get in the way and degrade the image quality of the fluorescent specimen. Moreover, some microscopists find it difficult to centre and focus the condenser, especially when there is only weak fluorescence or none at all. These handicaps do not occur in bright-field epi-fluorescence microscopy, where the objective acts simultaneously as condenser; there are therefore no problems with condenser centring, and a released optical system for dia-illumination may be used for combining epi-fluorescence microscopy with any desired transmitted-light technique (phase contrast, for instance).

With dia-illumination, the lower specimen layers facing the light source are subject to primary excitation, and the emitted fluorescent light must pass through the specimen before reaching the objective and is therefore partly absorbed and scattered. As a result, the fluorescence image becomes diffuse and less intense. This defect does not occur when the specimen is excited with epi-illumination since both excitation and observation of the specimen are undertaken on the same side. The fluorescence image is therefore brighter and its general quality is better, especially when thick histological sections are examined.

All these circumstances ensure that today—especially in view of the FITC method and the modern instrumentation available for it (dichroic interference beam-splitters, interference exciter filters)—bright-field epi-fluorescence micro-scopy has come to be generally preferred.

9.2. Immunofluorescence microscopy

Immunofluorescence is a special technique of fluorescence microscopy used for detecting antigens via *immune reactions*. Since the introduction of *fluorescent labelled antibodies* in laboratory practice in 1950 by Coons and Kaplan [9.3], this technique has been increasingly used for both research and routine investigations [9.4, 9.11–9.75], and is nowadays the most widely used method of fluorescence microscopy.

The principles upon which immunofluorescence is based were, of course, known earlier by microbiologists and immunologists, especially since the development of the antiglobulin assay in 1945, but Coons and Kaplan contributed most to the fluorescent visualization and localization of *antigen–antibody reactions* in biological specimens.

Human and animal tissues contain protein compositions, which are specific to a given species. Foreign proteins, especially those of viruses and bacteria, are hostile (pathogenic) to a given organism. Such foreign agents are referred to as *antigens* (AG). To defend itself against an antigen, the human body or other organism produces a specific protein, referred to as an *antibody* (AB),

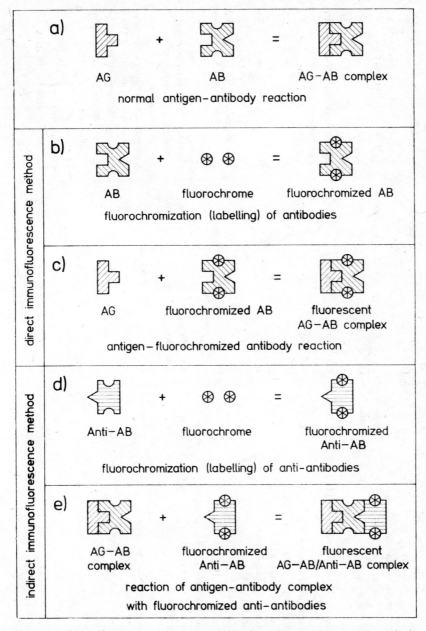

Fig. 9.4. Schematic diagram of direct and indirect immunofluorescence methods.

TABLE 9.1
Fluorochromes used in fluorescence microscopy [2.7, 9.76, 9.98, 9.107]*

Name of the fluorochrome	Spectral bands [nm]		Light wavelength maximally		Main field of application and/or suitable for
	Absorption (excitation)	Emission	absorbed [nm]	emitted [nm]	
Fluorescein isothiocyanate (FITC)	450–500	500–550	490	525	Immunology, detection of antigen-antibody reactions
Lissamine-rhodamine B (LRB)	530–570	550–700	568	597	As above
Tetramethyl-rhodamine-isothiocyanate (TRITC)	Green		550	580	As above
Dinaphtylaminesulfonic acid (DNAS)	UV				As above
Evans blue	Green				Red contrasting of FITC staining
Bisaminophenyloxidiazole (BAO), according to Ruch	UV	480–580	280	460	Cytology, DNA determination
Acriflavine	410–470	480–590	445	510	Cytology, determination of nucleic acids
Pararosaniline, according to Feulgen					Cytology, quantitative determination of small amounts of DNA
Formaldehyde-induced fluorescence (FIF), according to Falck	Violet		550	650	Cytology, determination of biogenic amines, especially catecholamine
Dansylchloride	UV				Lysines
Sulphaflavine	UV				Proteins, histones
Acridine orange	380–530	480–600	470	530	Cell nuclei, mucus, bacteria
Berberine sulphate	340–475	420–570	340, 425	472, 532	Cell nuclei, bacteria
Phosphine	425–505	490–585	465	530	Cell nuclei

Substance					Application
Coriphosphine O	400–500		470–660		Cell nuclei, cell walls, blood cells (lymphocytes, leucocytes), mucus, wood (cellulose), bacteria (especially diphteria)
Euchrysine	420–480		470–660		Blood cells (lymphocytes, leucocytes), mucus, wood (cellulose)
Thioflavine S	250–500 UV and green	365	410–510	445	Lymphocytes, leucocytes
Methyl green pyronine stilbene (MPS), according to Ploem	Blue				Blood differentiation by double staining
Aurophosphine G	Green				Mucus
Thiazine red R					Proteins
Auramine O	350–480	430	420–590	535	Tuberculosis, acid-resistant rods
Acridine yellow	350–500	440	460–600	508	Tuberculosis, acid-resistant rods
Primuline O	275–450	350	380–510	425	Cellulose
Tetracycline	390–420 Green		about 560		Bone tissue
Acid fuchsine					Osteons
Atebrin	400–450		500–590		Chromosome bands "drumsticks" (sex determination)
Quinacrine	Violet				As above
Quinacrine mustared	Violet				As above
Quinacrine-mustared-dihydro-chloride (QM)					As above
R-phycoerythrin	430–460 Blue	495	490–530 Yellow	576	Routine immunocytochemical analysis (see Proc. Roy. Micr. Soc., 23 (1988), 47)

* The above list is not complete; for further information the reader can consult Refs. [2.7] and [9.107].

directed against the antigen in order to neutralize it. An important factor in this defence mechanism, referred to as *immunity*, is that the body reacts against antigens and that there is always a specific antibody for every antigen. This reaction starts instantaneously if specific antibodies directed against a given antigen are already present in the organism at the beginning of the infection with this antigen. Such an organism is referred to as *immunized*. Roughly speaking, antibodies attach themselves to the antigens (Fig. 9.4a) and prepare them for the leucocytes, which dissolve and destroy them. If the organism has no antigen-specific antibodies, it becomes ill until it produces its own necessary antibodies or receives some via vaccination.

The desired antibodies can be separated from the serum of a suitable, that is an immunized, patient or animal. Smallpox antibodies, for instance, can be isolated from the serum of a smallpox-immunized patient.

Coons and Kaplan found a method of staining isolated antibodies with fluorochromes and determining, via fluorescence, the antigens against which the antibodies were directed. In general, the method depends on *labelling* (*fluorochromizing*) an antibody solution (Fig. 9.4b), applying it to the specimen containing the suspect (infected) material on a microscope slide, suitably rinsing the slide and placing it on the stage of a fluorescence microscope. The suspect material will fluoresce when and where the fluorochromized antibodies become attached (bound) to the antigens (Fig. 9.4c). This is a positive test. If no fluorescence occurs, the antibodies used were non-specific (not directed against the suspect antigens), and any unbound antibodies were washed out during the rinsing of the slide. In this case, the test is negative, and must be repeated with other labelled antibodies.

When this method was suggested by Coons and Kaplan, the initial problem was how to subject only the antibodies whose reaction we want observe to specific fluorochromization. After a great number of tests, Coons and Kaplan found that FITC (fluorescein isothiocyanate) was a specific fluorochrome of this kind, which labelled only antibodies contained in the specimen to be tested. At present, other specific fluorochromes are known, but FITC is the most universal. Table 9.1 gives an almost complete list of fluorochromes used in fluorescence microscopy.

The immunofluorescence test, which is described above and schematically shown in Fig. 9.4b and c, is known as the *direct antibody technique*. Its defect is that it requires each antibody to be labelled against a particular antigen, which is a rather time-consuming operation [9.76, 9.77]. To overcome this limitation, many versions of the original technique are now in use [9.32, 9.42]. Among these the so-called *indirect* or *sandwich antibody method* has become very popular in medicine. To avoid the fluorochromizing and storing of specifically labelled antibodies for each antigen, *anti-antibodies* are prepared in the laboratory and stored. A laboratory animal (a rabbit, for instance) is infected with human antibodies and immediately produces its own specific antibodies, which are directed gainst human antibodies. These anti-antibodies (Anti-AB) will now react with every human antibody complex as if it were an antigen. The Anti-AB can then be isolated, fluorochromized, and applied to the specimen slide. The test depends on proving whether an antigen-antibody reaction has taken place in the human body or not. If existing antigens have stimulated the production of specific antibodies and these have bound to the former while the latter, in turn, have coupled with Anti-ABs, there is fluorescence and the test is positive; if there were no existing antibodies, there is no fluorescence and the test is negative. Figures 9.4d and e illustrate schematically the mechanism of this antibody technique when a test is positive; a specific human antibody reacts with a hostile antigen and an AG/AB-complex occurs (Fig. 9.4a); a fluorochromized animal anti-antibody (Fig. 9.4d) reacts with the AG/AB-complex, and an AG/AB/Anti-AB-complex arises which fluoresces. Consequently, there must already have been an antigen in the specimen.

The advantages of the indirect antibody technique is twofold [9.77]. Since the fluorochromized anti-antibody is directed against the antibody of a species and not against the antigen in the specimen, it can be applied to a large variety of antibodies with a given primary antigen specificity. An antibody itself has several antigenic determinants, and the binding of fluorochromized anti-anti-bodies greatly enhances the sensitivity of the technique.

The third immunofluorescence antibody method which is also widely used is known as the *mixed antiglobulin* or *aggregation technique* [9.77]. It is based on the bivalency of antibody molecules. Thus, an antigen–antibody complex can bind an additional antigen added in the incubation-mixture if this has the same antigenic determinants as the first antigen. Consequently, a mixed AG/AB/AG-complex is formed, in which the newly added antigen can be recognized.

The performance of each antibody technique depends greatly on the quality of the specimen preparation as well as on the selection of exciter and barrier filters, sources of light, modes of illumination, and other aspects of fluorescence microscopy instrumentation dealt with in the following section. During the last two decads, immunofluorescence methods have been standardized by the introduction of distinctive fluorochromes, the unification of procedures for preparing specimens, and the availability of fluorescence microscopes which incorporate epi-illumination of the Ploemopak type [9.78–9.82]. It is worth noting that several international conferences (Florence 1967, London 1968, Stockholm 1970, Leiden 1974) have been devoted to the problem of standardization in fluorescence microscopy, especially in immunofluorescence.

9.3. Instrumentation of fluorescence microscopy

In general, any transmitted-light and/or reflected-light microscope can be adapted for secondary fluorescence microscopy by adding a suitable light source, exciter filters, and barrier filters. At present, however, leading manufacturers offer instruments which are designed especially for fluorescence microscopy. This practice is right because an advanced fluorescence microscope is of greater importance as an analytical tool than any other microscope. The design of fluorescence microscope must therefore take into account not only qualitative but also quantitative investigations (fluorometry and even spectrofluorometry).

Fundamental requirements relating to light sources, exciter and barrier filters, objectives, condensers, and other optical components of modern fluorescence microscopy will be discussed in the following subsections.

9.3.1. Light sources for fluorescence microscopy

In the past, ultraviolet light sources were required for fluorescence microscopy. Today this is not necessary, as the use of secondary fluorescence has become standard practice both in research and routine work. For the FITC, as well as for some other fluorochromes, tungsten halogen lamps are sufficient. However, the most popular light sources in fluorescence microscopy are high-pressure mercury lamps (HBO200 or HBO50). High-pressure xenon lamps may also be used. Recently, the attraction has become apparent of some types of laser beams as sources of highly monochromatic and intense light for sophisticated investigation both in biological and materials sciences.

High-pressure mercury lamps. Lamps of this kind are manufactured by several firms. The most popular are those available from Osram (HBO) and Philips (CS). Their counterparts are also manufactured in the USSR (Table 9.2). They have the discrete nature of a spectrum with a continuous background (see Fig. 1.13). The main source of luminous energy are the spectral maxima; most of them are in the ultraviolet (both near and far), violet, blue, and green parts of the spectrum. The principal maxima occur for the wavelengths $\lambda = 313, 334, 365,$

TABLE 9.2

Light sources (lamps) used in fluorescence microscopy and their basic parameters

Lamp type and producer	Power [W]	Luminous flux [lm]	Luminance at the centre of the arc or incandescent filament [Mnt = 10^6 cd/m²]	Approximate size of the arc or filament [mm]	Lamp life at the nominal supply [h]	Colour temperature at the nominal supply [K]
HBO50, Osram*	50	1700	300	0.6×1.2	100	—
HBO100, Osram*	100		1700 (average value 200)	0.25×0.25	200	—
HBO200, Osram*	200	9500	330	1.4×2.5	200	—
DRSh-100-Z, USSR	100		1000 (average value 220)	0.3×0.3	100	—
DRSh-250, USSR	250	12500	100	1.5×3.0	250	—
XBO75, Osram**	75		400	0.5×0.5	400	6100
XBO150, Osram**	150		150	0.5×2.2	1200	6000
XBO250, Osram**	250		260	0.7×1.7	1200	6100
XBO450, Osram**	450	2000	350	0.9×2.4	2000	6300
DKsSh-130, USSR	130		50		100	
DKsSh-200, USSR	200		90		500	
CSI250, Philips	250		150		200	3400
Halogen 12 V/100 W	100		40	2.3×4.2	50	3300
Tungsten 6 V/15 W***	15		10	1.8×1.8	100	2800–3000

* CS lamps from Philips are similar to HBO lamps from Osram.

** Xe lamps from Philips are similar to XBO lamps from Osram.

*** This lamp is a popular light source for ordinary bright-field microscopy; it is unsuitable for the excitation of fluorescence, and listed here only for comparison purposes.

406, 435, 546, and 578 nm. The first three are for UV-excitation, the next two for blue-, and the last for green-excitation. The lamps listed as HBO100, CS100, and DRSh100 are characterized by extreme luminance (brilliance) and small arc size, and are therefore suitable for epi-fluorescence microscopy. The HBO50 and HBO200, on the other hand, as well as their CS and DRSh counterparts, are more universal and suitable for both dia- and epi-excitation. In the HBO100 lamp, there is a greater stability of the luminous flux and it is therefore more suitable for microfluorometry. It is interesting to note, however, that the HBO50 and HBO200 lamps are of nearly equal luminance, though the arc size of the former is five times smaller than that of the latter. Hence, while both yield fluorescence images of nearly identical brightness, the HBO200 is less fastidious about its exact alignment and centring, especially when used for transmitted-light dark-field fluorescence microscopy. The HBO50 lamp, on the other hand, is perfectly adequate and more economical for epi-fluorescence microscopy [9.76].

High-pressure xenon lamps. These lamps are primarily available from Osram and Philips as the XBO and Xe models, respectively. Their construction and feeding are similar to those of the high-pressure mercury lamps; they do, however, have a different spectrum (Fig. 9.5), which manifests itself as a continuum of high intensity with a large irregular maximum between $\lambda \approx 800$–1020 nm. This spectrum is weak in the ultraviolet and relatively strong in the near infrared region.

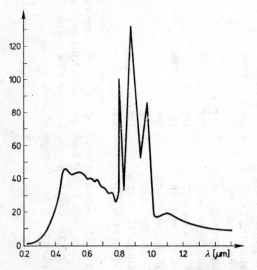

Fig. 9.5. Emission spectrum of a high-pressure xenon lamp.

The latter region is not useful for fluorescence microscopy and must be suppressed by a proper filter added to the main exciter filter. Unlike the high-pressure mercury lamp, the xenon lamp emits more light in the blue-violet region without the holes typical of the mercury spectrum (compare Fig. 1.13). The high-pressure xenon lamps are therefore readily used together with those fluorochromes whose main absorption bands occur in the holes of the mercury spectrum. The HBO lamps (and their counterparts) have no spectral lines within the range between $\lambda = 440$ and 540 nm, where there is a strong radiation continuum for the high-pressure xenon lamps. Moreover, the latter have a sufficiently stable luminous flux and are therefore recomended for fluorometry and spectrofluorometry.

Tungsten halogen lamps. Nearly two decads ago, there was no general agreement about the effectiveness of tungsten halogen lamps in fluorescence microscopy. Today, their usefulness is no longer questioned since a number of important fluorochromes, especially FITC, whose spectral excitation bands lie well within the blue and/or green spectral regions, actually do not require a UV source and the radiation intensively supplied by tungsten halogen lamps (12 V/100 W, for instance) is sufficient to excite satisfactory fluorescence [9.82–9.86]. Present-day microscopes for FITC fluorescence are fitted, as standard equipment, with halogen light sources and interference exciter filters, whose transmission spectrum is suitably adjusted to the emission spectrum of these sources.

Tungsten halogen lamps (known also as *iodine-quartz lamps*) belong to the group of incandescent filament light sources whose emission spectrum is continuous (see Fig. 1.12). The light output and colour temperature of any conventional filament lamp increase as the temperature of the incandescent filament is raised. The life of the filament is however shortened and blackening of the bulb is accelerated. Optimum operating conditions must therefore be a compromise between light intensity and the acceptable life of the lamp. In tungsten halogen lamps, on the other hand, the metal (tungsten) evaporated from the filament is bound by the halogen (iodine) vapour, and tungsten halide compounds are formed which re-deposit metal on the filament. A feed-back of this kind avoids blackening of the quartz bulb and permits both long life and high colour temperature (3300 K). In order to maintain this process, the bulb temperature must be at least equal to 250°C. Like a conventional filament, it has a negligible ultraviolet emission and a much increased emission in the blue and green regions of the spectrum (see Fig. 1.12). Compared with a high-pressure mercury lamp, its luminance is low (see Table 9.2) but sufficient for the FITC, which needs an excitation radiation with a peak wavelength equal to about 490 nm. In view of the hole for the mercury spectrum in this region, the tungsten halogen lamp is generally more convenient for FITC fluorescence microscopy than a high-pressure mercury lamp. Moreover, the latter has certain additional disadvantages: it cannot be switched on and off frequently (this may shorten the lamp's life or even cause its sudden failure); light output drops unavoidably during use; and it is relatively expensive. None of these defects apply to the tungsten halogen lamp.

There are several kinds of tungsten halogen lamp in use, but the most popular is the 12 V/100 W type. Its nominal life is 50 h, but this may be prolonged by applying a lower voltage. Figure 9.6 shows the relation between lamp life and

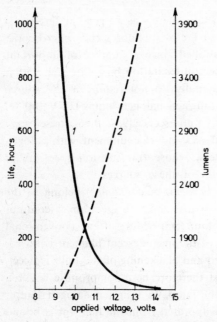

Fig. 9.6. Life (curve *1*) and emission (curve *2*) of the 12 V/100 W tungsten halogen lamp as a function of voltage.

voltage applied. Many fluorescent preparations can be observed at moderate magnifications, for which full lamp power (at 12 V) is not necessary. In fact, most microscopical observations may be performed at reduced voltage and in this case lamp life is significantly increased (150 h at 11 V or 1000 h at 9 V, for instance). However, the voltage cannot be reduced by more than 25%, otherwise halogen circulation may break down and bulb blackening will occur.

CSI lamps. These are metal-halide concentrated arc lamps, whose emission spectrum is approximately similar to that of high-pressure mercury lamps, except that it is almost without ultraviolet while having more intense maxima within the green spectral region (Fig. 9.7). Among light sources of this kind the best known is CSI250 lamp, which was initially designed for colour photomicrography [2.7] as its average spectral distribution of radiation resembles that of the sun. As far as fluorescence microscopy is concerned, this lamp is mainly suitable for exciting fluorochromes whose absorption band lies within the green region of the spectrum (for instance, lissamine-rhodamine B (LRB) or tetramethyl-rhodamine-isothiocyanate (TRIC)).

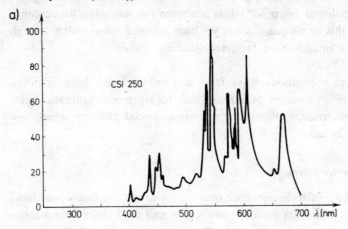

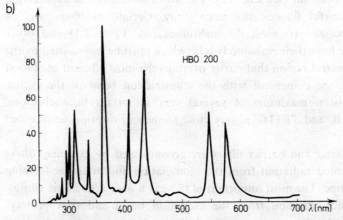

Fig. 9.7. Emission spectrum of the CSI250 lamp (a) compared with that of the HBO200 lamp (b).

Flash-lamps. An off-recurring feature of fluorescence microscopy is the low intensity of emitted light compared to the amount of light absorbed by molecules to promote them to an excited state. Photographic exposures of fluorescence images are therefore long and can vary from seconds to several minutes or more. During exposure, the fluorescence colours may alter and/or fade, thus making it impossible to obtain accurate colour photomicrographs. A further problem arises when fluorescent motile microobjects (vitally fluorochromized cells, for instance) must be photographed.

To overcome these limitations, Young used a flash-lamp for taking colour photomicrographs of fluorescent living cell cultures vitally stained with euchrysine [9.88]. A 100 watt halogen lamp served as a source for screening fluorescence exitation. However, this technique has not yet been adopted more widely though excellent fluorescence images have been recorded by Young.

Lasers. Recently both continuous wave (cw) and pulsed lasers have acquired increasing importance as sources of intense and highly monochromatic light. Laser fluorescence microscopy, however, presents a special problem which will be discussed separately later.

9.3.2. Exciter and barrier filters

In general, spectral filters for fluorescence microscopy are of the *broad pass-band* type; the exciter filter transmits a short-wave region and the barrier filter a long-wave region of the spectrum (see Fig. 9.2). The art of the correct filter selection is decisive for successful fluorescence microscopy. Certain problems are, of course, involved because: (i) most fluorochromes (see Table 9.1) and other fluorescent substance have their emission bands (which must be transmitted by the barrier filter) in a spectral region that partly overlaps the most efficient excitation range (which should be coincident with the transmission band of the exciter filter); (ii) the emission maximum of several very important fluorochromes, including FITC, LRB, and TRITC, is very close to their absorption (excitation) maximum.

As a rule, the exciter and barrier filters are accompanied by *excluding filters* that eliminate unwanted radiation from the illumination and/or image-forming space of the microscope. The most important of these is a *heat filter*. This eliminates infrared (thermic) radiation from the excitation beam and thus protects the optical components of the microscope and the specimen against overheating. Moreover, it reduces the quenching of fluorescence by far red and infrared waves that can residually pass through most ultraviolet or even blue exciter filters. In the past, liquid filters were largely used as heat-excluding filters. Present-day fluorescence microscopes are usually fitted with an interference heat filter, which is installed between the light source and exiciter filters.

Another excluding filter is an *UV-filter* that should be installed just above the microscope objective. Its function is (i) to protect the microscopist's eyes against ultraviolet radiation, and (ii) to eliminate any possible autofluorescence of barrier filters and/or other optical elements in the image-forming space of the microscope.

Filters for fluorescence microscopy (as well as other spectral filters) are characterized by the shape of the *transmission curve*. Terms like "short-wave pass", "long-wave pass", "wide-band", "narrow-band", "cut-off", etc. always refer to the shape of the filter transmission curve (Fig. 9.8), which is usually a plot of transmittace (τ) vs wavelength (λ). *Absorbance* (or *optical density*) D, defined as

$$D = \log \frac{1}{\tau} = -\log \tau, \tag{9.2}$$

is frequently used instead of transmittance τ, which is the ratio of the light intensity (I_t) transmitted by a filter to the incident intensity (I), i.e., $\tau = I_t/I$.

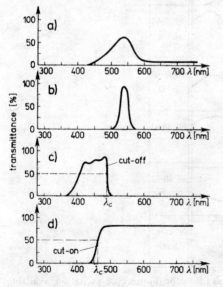

Fig. 9.8. Basic types of filters for fluorescence microscopy: a) wide-band pass filter, b) narrow-band pass filter, c) short-wave pass filter, d) long-wave pass filter.

If transmittances τ_1, τ_2, ..., τ_m of particular filters are known, we can calculate the resultant transmittance of a combination of two or several filters from the following relation:

$$\tau = \tau_1\tau_2 \ldots \tau_m = \prod_{i=1}^{m} \tau_l. \qquad (9.3)$$

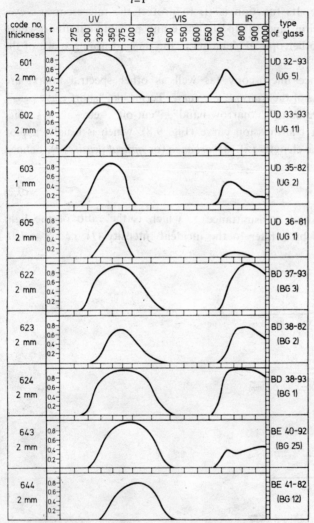

Fig. 9.9. Spectral pass bands of several colour-glass filters for fluorescence excitation, available from VEB Jenaer Glaswerk Schott und Gen. (according to Wunderlich W., Lumineszenz- und Ultraviolett-Fotografie, *Bild und Ton*, **31** (1978), 261–265).

Similarly, if absorbances of single filters are known, we can calculate the resultant absorbance of a filter combination from the formula below

$$D = D_1 + D_2 + \ldots + D_m = \sum_{i=1}^{m} D_i. \tag{9.4}$$

The resultant transmittance is therefore a product of the individual transmittances, while the resultant absorbance is a sum of the individual absorbances.

Wide-band pass filters. The first filters of this kind (Fig. 9.8a) were *liquids filters* (dyed solutions in cuvettes). Next, *gelatine* and *solid-colour glass filters* were developed, and now *wide-band interference filters* are also produced. The most popular are, however, those made of Schott colour glasses. Figures 9.9 and 9.10 show the transmission curves of several such filters used for fluorescence excitation with UV and/or blue light. As can be seen, they have a residual maximum of transmission in the far red and infrared, which must be suppressed by an excluding

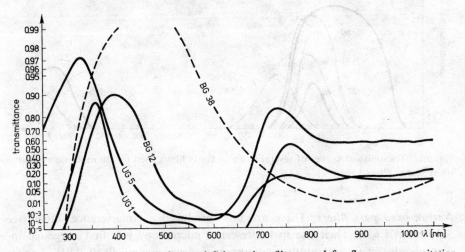

Fig. 9.10. Transmission curves of several Schott glass filters used for fluorescence excitation (thickness of filters 1 mm).

filter (BG 38, for instance). The plotted curves represent a glass thickness of 2 mm in Fig. 9.9 and 1 mm in Fig. 9.10. The latter expresses transmittance in terms of a logarythmic scale, which presents low values of transmittance better than a decimal scale (Fig. 9.9). In practice, the same type of glass exciter filter is often used in several thicknesses, e.g., BG 12/4 mm + BG 38/2.5 mm or BG 12/3 mm +

+BG 38/4 mm. Increasing thickness makes the pass-band narrower and *intrinsic transmittance* of the filter decreases according to Lambert's law (see Subsection 1.2.2 in Volume 1 and Eqs. (1.21)). The term "intrinsic" (or internal) "transmittance" means that the correction for glass surface reflection has been made for spectral curves representing the filters.

Until recently standard use was made of UG 1 and/or UG 5 filters for UV excitation, of BG 3 for violet excitation, and of BG 12 for blue excitation.

Glass filters for fluorescence microscopy are also produced by the Corning Glass Company (USA). Figure 9.11 represents several filters made by this manufacturer for UV, violet, and blue excitations. For some special purposes, these glass filters are accompanied by a selected Kodak Wratten gelatine filter. In general, though gelatine filters find less use in fluorescence microscopy than glass filters.

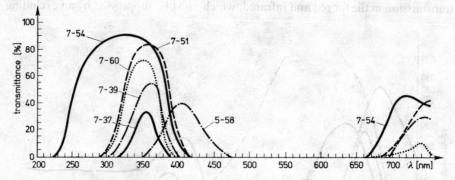

Fig. 9.11. Transmission curves of several Corning Glass filters used for fluorescence excitation (thickness of filters 1 mm).

Narrow-band pass filters. These are now produced as interference filters (see Subsection 1.6.4). Their use in fluorescence microscopy was first suggested independently by several authors almost two decades ago [9.89–9.91]. Exciter interference filters can be optimally adapted to the specific absorption (excitation) band of the substance under test. Moreover, their transmission curves may be adjusted so that the peak transmittance of the filter is coincident with the exciting radiation to be selected from the single emission line of a high-pressure mercury lamp (HBO). Similarly, a barrier interference filter can be optimally adapted to the specific emission band of a fluorochrome or other fluorescent substances. Figure 9.12 shows the typical spectral characteristics of a few narrow-

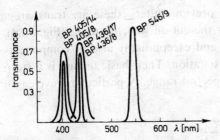

Fig. 9.12. Spectral characteristics of several narrow-band pass interference filters available from C. Zeiss Oberkochen [9.76].

band pass filters produced by C. Zeiss Oberkochen. The numbers that follow the acronym BP refer to the peak wavelength and half bandwidth in nm.

Short-wave pass filters. These are transparent for light wavelengths that are shorter than the cut-off wavelength (Fig. 9.8c). Filters of this kind are at present manufactured as multidielectric interference filters. Their characteristic feature is a steep slope of the transmission curve at the long-wave side of the spectrum. A particular wavelength λ_c for which transmittance at the slope falls to 50% is referred to as the cut-off wavelength.[1]

Short-wave pass filters are used as exciter filters of high performance. Figure 9.13 shows the spectral characteristics of several filters of this kind produced by C. Zeiss Oberkochen. The numbers that follow the acronym KP refer to the cut-off wavelengths expressed in nm.

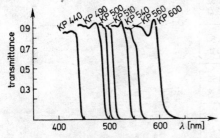

Fig. 9.13. Spectral characteristics of several short-wave pass interference filters available from C. Zeiss Oberkochen [9.76].

[1] For a standard interference filter, the cut-off (or cut-on) wavelength is usually defined as that whose transmittance falls to (or reaches) 5% of the peak transmittance (see Subsection 1.6.4).

Long-wave pass filters. In contrast to the previous filters, these are transparent for light wavelengths that are longer than the cut-on wavelength λ_c (Fig. 9.8d). They are generally used as barrier filters and exceptionally also as components of a filter combination for fluorescence excitation. Their basic feature is a steep increase of transmittance within a short spectral range. A particular wavelength

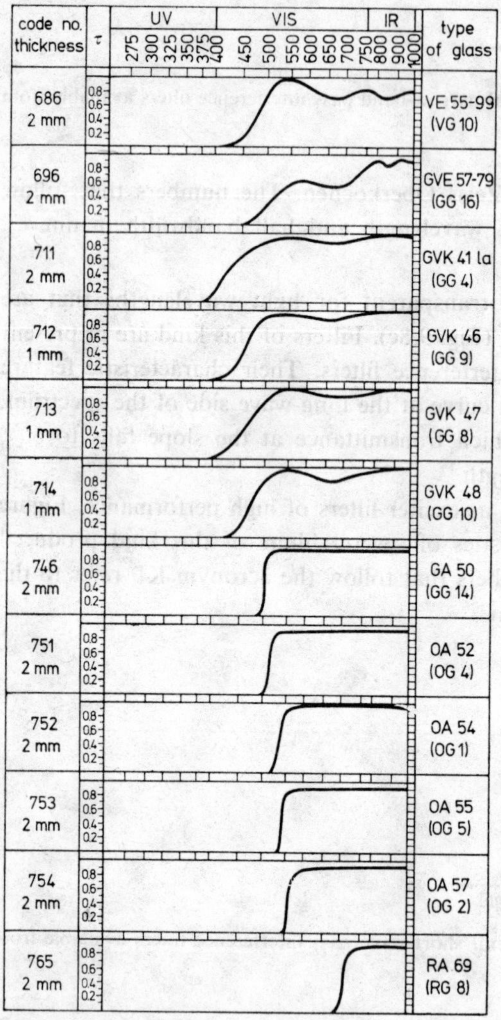

Fig. 9.14. Spectral pass bands of colour glass barrier filters available from VEB Jenaer Glaswerk Schott und Gen. (according to Wunderlich W., Lumineszenz- und Ultraviolett-Fotografie, *Bild und Ton*, **31** (1978), 261–265).

for which transmittance reaches 50% is known as the cut-on wavelength. They exit as either colour glass (Figs. 9.14 and 9.15) or multidielectric interference filters (Fig. 9.16). The cut-on of the latter is steeper than that of the former

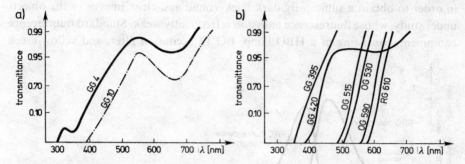

Fig. 9.15. Transmission curves of several barrier glass filters: a) green-yellow filters free from autofluorescence, commonly used together with the exciter filters UG1 and UG5, b) long-wave pass filters with a steep cut-on.

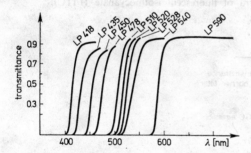

Fig. 9.16. Characteristics of long-wave pass interference filters available from C. Zeiss Oberkochen. The numbers that follow the acronym LP refer to the cut-on wavelengths expressed in nanometres.

Optimum filters for FITC. As was mentioned earlier, FITC is the most universal fluorochrome for immunofluorescence microscopy. Its role greatly increased when special interference filters were first developed by Rygaard and Olsen [9.91–9.93], refined by other researchers and then commercially introduced by E. Leitz Wetzlar [9.94]. This fluorochrome effectively absorbs blue light of the wavelengths $\lambda = 450$ to 500 nm with a maximum at $\lambda = 490$ nm. For its excitation, the light intensity should therefore have a maximum also at $\lambda = 490$ nm. The fluorescent light emitted by the FITC is green-yellow with a maximum intensity for $\lambda = 525$–530 nm (see Table 9.1 and Fig. 9.17). Correspondingly, the exciter filter should be highly transparent for λ between 450 and 495 nm and the barrier filter for λ between 510 and 550 nm. Moreover, this latter filter must cut-on completely the excitation light of wavelengths below 500 nm in order to obtain a sufficiently dark background and clear images of the objects under study, whose fluorescence emission is frequently weak. Standard fluorescence equipment, consisting of a HBO lamp, BG 12 excitation filter, and yellow-green

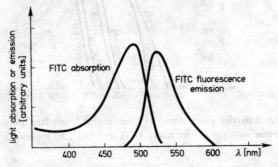

Fig. 9.17. Absorption and emission spectra of fluorescein isothiocyanate (FITC).

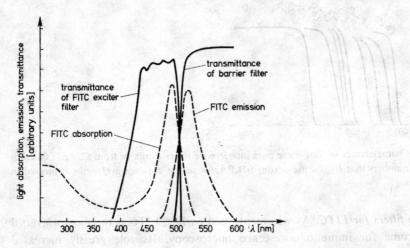

Fig. 9.18. Transmission curves of optimal exciter and barrier filters for FITC.

barrier filter is not suitable for this purpose. The HBO lamp does not have emission lines within the spectral range between 450 and 495 nm but a radiation continuum, which though strong enough, is, however, suppressed by the BG 12 filter, whose transmittance is low (see Fig. 9.10) within the spectral range of an effective

excitation of the FITC fluorochrome. All modern systems for FITC immunofluorescence microscopy therefore employ tungsten halogen lamps and custom-
designed multidielectric interference filters. The typical transmission curve
of an interference filter for FITC excitation is shown in Fig. 9.18. This filter can
actually be classified as a wide-band pass filter with high transmittance and very
steep cut-off on both sides of the transmission curve.

Another effective solution of the problem of FITC fluorescence microscopy
is shown in Fig. 9.19. The required radiation for FITC excitation is produced by
two interference filters (manufactured by C. Zeiss Oberkochen): one of them is
the KP 490 short-wave pass filter and the other the LP 450 long-wave pass filter.
A combination of this kind produces the same result as that shown in Fig. 9.18.

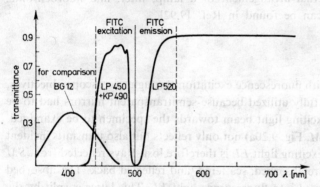

Fig. 9.19. Combination of short-wave (KP 490) and long-wave (LP 450, LP 520) filters for
FITC fluorescence microscopy.

Another long-wave pass filter, the LP 520, is recommended as a barrier filter
[9.76]. Thanks to the enormous progress in interference filter technology, certain
manufactures (C. Zeiss Oberkochen, for instance) offer several different filter
combinations for FITC fluorescence microscopy which are specifically adapted
for various possible FITC reactions. Highly efficient interference filters are
also offered for fluorescence studies based on other fluorochromes (LRB, TRITC).

Rygaard and Olsen have suggested that the *weighted transmission factor*
(*WTF*) could be used to assess the performance of interference exciter filters.
This factor is defined as [9.92]

$$WTF = \int_{\lambda_1}^{\lambda_2} \frac{\alpha(\lambda)\tau(\lambda)}{\lambda_2 - \lambda_1} d\lambda, \tag{9.5}$$

where $\alpha(\lambda)$ is the spectral distribution of the absorption coefficient of the substance (fluorochrome) to be examined, $\tau(\lambda)$ the filter transmission curve, and $\lambda_2 - \lambda_1$ the wavelength range within which the effectiveness of the filter is tested. This definition is valid for a light source that gives the same intensity of light throughout the entire spectral range considered $(\lambda_2 - \lambda_1)$. In particular, λ_1 and λ_2 can be selected as the cut-off and cut-on wavelengths, respectively. Given a light source with an equal distribution of energy in the range $\lambda_2 - \lambda_1$, Eq. (9.5) can yield quantitative information with two desirable properties: (i) the higher the *WTF* the stronger the fluorescence, (ii) two filters that have the same *WTF* yield the same fluorescence intensity.

It is, of course, possible to test an exciter filter for efficiency by practical measurements on an actual arrangement of a lamp, filter, and fluorochrome. The proper procedure can be found in Ref. [9.92].

9.3.3. Dichroic mirrors

In the first experiments with fluorescence excitation through microscope objectives, its advantages were not fully utilized because semitransparent mirrors had to be used for reflecting the exciting light beam towards the specimen to be examined. A mirror of this kind (*SM*, Fig. 9.20a) not only reflects but also transmits incident light; about 50% of the exciting light *EL* is therefore lost. Rays reflected from *SM* strike the specimen *O*, are absorbed, scattered, and reflected back. The absorbed rays are partly transformed into fluorescence light *FL*. This latter is split by the semitransparent mirror *SM*, and only half of the fluorescent light reaches the observer, while the other half is reflected by *SM* and lost. This loss is especially undesirable when weakly fluorescent specimens are examined. Moreover, the

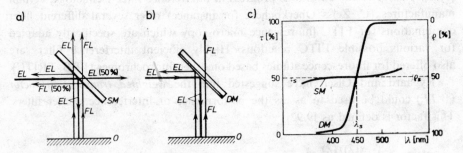

Fig. 9.20. Working principle of a dichroic mirror (b) in comparison with a conventional semi-transparent mirror (a) and their spectral characteristics (c).

excitation rays that are scattered or reflected back by the object O are partly (as much as 50%) transmitted by SM and form an unwanted background.

All these defects can be overcome by using *dichroic mirrors* (Fig. 9.20b), also referred to as *chromatic beam-splitters* or *dichromatic beam-splitters*. The first who applied them to fluorescence microscopy was Brumberg [9.5, 9.6]. His ideas were then embodied in practice at the Optical-Mechanical Works in Leningrad (LOMO), from whom fluorescence epi-illuminators (OSL-1, OI-17, OI-18) and an epi-fluorescence microscope (ML-2) with dichroic mirrors were first commercially available. Now all modern fluorescence microscopes with epi-illumination are fitted with these mirrors.

The dichroic mirror (DM, Fig. 9.20b) is a special multidielectric interference filter slanted at 45° to the incident illuminating beam, which reflects short-wave light and transmits long-wave light. A particular wavelength λ_s (Fig. 9.20c), for which the reflectance ϱ_s and transmittance τ_s are identical ($\varrho_s = \tau$), is known as the separating wavelength between reflection and transmission. For $\lambda < \lambda_s$, the reflectance ϱ suddenly increases and reaches a maximum; simultaneously the transmittance τ rapidly drops to an extremely small value. For $\lambda > \lambda_s$,

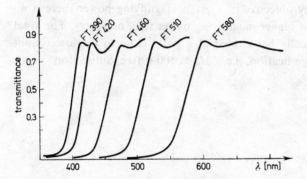

Fig. 9.21. Transmission curves of dichroic mirrors available from C. Zeiss Oberkochen [9.76].

the situation is reversed—τ suddenly increases while ϱ decreases. Consequently, the dichroic mirror clearly separates fluorescence light from excitation radiation. The latter reaches the specimen O (Fig. 9.20b) with no loss of intensity at the mirror DM, while the fluorescence light by-passes the mirror completely—as long as λ_s is set between the maximum of excitation (absorption) and the fluorescence band. The technology of dichroic mirrors is now so well established that the separating wavelengths λ_s can be selected at will at any point of the spectrum (Fig. 9.21; see also Refs. [9.95]–[9.97]).

9.3.4. Objectives and condensers for fluorescence microscopy

All existing microscope objectives are suitable for fluorescence microscopy using violet, blue, and green excitation. UV excitation, however, requires special quartz condensers for dia-fluorescence and special UV-transparent objectives for epi-fluorescence if UV excitation is carried out through the objective.

In general, the brighter the image of fluorescent objects, the higher the performance of a fluorescence microscope. The image brightness B' is directly proportional to the squared numerical aperture A of the objective and inversely proportional to the squared total magnifying power of the microscope, i.e.,

$$B' \propto \frac{A^2}{M^2}. \tag{9.6}$$

From this it follows that objectives with the highest possible numerical apertures are preferable for fluorescence microscopy. Apochromates or planapochromates are therefore particularly suitable for all fluorescence techniques with violet, blue, or green excitation, in view of their essentially higher numerical aperture compared to achromats and planachromats (see Figs. 2.48 and 2.49). Formula (9.6) also shows that for any objective of a given magnifying power there is no advantage in using oculars of higher magnifying power than necessary. The total magnification M should usually be selected at a point close to the lower limit of the range of useful magnification, i.e., $M_u \approx 500A$ (see Subsection 3.6.8).

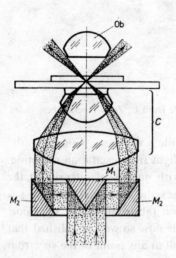

Fig. 9.22. Highly efficient dark-field condenser manufactured by Polish Optical Works, Warsaw.

Although nowadays dia-fluorescence is less popular than epi-fluorescence, the former—especially its dark-field version—still has its applications and will certainly continue to be used. This technique, however, requires dark-field condensers of high luminous efficiency. A conventional cardioid or paraboloid condenser (see Fig. 4.6) incorporates a circular opaque screen which masks the central region of the illuminating beam and thus greatly reduces the total amount of energy of the excitation radiation. To overcome this defect, highly efficient dark-field condensers were developed for dia-fluorescence microscopy. Two of these, the K5C model (developed in the Central Optical Laboratory, Warsaw) and the "superwide dark-field condenser" (manufactured by Tiyoda, Japan), are shown in Figs. 9.22 and 9.23; their basic parameters are listed in Table 9.3.

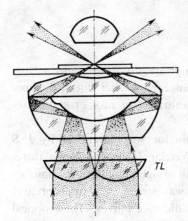

Fig. 9.23. Superwide dark-field condenser manufactured by Tiyoda (Japan).

TABLE 9.3

Optical parameters of the dark-field condensers shown in Figs. 9.22 and 9.23

Parameter	Condenser	K5C (PZO)	SDC (Tiyoda)
Internal numerical aperture*		0.95	1.2
External numerical aperture*		1.27	1.4
Immersion liquid		oil	oil
Diameter of accessible light beam		12 mm	31 mm
Maximum numerical aperture of objectives to be used		0.95	1.2

* See Subsection 2.4.4 for definition.

As can be seen, the K5C model uses a bright-field condenser (C, Fig. 9.22) which is transformed into a dark-field version by means of two conical mirrors M_1 and M_2. The Tiyoda system consists of a cardioid dark-field condenser fitted with a toroidal lens (TL). Both systems transmit the entire luminous flux without occlusion of its central region.

As was mentioned earlier, the epi-fluorescence microscopy for general use is of the bright-field type. Fluorescence excitation is carried out through the microscope objective, which for excitation light acts as a bright-field condenser. This latter function also requires the use of objectives with highest possible numerical aperture. Hight-power objectives are therefore more suitable for epi-fluorescence than low-power objectives. The converse holds good for dia-fluorescence microscopy.

9.3.5. Ploemopak epi-illuminator

This illuminator is now standard equipment for high quality fluorescence micro-scopes. It features a wide range of easily interchangeable exciter filters, dichroic mirrors, and barrier filters which are especially valuable for *two-* or *multiwave-length fluorescence microscopy*.

The first version of the Ploemopak epi-illuminator was developed by J. S Ploem (Fig. 9.25) and manufactured by E. Leitz Wetzlar. It incorporates a turret with four dichroic mirrors placed at 45° to the path of light. Each of these mirrors is associated with a barrier filter whose cut-on wavelength λ_c is appropriately matched to the separation wavelength λ_s of the dichroic mirror. These optical components can be combined as follows: TK 400 + K 400, TK 455 + K 460, TK 495 + K 495, and TK 580 + K 590.[2] Glass filters (UG 1, BG 3, BG 12, and others) as well as interference filters of the short-wave pass type are used as exciter filters. FITC-optimum exciter and barrier filters are installed for FITC fluorescence (in addition to the combination TK 495 + K 495).

Figure 9.24 shows the Ploemopak 2 epi-illuminator, which is an improved version of the above Ploemopak 1. The main difference between them is that the later model has a revolving disc which contains not only the dichroic mirrors and barrier filters but also the exciter filters. Moreover, the Ploemopak 2 is equipped with twelve interchangeable combinations of exciter filters, dichroic mirrors, and barrier filters (see Table 9.4), though only four combinations can

---

[2] TK and K are respectively acronyms for the dichroic mirrors and long-wave pass inter-ference filters offered by E. Leitz Wetzlar.

be installed simultaneously in the revolving disc. However, the required combination can be obtained by the user himself without difficulty. An important advantage of this epi-illuminator is that all filter combinations can be arranged next to each other and swung together into the path of light. In particular, the filter combinations required for the observation of double staining (two-wavelengths) fluorescence can be installed side by side. The epi-illuminators Ploemopak 1 and Ploemopak 2 are described in greater detail in Refs. [9.98]–[9.101].

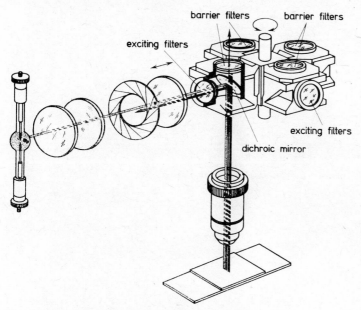

Fig. 9.24 Fluorescence epi-illuminator Ploemopak 2.

The Ploemopak epi-illuminators are offered by E. Leitz Wetzlar as fluorescence equipment for different microscope (Orthoplan, Ortholux II, Dialux, SM Lux, Diavert) manufactured by this firm. Similar illuminators are also available from other firms. For instance, the epi-condensers IIIRs and IVF (produced by C. Zeiss Oberkochen) are among devices of this category. The IIIRS model is for use with large research microscopes (Universal, Photomikroskop III) and the IVF with standard microscopes. Another example is the 2070H or 2070C Vertical Illuminator manufactured by the American Optical Corporation as equipment for the H10 and H20 Fluorstar microscopes for incident-light fluorescence. The 2070H model is fitted with a tungsten halogen lamp (12 V/100 W), while the 2070C model uses a high-pressure mercury lamp (50 W).

Fig. 9.25. Johann S. Ploem

TABLE 9.4

Combinations of filters and dichroic mirrors for the Ploemopak 2 epi-illuminator [9.100]*

Combination	Exciter filters	Dichroic mirrors	Barrier filters	Especially suitable	
				for fluorochromes	with excitation light source
A	2×2 mm UG1	TK 400	K 430	BAO, DANS	HBO
B	2×3 mm BG3	TK 455	K 470	QM, tetracycline	HBO
C	3 mm BG3+S 505	TK 455	K 460	Biogenic amines	HBO
D	3 mm BG3+KP 425	TK 455	K 460	Biogenic amines	HBO
E	3 mm BG3+AL 436	TK 455	K 490	QM	HBO
G	3 mm BG12	TK 510	K 515	Acridine orange, tetracycline	HBO
H	2×KP 490 (= KP 500)	TK 510	K 515	Acridine orange, FITC, FDA, tetracycline, QM	XBO or TH, HBO
I	2×KP 490+1 mm GG 455	TK 510	K 515	FITC, FDA	XBO or TH
K	2×KP 490+2 mm GG 475	TK 510	K 515	FITC, FDA	XBO or TH
L	2×KP 490+2 mm GG 475	TK 510	K 515+S 525	FITC, FDA	XBO or TH
M	2 mm BG 36+S 546	TK 580	K 580	LRB, Feulgen stain, methyl green pyronine, TRITC	HBO
N	2 mm BG 36+KP 560+K 530	TK 580	K 580	LRB, Feulgen stain, methyl green pyronine, TRITC	HBO

* KP—cut-off interference filter, S and AL—selective interference filters, HBO—high-pressure mercury lamp HBO100 or HBO 200, XBO—high-pressure xenon lamp XBO75 or XBO150, TH—tungsten halogen lamp 12 V/100 W.

It is worth noting here that an excellent example of the modern immuno-fluorescence technique was recently presented by W. Stöcker who used the Leitz incident-light microscope Dialux 20 equipped with a HBO 50 W lamp, a relatively simple set of exciter and barrier filters (BP 450–490 nm; RKP 510 nm and LP 520 nm), and oil immersion objectives 10/0.45, 25/0.75 and 100/1.32 (see *Leitz Mitt. Wiss. u. Techn.* **9** (1987), 36–44). The Leitz-Vario-Orthomat was used for recording the fluorescent images of different cells and tissues.

9.3.6. Other devices and facilities

The present-day instrumentation of fluorescence microscopy is highly differentiated and divided—more obviously than is the case with other microscopical tech-niques—into routine and research microscopes [9.102–9.111]. The former are frequently relatively simple and therefore inexpensive, but have a very limited field of application. An example is shown in Fig. 9.26, which illustrates a dia-fluor-

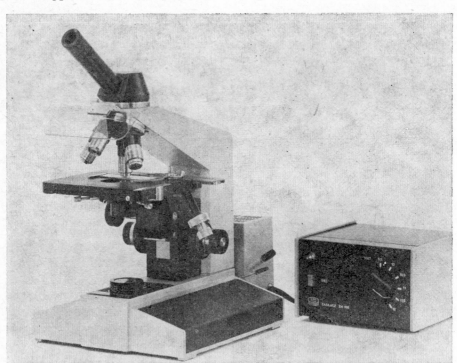

Fig. 9.26. An inexpensive fluorescence microscope for routine work—the Biolar FR (photo by courtesy of PZO, Warsaw).

TABLE 9.5

Filters installed in the Biolar FR fluorescence microscope (Fig. 9.26)

Exciter filters	Barrier filters	Especially suitable	
		for fluorochromes	with light source
FIK 495*	OG 4/2 mm	FITC, acridine orange	TH 12 V/100 W
FIK 570	OG 3/2 mm	LRB	TH 12 V/100 W

* FIK—cut-off interference filter, TH—tungsten halogen lamp.

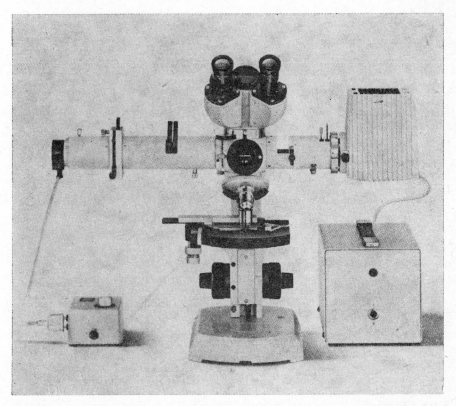

Fig. 9.27. Fluorescence microscope Standard IFP for routine diagnosis (C. Zeiss Oberkochen).

escence microscope constructed in the Central Optical Laboratory, Warsaw [9.104], and manufactured by PZO. This instrument contains a 12 V/100 W tungsten halogen lamp, a highly efficient dark-field condenser (K5C, Fig. 9.22), and only two combinations of filters (see Table 9.5); it is mainly intended for the detection and identification of syphilis (FTA test) and tuberculosis (according to Adamczyk procedure).

Another example is shown in Fig. 9.27, which illustrates the Zeiss (Oberkochen) Standard IFD microscope for routine immunofluorescence diagnosis. This original instrument incorporates a comparison system which facilitates the speedy identification, for instance, of toxoplasmosis. About two thirds of the field of view is occupied by the fluorescent image of the specimen under test, while the remaining field is divided into two segments showing transparent photomicrographs of positive and negative toxoplasmosis.

9.4. Fluorescence combined with other microscopical techniques

Fluorescence microscopy combined with standard bright-field, dark-field, or phase contrast observation is often useful in biological and medical research. The alternating or simultaneous display of the fluorescence and other images of the same specimen makes it possible, in particular, to exactly localize fluorescing objects, e.g., specifically fluorochromized antigens in cell cultures and tissue sections. Such a combination poses no problem if the fluorescence excitation is by radiation from an epi-illuminator. The use of any conventional bright-field, dark-field, or phase contrast condenser for transmitted light is then possible. The universal fluorescence microscopes, such as shown in Fig. 9.28 or 9.29, are especially suitable for this purpose. Apart from that, a microscope of this kind permits the specimen to be excited through both the condenser and the objective. This is especially useful when the specimen under study is relatively thick and weakly fluorescing.

The above-mentioned combination of fluorescence with other techniques, and particularly with the phase contrast method, is less useful if fluorescence excitation is only in transillumination through the condenser as the annular diaphragm for the phase contrast method greatly reduces the intensity of the exciting light. Several systems have been devised to overcome this limitation [9.104, 9.112–9.114]. A well known one is that developed by Gabler and Herzog [9.112] and manufactured by C. Reicher Wien. This consists of a double illuminator and special phase-fluorescence condenser which incorporates annular

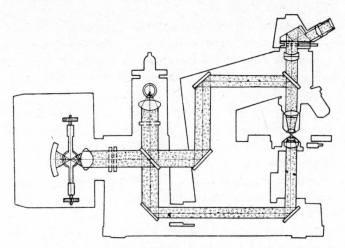

Fig. 9.28. Pathway of light in a fluorescence microscope for research work (Fluoval, VEB Carl Zeiss Jena).

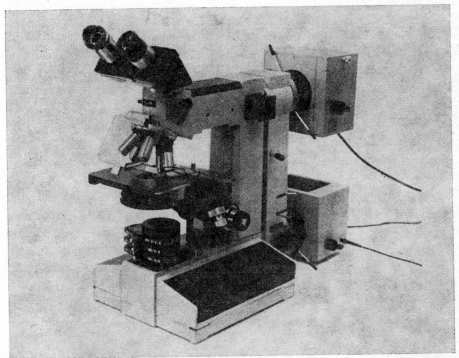

Fig. 9.29. Universal fluorescence microscope Fluar (prototype) constructed in the Central Optical Laboratory, Warsaw.

diaphragms made of dark glass (e.g., Schott UG1 filter) transparent to the exciting light (near ultraviolet and blue-violet). The phase contrast image is produced by visible light from a separate low-voltage illuminator which passes through a clear condenser annulus conjugate with the phase ring. The fluorescence image is excited by short-wavelength light emitted by a high-pressure mercury lamp and transmitted to the specimen by both the clear annulus and the remaining filter-covered area of the condenser diaphragm. A 45° mirror in the light path partially reflects the light beam from the low-voltage illuminator. In order to increase the colour contrast between fluorescent and non-fluorescent parts of the image, a red filter is sometimes inserted into the beam producing the phase contrast image. An example of the performance of this system is shown by photo-micrographs in Fig. 41 in Ref. [5.11].

The author of this book has combined dia-fluorescence microscopy with phase contrast by using annular condenser diaphragms made of thin interference films instead of a dark glass filter [9.114]. The use of such diaphragms (Fig. 9.30b) has some advantages because the spectral transmission curve of an interference

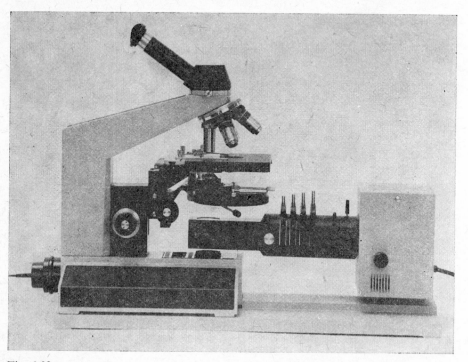

Fig. 9.30a

b)

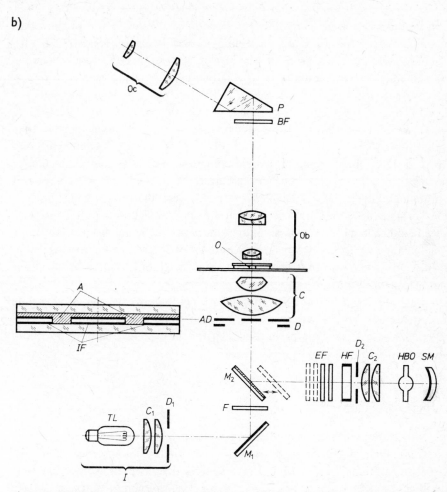

Fig. 9.30. Dia-fluorescence microscope Biolar FL (a) and its optical system (b) combined with the phase contrast technique in transillumination: *I*—conventional illumination, *TL*—tungsten lamp (6 V/15 W), C_1 and C_2—collectors, D_1 and D_2—field diaphragms, M_1—mirror, *F*—colour glass filter (preferably orange or red used for simultaneous fluorescence and phase contrast microscopy), *SM*—spherical mirror of the illuminator for fluorescence excitation, *HBO*—high-pressure mercury lamp (50 W), *HF*—heat filter, *EF*—exciter filters, M_2—mirror partially transparent (or dichroic mirror), *D*—aperture diaphragm (iris), *AD*—annular diaphragm for phase contrast, *IF*—short-wave pass interference filter, *A*—clear annulus, *C*—condenser, *O*—object under study, *Ob*—phase contrast or ordinary objective, *BF*—barrier filter, *P*—reflecting prism, *Oc*—ocular.

filter can be properly shaped, and fluorescence light of wavelengths much closer to the exciting wavelengths are observed. Experiments with this system show that for simultaneous phase contrast and fluorescence imaging (see Fig. XLVIb) negative phase objectives, such as described in Subsection 5.5.2, are preferable.

Essentially different equipment for dia-fluorescence and phase contrast has been developed by VEB Carl Zeiss Jena. In this, an image-transfer sub-condenser system is applied, which projects the image of a clear annulus through the partially transparent 45° mirror into the front focal plane of the condenser [5.27, 9.104].

Geissinger *et al.* [7.18–7.20] suggest that epi-fluorescence combined with Nomarski DIC microscopy in transillumination can be a useful technique especially suitable for resolving the fluorescence image and non-fluorescent surrounding tissue structures. Evidence for this were their experiments on un-stained or stained sections, smears, and wet mounts, and on fluorochromed sections and cell culture monolayers. On some specimens this technique is even preferable to phase contrast, especially where preparations of a certain thickness are concerned, since it can resolve structures that are invisible or obscured by the phase contrast system.

It was also found advantages to combine epi-fluorescence microscopy with transmitted-light bright-field microscopy using polarized white light [9.115]. This combination is very useful, in particular for the examination of thin sections of bone tissues.

9.5. Laser fluorescence microscopy

In this application, the laser is used as an excitation light source to cause secondary fluorescence in a fluorochromed specimen, or primary fluorescence in a non-stained preparation [9.116–9.124]. Light sources described in Subsection 9.3.1 radiate either as quasi-black body (e.g., high-pressure xenon or tungsten halogen lamps) or at peak wavelengths with a continuous background (e.g., high-pressure mercury lamps). The average power output of these sources can be quite high, but the spectral brightness or power per wavelength band is only a small fraction of the full power. Since fluorochromes effectively absorb light in relatively narrow spectral bands (5–100 nm), the useful excitation energy from these lamps is low and long periods of illumination are necessary for exact visual monitoring or photographing fluorescence images. During these lengthy excitation periods, often lasting several minuts or even longer, a fluorochrome may completely photodecompose (near UV light is the spectral region most often responsible

for fluorochrome decomposition). This limitation can be overcome if a laser
is used, e.g., a cadmium or argon laser, since its spectral brightness is very high.
Moreover, laser light can be higly monochromatic and is therefore easy to separate
from the fluorescent light.

Another advantageous feature of a laser source of light is the parallelism
of its radiation beam. This feature was used by Wayland [9.117] for laser stimulation of fluorochromes in intravital microscopy. His experimental apparatus
(Fig. 9.31) incorporates a *LWD* mirror objective Ob, transfer lens *TL*, dichroic
mirror *DM*, silicon intensifier target camera *SIT* with a barrier filter *BF*, and
silicon vidicon camera *SV*. For epi-fluorescence, the laser beam (*EF*) is reflected
onto the object *O* (tissue) by the fully reflecting plane mirror *M* mounted below

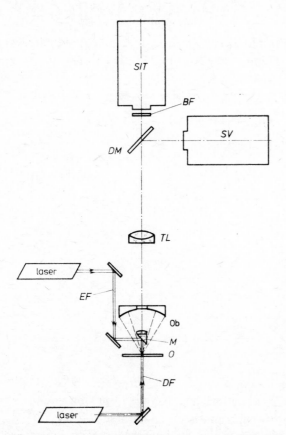

Fig. 9.31. Laser fluorescence microscopy according to Wayland [9.117] (see text for explanation).

the convex mirror of the objective Ob at 45° to the optic axis. Another laser beam (*DF*) for dia-fluorescence transilluminates the object and is reflected out of the field by *M*. This double illumination permits high energy excitation of fluoro-chromes with greatly reduced laser light reaching the imaging system. By splitting the image-forming light by a suitable dichroic mirror *DM*, a dark-field image of the entire field of view of the objective Ob can be formed on the camera *SV* and an image of the fluorescent areas on the camera *SIT*, to permit display of both images on side-by-side monitors.

A valuable feature of lasers is the possibility of focalizing their beams with diffraction-limitted divergence permitting a spatial resolution in the μm-range. The laser may thus be considered an ideal light source for the excitation of single living cells. Many studies of this kind have been published [9.118–9.120, 9.122–9.125].

It is a well known fact that the fluorescence intensity of fluorochromes decays in the course of continuous excitation (curve *A* in Fig. 9.32). This phenomenon

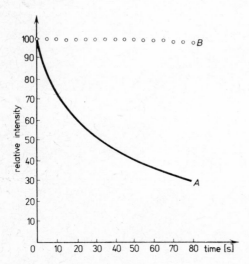

Fig. 9.32. Fluorescence decay.

is a consequence of the fluorochrome photodecomposition mentioned above, also referred to as fluorescence photobleaching. Another situation occurs if fluorescent emission is produced by short pulses of light (curve *B* in Fig. 9.32); the bleaching effects are now reversible, that is, initial fluorescence intensity is recovered after short null periods which separate very short, say, ms-pulses of light such as those produced by pulse lasers. This is the basis of a relatively new technique, known as *fluorescence photobleaching recovery* (FPR), for studying the motion of specific components in a complex mixture [9.124, 9.125]; pulse excitation is also a standard procedure in modern cytofluorometry (see Chapter 14).

All lasers that emit short-wavelength light are suitable for fluorescence microscopy (see Table 1.5). Argon or helium–cadmium cw-lasers and pulsed dye laser (optically pumped by nitrogen laser) are the most frequently used. Since the early 1970s, a number of laser fluorescence microscopes have been available. For instance, in 1973 *The Laser Weekly* (May 21) announced that a group of researchers was teaming up a UV-laser and a computer in a cancer research project at the University of Rochester's Medical Center. Detection of cancer cells involves staining the cells with a fluorochrome that emits green light under UV excitation. The amount of fluorescence light emitted by the cell nucleus is related to the amount of DNA (deoxyribonucleic acid) in the nucleus; a raised level of DNA is evidence of abnormality. To supplement the test, the ratio of the diameter of the nucleus to the diameter of the cell is measured at the same time. This ratio can vary and its variations, together with the abnormality of DNA, make it possible to distinguish between normal and abnormal types of cells. The system incorporates a computer on line with the microscope and is capable of screening hundreds of cells a second. There may be one abnormal cell in 10,000, and that is the cell that requires the attention of an experienced cytologist. The system not only detects abnormal (cancer) cells, but also enables one to recognize atypical cells classified between normal and malignant cells. It was originally devised for detecting early states of cervical cancer. Although the Papanicolaou smear is still the prime method of detecting this cancer, the above system promises to extend preliminary screening tests to many more women.

Another new departure is a laser fluorescence exciter announced by Liconix in 1975. This exciter can be used with clinical and research microscopes that accept a high-pressure mercury lamp. It uses a helium–cadmium laser and offers both full-field and selective spot illumination in the UV ($\lambda = 325$ nm) and deep blue ($\lambda = 441.6$ nm) spectral regions. This illumination is effective for exciting secondary fluorescence of a few popular fluorochromes (see Table 9.1) and primary fluorescence of cellular molecules. The capacity to focus intensely on a spot of several μm diameter permits close inspection of ultra-weak fluorescence under normal external lighting.

It is, however, important to note that laser fluorescence excitation, especially pulsed excitation, is more useful in quantitative than qualitative fluorescence microscopy. For further information regarding laser fluorescence microscopy, the reader is therefore referred to Volume 3, especially to Chapter 14 (micro-fluorometry and flow cytometry).

9.6. Total internal reflection fluorescence microscopy

Microscopy of this kind is quite new and was initiated by Axelrod in 1981 [9.126]. Its underlying principles are based on the phenomenon of evanescent waves (see Subsection 1.5.2), while its instrumentation is similar to that used in reflection contrast microscopy as initiated by Ambrose (see Chapter 8 and Refs. [8.7], [8.8]). When totally internally reflected in a transparent solid at its interface with liquid (Fig. 9.33), the excitation light beam (preferably laser beam) pen-etrates only a short distance into the liquid (see Eqs. (1.69)) as a surface electro-magnetic field, called the *evanescent wave*. This latter can selectively excite fluoro-chromed molecules in the liquid on the cell surface or submembrane structure at cell-substrate contract regions. The technique of total internal reflection fluor-escence (TIRF) was used by Axelrod *et al.* [9.126] to examine the cell–substrate

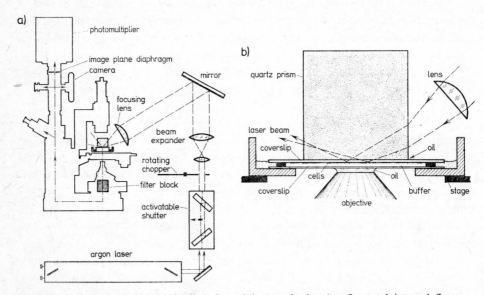

Fig. 9.33. Optical system (a) and detailed view of the sample chamber for total internal fluor-escence microscopy (b) according to Axelrod *et al.* [9.126].

contact regions of primary cultured rat myotubes with acetylcholine receptors fluorochromed with bungarotoxin and human skin fibroblasts fluorochromed with a membrane-incorporated fluorescent liquid. With only minor modification of the optical system, TIRF microscopy can also be combined with the fluorescence photobleaching recovery technique (mentioned in the preceding section and discussed in greater detail in Chapter 14) and also with correlation spectroscopy to measure the chemical kinetic binding rates and surface diffusion of fluorescent biological molecules (e.g., serum proteins, immunoglobulins, hormons) adsorbed at equilibrium at a solid–liquid interface.

In their review article [9.126], Axelrod *et al.* pointed to several advantages of TIRF microscopy. Firstly, this technique greatly reduces fluorescence from cytoplasmically internalized labelling and cellular debries as well as primary fluorescence in thick cells while intensifying the fluorescence of membrane regions close to the substrate. Consequently, it is possible to detect lower concentrations of fluorochromed membrane receptors than would otherwise be the case. Secondly, examination of the submembrane structure of the cell-substrate contact in thick cells is facilitated when compared with studies using conventional reflection contrast microscopy. Thirdly, by varying the angle of light incidence, the topography of the cell membrane facing the substrate can be reconstructed. Fourthly, reversibly bound fluorescent ligands on membrane receptors can be visualized without exciting background fluorescence from unbound ligands in the bulk solution. Certain receptors might therefore be studied without the necessity of blocking them by irreversible antagonists.

This new technique will undoubtedly lead to further applications in cell biology, especially when used together with reflection contrast microscopy.

9.7. Surgical fluorescence microscopy

There is no need to justify the importance of diagnostic or surgical fluorescence microscopy in the clinical diagnosis of certain afflictions of the human body, especially of malignant tumours. However, progress in this field has been slow, the sole examples of new sophisticated instruments being the *fluorescence recto-microscope* MLK-1 [9.127] and the fluorescence contact microscope OLK-2 [9.128, 9.129] developed by a group of Russian researchers. These incorporate special objectives whose working distance is equal to zero; consequently, the object under study (living tissue, human or animal organ) is in direct contact with the front lens of the objective.

Figure 9.34 shows the optical system of the rectomicroscope, whose objective tube is ended in a contact objective Ob of magnifying power 11.6×. This part is made of sterile material because it is directly introduced into the rectum or other organs of the patient. Epi-excitation of fluorescence is carried out through the objective tube. The epi-illumination system IS incorporates two sources of light: a tungsten halogen lamp (9 V/75 W) and a flash-lamp whose pulse duration is equal to 2.5 ms. The former is used for visual observation of fluorescence images and the latter for photomicrography. Exciter filters (*EF*) and barrier filters (*BF*) as well as dichroic mirror (*DM*) are primarily adapted to secondary fluorescence of living tissues fluorochromized with an acridine orange solution of concentration 1:5000. The basic technical parameters of this instrument (manufactured by LOMO, Leningrad) are given in Table 9.6.

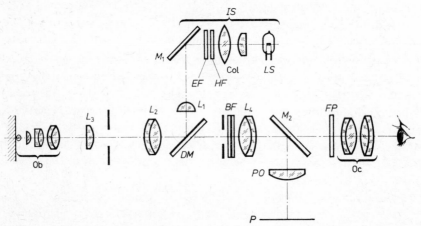

Fig. 9.34. Optical system of the contact fluorescence microscope MLK-1; *IS*—epi-illuminator, *LS*—light source, Col—collector, *HF*—heat filter, *EF*—exciter filter, M_1 and M_2—mirrors, L_1 to L_4—lenses, *DM*—dichroic mirror, Ob—contact objective, *BF*—barrier filter, *EP*—focal plate, Oc—visual ocular, *PO*—projection (photographic) ocular, *P*—plane of photographic film.

TABLE 9.6
Basic optical parameters of the MLK-1 contact fluorescence microscope (Fig. 9.34)

Name of the parameter	Magnitude
Total magnifying power: (a) visual	93× with 8× ocular
	174× with 15× ocular
(b) photographic	33×
Field of view in the object plane	0.65 mm
Spectral range of excitation	400–440 nm
Spectral range of fluorescence	440–700 nm

The OLK-2 microscope is intended for the study of thick fragments of tissues and small animals installed on the microscope stage. The optical system of this instrument is similar to that shown in Fig. 9.35, which represents a simple contact microscope developed by Brumberg and his co-workers [9.129] and intended for anatomists. It incorporates contact epi-objectives of different magnifying power (Fig. 9.36) corresponding to that of conventional microscope objectives. A dark-

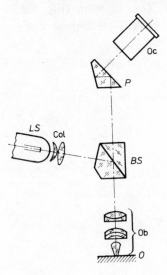

Fig. 9.35. Optical system of a contact microscope [9.129]; *LS*—light source, Col—collector, *BS*—polarizing interference beam-splitter, Ob—contact objective, *O*—object under study, *P*—reflection prism, Oc—ocular.

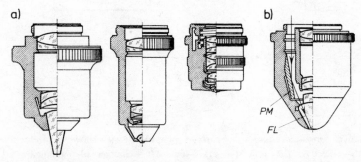

Fig. 9.36. Contact epi-objectives: a) bright-field systems, b) dark-field system (*PM*—paraboloid mirror, *FL*—front lens) [9.128, 9.129].

field version such as shown in Fig. 9.36b is more suitable for contact epi-fluorescence than bright-field models (Fig. 9.36a).

A medical fluorescence microscope quite different from those described above
has also been developed in the Central Optical Laboratory, Warsaw [9.130].
This is based on the colposcope, which enables one to observe an object stereoscopically. Originally, two slightly different prototypes of this instrument were
constructed and carefully tested in clinical practice. One of them (Fig. 9.37)
contained a high-pressure mercury lamp of moderate power (HBO50) or a tung-

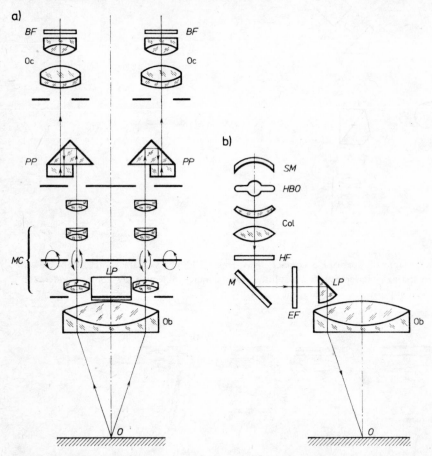

Fig. 9.37. Optical system of fluorescence colposcope: a) front view, b) side view showing epi-illuminator; O—object, Ob—objective, LP—reflecting prism-lens, MC—changer of magnification,
PP—Porro prisms, Oc—oculars, BF—barrier filters, EF—exciter filter, M—mirror, HF—heat
filter, Col—collector, HBO—high-pressure mercury lamp (50 W), SM—spherical mirror.

sten-halogen lamp (12 V/100 W), while the other was fitted with a high-pressure mercury lamp of high power (HBO 200). This latter lamp was installed on the microscope (colposcope) base from which excitation light was guided via a bundle of optical fibres to the optical system of the epi-illuminator. The prototype with the HBO 200 lamp and fibre illumination optics was judged to be less useful than that equipped with the HBO 50 lamp. The latter has consequently been selected for commercial production by the Polish Optical Works (PZO).

This fluorescence colposcope is fitted with several combinations of exciter and barrier filters suitably adapted for use with the most popular fluorochromes (see Table 9.7), although the standard preference is for acridine orange in fluoro-chromizing cells and tissues in vivo. Its basic optical parameters are as follows: working distance—ca 200 mm, maximum magnifying power—50 ×, minimum magnifying power—3.2 ×, field of view in the object plane—from 80 mm to 5 mm, respectively for minimum and maximum magnification.

TABLE 9.7

Filters installed in the fluorescence colposcope shown in Fig. 9.37

Exciter filter*	Barrier filters	Especially suitable	
		for fluorochromes	with light source
BG 3/2.5 mm	GG 475/2 mm	QM, tetracycline	HBO50
BG12/2.5 mm	OG 515/2 mm	Acridine orange, tetracycline	HBO50
BG 38/2 mm+		Rhodamine, TRITC,	
+S 546	OG 590/2 mm	Feulgen stain	HBO50
FIK 495	GG 515/2 mm	Acridine orange, FITC	Tungsten-halogen 12 V/100 W
FIK 570	OG 590/2 mm	Rhodamine, LRB	Tungsten-halogen 12 V/100 W

* S—selective interference filter, FIK—cut-off interference filter.

At present, the main field of application of this instrument is gynaecology (detection of early states of cervical cancer [9.131]). Its usefulness has also been recognized for otolaryngology.

Certain aspects characteristic of the instrumentation of surgical fluorescence microscopy occur also in the "intravital microscopy" developed mainly by H. Wayland. In his review article [9.117], Wayland discusses, among other things, the problems involved in measuring microvascular organization and macromolecular diffusion by means of fluorescent tracers.

9.8. Fields of applications of fluorescence microscopy

Biology and medicine are the main fields where fluorescence microscopy is widely used in both research and routine work [9.11–9.162] though techniques of primary or even secondary fluorescence are important in the materials sciences and may prove valuable for certain problems in chemistry [9.163–9.171].

9.8.1. Applications in biology and medicine

Medical research and diagnostics are where fluorescence microscopy has its widest application [9.132, 9.133]. The field includes, in particular, cell biology [9.138–9.140], microbiology [9.137], bacteriology [9.138–9.140], hematology [9.141–9.144], genetics [9.145–9.149] (see also Fig. XLVII), histology, with particular reference to bone tissues [9.150–9.154], cyto- and histochemistry [9.155–9.159], and botany [9.160–9.164].

In immunofluorescence microscopy alone [9.11–9.75], many antibody techniques are now routine research and clinical practice. Immunofluorescence microscopy covers such wide fields as bacteriology, virology, mycology, parasitology, histo- and cytochemistry. It is also useful for the diagnosis of certain autoimmunodiseases (e.g., lupus erythematodes visceralis, struma Hashimoto, myxoedema, rheumatism). The most popular fluorochromes, such as FITC, acridine orange, auramine, rhodamine, LRB, TRITC, QM, atebrine, are regularly used for the detection and identification of [9.42, 9.76]: antinuclear antibodies, Bacillus anthracis, bacterial meningitis, Blastomyces dermatitis, Bordetella pertussis, Brucella (B. abortus, B. suis, B. melitensis), Candida albicans (and other Candida species), Cladosporium carrionii, Cladosporium bantianum, Corynebacterium diphtheriae, cryptococcosis (Cryptococcus neoformans), Diplococcus pneumoniae, Escherichia coli, Erysipelothrix insidiosa, Fungi of various origin, Hemolytic streptococcus, Haemophilus influenzae, Histoplasma capsulatum, infectious mononucleosis, Klebsiella, Leptospira, Listeria monocytogenes, malaria (Plasmodium vivax, Plasmodium malariae, Plasmodium falciparum, Plasmodium ovale, Plasmodium immaculatum), Malleomyces (M. mallei), Mycobacteria, Neisseria gonorrhoea, Neisseria meningitidis, Nocardia asteroides, Pasteurella pestis, Pasteurella tularensis, Pseudomonas pseudomallei, rabies, Rickettsia, Salmonella, schistosomiasis (Schistosoma mansoni), Sporotrichum Schenckii, Shigella, Staphylococcus aureus, Streptococcus, syphilis (Treponema pallidum), toxoplasmosis, trichinosis, tuberculosis. Moreover, it is possible to diagnose almost all viral diseases by using the appropriate FITC fluorescence methods.

Immunofluorescence microscopy and especially the antibody techniques have gained their great present-day importance thanks to development of suitable flurochromes, highly effective exciter and barrier filters, dichroic mirrors, and other accessories. It is, however, worth noting here that since the beginning of the 1980s, colloidal gold has been increasingly used as a cytochemical marker, not only in its traditional domain of electron microscopy but also in light microscopy. Nowadays, immuno-gold reagents have become a competitive alternative to fluorochromes and immunfluorescence microscopy [9.172] and are commercially produced for both electron and light microscopy. A silver enhancement kit has also recently been introduced for use with the immuno-gold technique. It appears that immuno-gold and silver staining is much more sensitive than the PAP method (see *Proc. Roy. Micr. Soc.*, **20** (1985), 201).

9.8.2. *Applications in materials sciences and chemistry*

Unlike biologists, chemists and physicists, or researchers in the materials sciences are mainly interested in primary fluorescence, though secondary fluorescence techniques may prove valuable for polymer or ceramic components as well as for certain manufactured products such as food, paper, textiles, etc. In particular, fluorescence microscopy is a valuable tool for mineralogists [9.165, 9.166], petrographers [9.167], geochemists [9.169], coal petrologists [9.170], and in the field of semiconductor materials [9.171].

*
*   *

In this chapter, the emphasis has been placed on qualitative methods in fluorescence microscopy, which only covers half the field. The other half, i.e., quantitative fluorescence microscopy, is classed among photometric methods and belongs therefore to Volume 3 (see Chapter 14), which is devoted in its entirety to the most significant quantitative techniques and measuring procedures of light microscopy.

Finally, mention should also be made of the fluorescent activated cell sorting technique (see *Proc. Roy. Micr. Soc.*, **16** (1981), 256). It started in the mid 1960s, its commercial version appeared in the early 1970s, and today is widely used in many laboratories around the world to current medical research.

10. Ultraviolet and Infrared Microscopy

Visible light covers the wavelengths of electromagnetic radiation from about 400 to 700 nm.[1] Wavelengths (λ) shorter than 400 nm lead up to the ultraviolet spectrum (UV) and those longer than 750 nm—to the infrared spectrum (IR). The lowest limit for the useful UV region is approximately at $\lambda = 100$ nm. Below 100 nm air strongly absorbs invisible light and this region is known as the *vacuum ultraviolet*, which adjoins X-rays.

The long-wavelength limit for IR radiation is approximately at 1000 μm. The entire IR spectrum is divided into several regions. Light sources also emit what is known as *actinic radiation*, which adjoins the long-wave visible red light and has the ability to produce chemical changes in many substances. The actinic infrared is followed by the radiation of hot bodies with a temperature of about 500–700 K. The next spectral band lies near wavelengths of 9000 nm; this is the region which also includes the radiation emitted by the human body. Longer IR waves are usually considered as *thermal radiation*. Near-infrared radiation produces a sensation of heat.

In general, UV or IR microscopy is useful for the study of materials that are uniformly transparent or opaque in the visible spectrum, but have selective absorption and/or transmission bands in the UV or IR spectral regions.

10.1. Observation of UV and IR images

The conversion of UV and IR images to visible images is accomplished by photography, fluorescence, photoelectronic detection, TV-reception, or by using image converters [10.1–10.8].

Commercially available photographic emulsions for scientific research, such as Agfa-Gevaert Scientia films and plates, are suitable for recording UV images

---

[1] There are no precisely defined limits for the visible spectrum. Some authors accept a short-wavelength limit of 380 nm or even 360 nm and a long-wavelength limit of 760 nm or even higher.

within the spectral range qualified as the *near-ultraviolet* (300–400 nm). On the other hand, photomicrography in the IR region requires specially sensitized emulsions such as Scientia emulsion 52A86, whose long-wave limit of spectral sensitivity reaches ca 950 nm. Only some spectrographic emulsions have the upper spectral limit at maximum 1300 nm. For all practical purposes, the photomicrographic spectral range generally ends at 900 nm.[2]

Ultraviolet radiation whose wavelength is shorter than 300 nm and infrared radiation whose wavelength is longer than 900 nm are unsuitable for direct photographic recording and require other receptors. One of these is a *luminescent screen* that emits visible light when excited by UV radiation, according to the Stokes law. The amount of visible light emitted by the screen is proportional to the amount of excitation radiation. A chemical substance (*luminophor*) covering a substrate (glass plate or other transparent material) is responsible for the emission of visible light of a given colour. If the luminescence decays rapidly once the excitation radiation is removed, the luminophor is said to be *fluorescent*. If the luminescence decays slowly, on the other hand, the luminophor is said to be *phosphorescent*. In order to record infrared radiation, use is made of luminescent substances known as *anti-Stokes luminophors*, which have the ability to emit light of a wavelength shorter than that of incident radiation. An example of IR microscopy using a luminophor of this kind can be found in Ref. [10.7]; infrared radiation of 1.5 μm wavelength is transformed into visible light of 0.66 μm wavelength.

At present, the most common detectors of UV and/or IR radiation function by electronic or electric means. These devices are transducers that convert the radiation received into an electronic or electric signal, enabling the data to be processed, stored (nowadays most frequently in digital form), and of course displayed on a cathode-ray tube.

UV and IR cameras have also been developed. Some of them use a small area transducer and scan the object being studied. The others use a large photosensitive surface on which the optical image is projected and scanned by an electron beam. Among these devices are TV cameras and also the vacuum pyricon tube whose spectral range lies between 2 and 25 μm and maximum sensitivity between 8 and 14 μm [10.36]. Another type of IR camera, operating between

---

[2] Problems of UV and IR photomicrography are discussed in detail by R. P. Loveland [2.7]. See also Ref. [10.17] for information concerning the IR photomicrography of biological objects.

8 and 14 μm [10.5], incorporates a liquid crystal film. The image is obtained with a scanning polarimeter and displayed on a TV screen.

The most popular devices, however, are *image converters* without a scanning procedure, which can be used in the infrared, ultraviolet, and visible spectral regions. Figure 10.1 shows a typical image converter suitable for microscopy. This is an electron tube that can produce a visible replica of an image formed on its cathode by invisible radiation. Electrons ejected from the cathode by incident rays are accelerated by the electric field of the anode and focused upon a fluorescent screen. The total number of ejected electrons is proportional to the spectral efficiency of the cathode and to the intensity of incident radiation. An image converter of this kind functions in the UV, VIS, and IR spectral regions (Fig. 10.2). The required spectral band, e.g., near-ultraviolet or near-infrared, is selected by using an appropriate light source and filters.

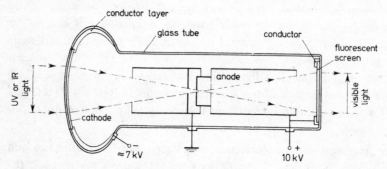

Fig. 10.1. Schematic diagram of an image converter.

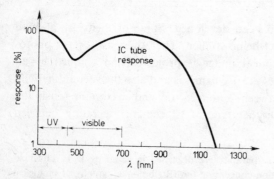

Fig. 10.2. Spectral response of a typical image converter.

The basic components of an UV or IR microscope with an image converter are shown in Fig. 10.3. The primary (real) image of a transilluminated or epi-illuminated object is formed by the objective on the photocathode of the converter. Electrons emitted from a photocathode area covered by this image form the secondary image on the converter fluorescent screen. Note that the secondary image is reversed with respect to the primary one, thus the visible image is upright with respect to the object; it is viewed through an ocular or photographed. Obviously, appropriate spectral filters (F_1, F_2) are used in the illumination systems. Sometimes, studies in polarized light are useful, thus the polarizers (P_1, P_2, P_3) for UV and/or IR radiation (see Section 1.4) must be employed.

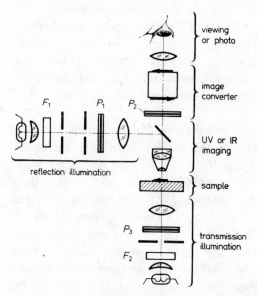

Fig. 10.3. Basic components of an UV or IR microscope with an image converter.

It is worthwihle mentioning the Wild M520 infrared image converter, which can be attached to various Wild stereomicroscopes and Leitz microscopes as part of an extensive modular system. This highly versatile instrument makes different structures visible in the near-infrared range (760 to 1200 nm). Moreover, it also makes photomicrography possible on any film material via the image conversion system or directly in infrared rays of wavelength 700 to 900 nm on standard commercially available infrared films. The exposure can be focused accurately in every case with binocular viewing.

10.2. UV microscopy and its applications

An ultraviolet microscope is intended for use with light wavelengths shorter than 380 nm. With the exception of visual ocular (Fig. 10.3), its optical components must be made of quartz, calcium fluoride, lithium fluoride, and/or other materials transparent in the UV spectral region, as optical glass strongly absorbs light of wavelengths shorter than 380 nm.

A. Köhler developed the first ultraviolet microscope in about 1904 [10.1]. His original concept was formulated as part of an attempt to improve resolution by making use of the short wavelength of UV radiation. Today, ultraviolet rays have been replaced in this respect by electron microscopes, while UV microscopes are used for the detection and study of the selective absorption of various UV bands by the specimen. A microscope based on Köhler's design was manufactured by the Zeiss factory. The objectives were corrected for only one wavelength ($\lambda = 257$ nm or 275 nm) and made of quartz. They are referred to as *monochromats*. Photographic materials or a fluorescence screen were used to render the UV image visible. Further details regarding this microscope may be found in Ref. [2.28], pp. 346–350.

Today, microscopes for ultraviolet imaging are equipped with an image converter. An instrument of this type is the MUF-6 ultraviolet microscope manufactured by LOMO (Figs. 10.4 and 10.5) and intended mainly for visual observation and photographic documentation of biological specimens both in UV and VIS light within the spectral range from 250 to 700 nm. The objectives are of three types: dioptric (quartz–fluoride) for UV radiation, catoptric for UV and visible light, and dioptric solely for visual observation and microphotography in visible light. Their magnifying powers are from $10\times$ to $125\times$. Some of these objectives are dry systems while others are water and glycerine immersion systems. Conventional oculars of magnifying power $4\times$ to $15\times$ are used for visual observation via an image converter, while quartz oculars of magnifying power $3\times$ to $10\times$ are available for photomicrography directly in UV light. Further information on this versatile instrument is given in Table 10.1, and more details regarding its construction and operating principles may be found in Ref. [2.8].

Another UV microscope manufactured by LOMO (MUF-5) can be classified as a microspectrophotometer for quantitative studies of biological specimens within the spectral range from 250 to 650 nm. However, this instrument may also be used for visual observations in UV rays converted into visible light via a fluorescent screen.

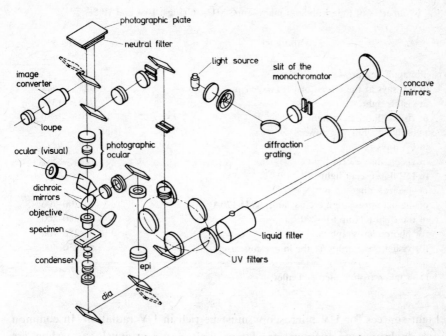

Fig. 10.4. Optical system of the UV microscope MUF-6 (LOMO, Leningrad).

Fig. 10.5. The UV microscope MUF-6 whose optical system is shown in Fig. 10.4.

TABLE 10.1

Optical parameters of the ultraviolet microscope MUF-6 (Figs. 10.4 and 10.5)*

Parameter	Value
Visual magnifying power	
in UV rays at the screen of the image converter	$28 \times - 948 \times$
in visible light	$40 \times - 1350 \times$
in visible fluorescent light	$40 \times - 630 \times$
Photographic magnifying power	
in UV rays	$28 \times - 948 \times$
in visible fluorescent light	$28 \times - 853 \times$
in UV fluorescent light	$28 \times - 440 \times$
Useful spectral range (in nm)	
of the high-pressure mercury lamp SVD-120A	200–700
of the ribbon lamp SI8-200	380–800
for photomicrography	250–700
for visual observation at the image converter	250–400

* Data taken from a factory leaflet.

Light sources for UV microscopy must be rich in UV radiation. In common use are high-pressure mercury arc lamps with a quartz envelope, carbon arc sources, and cadmium spark lamps. The lamp must be used with appropriate spectral filters or with a grating (or prism) monochromator which isolate the particular region of the spectrum required. This problem is largely discussed by Loveland in his monograph [2.7].

Ultraviolet microscopy has its main use in the biological sciences [10.9–10.11]. Many biological substances that are transparent in visible light show differing absorption in various parts of the ultraviolet spectral region (Fig. 10.6) and can therefore produce contrast without the need for staining. Living cells or tissues which are entirely colourless in visible light manifest themselves in ultraviolet light as selectively stained specimens. At 260 nm, for instance, ultraviolet absorption coincides with object areas rich in nucleic acids which would require basic stains for their presence to be indicated in visible light. Mention should be made here of a number of photomicrographs taken by R. Barer (see *Proc. Roy. Micr. Soc.*, **19** (1984), 228) showing living locust spermatocytes by means of UV microscopy using a Cooke quartz monochromatic objective of 1.25 numerical aperture (glycerine immersion), corrected for the wavelength of 254 nm. The UV image was projected onto the photosensitive plate of an EMI UV-sensitive vidicon TV camera, and the photomicrographs were recorded from a TV monitor screen at an exposure time of 0.2 second. The absorption of UV radiation at 250 nm mentioned here enables us to show clearly many details of chromosomes in division by means of this technique, which can be considered as an alternative to the use of an instrument with an image converter.

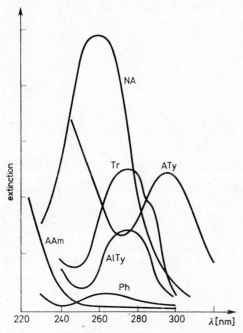

Fig. 10.6. Ultraviolet absorption bands of several biological substances: NA—nucleic acids Tr—tryptophan, ATy—acid tyrosine, AlTy—alkalic tyrosine, Ph—phenylalanine, AAm—aliphatic amino acids.

Occasionally, there are advantages in using a combination of UV microscopy with the phase contrast method (see Ref. [5.11] and the papers cited there). Ultraviolet phase contrast systems are very similar to standard phase contrast devices for visible light, the only difference being that the phase plate is optimized for UV radiation. A visible light phase contrast attachment for ultraviolet microscopes may also be found useful since it helps in the location of intracellular structures as a prelude to quantitative microphotometry, both in ultraviolet radiation and visible light.

10.3. IR microscopy and its applications

In the main, IR microscopy deals with the study of objects either illuminated by or emitting near-infrared radiation of wavelength ranging from 750 to 1200 nm. For this spectral range, systems with image converters are the most popular. Infrared microscopes of this kind are manufactured by several leading firms. Figure 10.7 shows the optical system of the near-infrared microscope MIK-4 manufactured by LOMO. This is an universal instrument for both dia- and epi-

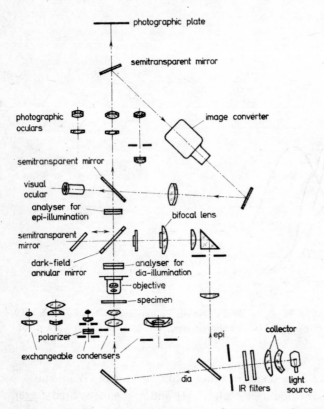

Fig. 10.7. Optical system of the IR microscope MIK-4 (LOMO, Leningrad).

microscopy with bright-field and dark-field illumination. In transillumination, it is also possible to use linearly polarized infrared radiation and visible light. The MIK-4 microscope (Fig. 10.8) is equipped with several conventional objectives of magnifying power ranging from 3.3× to 95×. For epi-microscopy, use is also made of a special immersion catadioptric epi-objective $75 \times /1.0$. The basic optical parameters of this instrument are listed in Table 10.2, while a more detailed description of its operation may be found in Ref. [2.8].

Fig. 10.8. The IR microscope MIK-4 whose optical system is shown in Fig. 10.7.

Another infrared microscope manufactured by LOMO is shown in Fig. 10.9. This is a monoobjective stereoscopic system which uses two image converters installed in the tubes of a binocular head. A stereoscopic effect is achieved by reimaging the exit pupil of the objective to an easily accessible plane and dividing the secondary pupil into two halves. Table 10.3 illustrate the optical features of this microscope. The LOMO in Leningrad also offers some very simple infrared devices which can by attached to conventional dia- and epi-microscopes [2.8].

The Polish Optical Works also manufactures infrared microscopes, and an example of one is shown in Fig. 10.10. This is a conventional stereoscopic microscope, to which a binocular head with two image converters is attached. The optical features of this instrument are listed in Table 10.4.

The Polyvar microscope, offered by Reichert-Jung (see *Proc. Roy. Micr. Soc.*, **19** (1984), 155) has recently been equipped with devices for IR microscopy,

TABLE 10.2

Optical parameters of the infrared microscope MIK-4 (Figs. 10.7 and 10.8)*

Parameter	Value
Spectral range, in nm	750–1200
Visual magnifying power	
in transmitted IR rays	$44\times-5260\times$
in reflected IR rays	$94\times-5517\times$
in transmitted visible light	$21\times-1080\times$
in reflected visible light	$45\times-1425\times$
Magnification for conoscopic observation	$3\times$
Magnifying power of the optical system by means	
of which the screen of the image converter is observed	$6\times-33.6\times$
Photographic magnifying power	
with photomicrographic attachment	
in transmitted IR rays	$22\times-657\times$
in reflected IR rays	$46\times-684\times$
in transmitted visible light	$10\times-250\times$
in reflected visible light	$22\times-332\times$
without photomicrographic attachment	
in transmitted IR rays	$12\times-261\times$
in reflected IR rays	$26\times-275\times$
in transmitted visible light	$16\times-540\times$
in reflected visible light	$36\times-712\times$
Numerical apertures of condensers	
for bright-field and oblique illumination	0.3 and 1.2
for examination in polarized light	0.22 and 0.85
for bright-field and dark-field illumination	0.6–0.7

* Data taken from a factory leaflet.

although it should be noted that the instrumentation for near-infrared microscopy is along "do-it-yourself" lines [10.13, 10.14]. This also holds good for the middle- and far-infrared microscopes. The latter, however, are more complicated in their construction than the former and usually function as scanning systems (see Section 12.2).

The applications of infrared microscopy are more varied than those of ultra-violet microscopy, although it was long neglected by biologists. However, this situation has recently changed and today infrared microscopy is a useful tool not only for researchers of solid-state materials but also in the biological sciences [10.13–10.17]. The main attribute of near-infrared light is the low energy of its

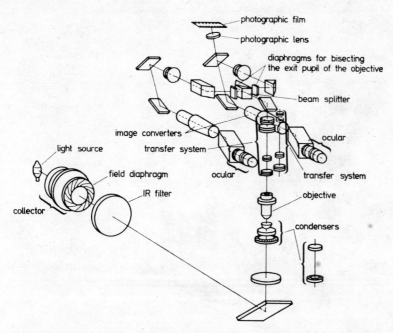

Fig. 10.9. Optical system of the IR stereoscopic microscope MIK-3 (LOMO, Leningrad).

TABLE 10.3

Optical parameters of the stereoscopic infrared microscope MIK-3 (Fig. 10.9)*

Parameter	Value
Visual magnifying power in transmitted IR rays	25×–272× or
	50×–544×
Photographic magnifying power	6×– 60× or
	12×–120×
Magnifying power of the telescopic loupe	8×
Numerical aperture of the condenser for	
bright-field and oblique illumination	0.3 and 1.2
Numerical aperture of the condenser used together with the 3.7×	
objective	0.1
Spectral range, in nm	750–1200

Data taken from a factory leaflet.

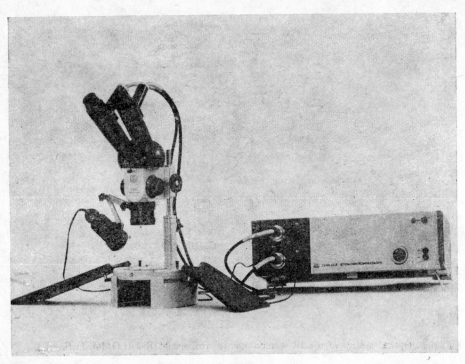

Fig. 10.10. The IR stereoscopic microscope Minfra (photo by courtesy of PZO, Warsaw).

TABLE 10.4

Optical parameters of the infrared stereoscopic microscope Minfra (Fig. 10.10)

Parameter	Value				
Spectral range, in nm				700–1200	
Total magnifying power	4.5×	7.2×	11.5×	18.0×	28.8×
Resolving power, in μm	80	55	40	25	20
Field of view in the object plane, in mm	32	20	12	8	5
Working distance, in mm				100	

photons, so that living cells are not killed or caused to mutate, as is often the case with ultraviolet radiation. In particular, the near-infrared microscope allows many studies to be performed on photosensitive microorganisms, for example accurate measurements can be made of the velocity of chloroplast movement

[10.18]. Infrared microscopy is also a suitable method for the study of neurosecretory cells [10.16].

The infrared microscope is also a useful tool for mineralogists and paleontologists (see, e.g., *Handbook of Paleontological Techniques*, W. H. Freeman and Co., San Francisco 1965) since infrared rays show up distinctly the structure and carbonization of peats, lignites, coals, and other materials. Their most important advantage in paleontological studies lies in the higher transparency of organic components in infrared light. Consequently, thicker sections can be examined than is possible in visible light microscopy.

The infrared microscope makes possible the rapid and convenient examination of forensic materials such as documents, for alterations or obliterations, fingerprints, gunpowder traces, and other particles.

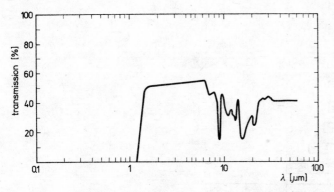

Fig. 10.11. Spectral transmission curve of a 2.5 mm thick silicon plate [10.13].

However, the widest field of application of IR microscopy is solid-state physics and especially semiconductor technology [10.19–10.34]. Semiconductor materials such as silicon (Fig. 10.11), germanium, gallium arsenide, indium arsenide, and many others (Fig. 10.12) are opaque to visible light but transparent for infrared radiation. It is therefore possible to detect subsurface faults that would have a deleterious effect on the performance of semiconductor devices. Infrared light can also be linearly polarized to help in detecting unwanted strains in semiconductor materials, epitaxial layers, mounted devices, etc. Microwave transistor chips, deformities in the die-attach bond, precipitates, inclusions, strain configurations, internal flows in the material, faulty processes of diffusion or metallization are among faults that can be effectively detected and localized with the aid of an infrared microscope operating both in transmission and reflection.

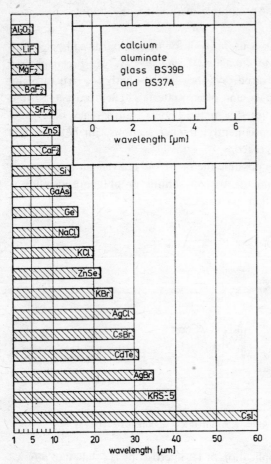

Fig. 10.12. Diagram showing the IR transmission ranges of various materials (specification taken from Ref. [10.19], but for calcium aluminate glass from a catalogue of Pilkington).

10.4. Thermal microscopy

Thermal radiation is classified as infrared emission originating in the thermal motion of the atoms or molecules of a given material. The process of producing a visible image of the temperature distribution in an object at a microscopic level is the subject of thermal microscopy. In fact, this specialized technique is concerned with the non-contact measurement of temperature by means of recording infrared radiation within a large spectral region. A photodetector is therefore used which has to record not only the near-infrared, but also radiation of wavelengths greater than 2 μm. The image converters normally used in near-infrared microscopy are unsuitable for thermal microscopy, where they are commonly replaced by small-area detectors based on InSb [10.26], InAs [10.29], HgCdTe, or TGS [10.35] sensors. These need to scan the object field and have in most cases to be cooled to a low temperature because of their instability, caused by the ambient temperature. These disadvantages do not apply to the vacuum pyricon tube [10.36] or the liquid crystal camera [10.5], which operate near room temperature.

A number of thermal microscopes are described in the literature; here two examples will be presented. One of them is the Infrared MicroScanner, Model

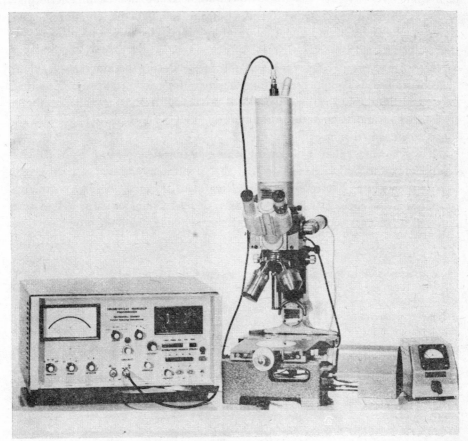

Fig. 10.13. The pyrometric thermomicroscope PMP-1 (photo by courtesy of A. Nowakowski [10.35]).

RM-50, developed by Barnes Engineering Co. (Stanford, USA). This combines the capabilities of an infrared scanner and a microscope to produce finely detailed thermal images (maps) of objects. The thermal image is continuously displayed in real time on a cathode ray tube and can be recorded by a built-in camera. Each frame is produced in 1 second and contains 4096 temperature measurements. Over-all temperature range of this instrument is from ambient to 200°C or even more. A variety of image displays is provided. Conventional analog display shows the coldest thermal details as black and the hottest white, with intermediate gray tones representing the temperatures between these extremes. Superimposed on the image are numbers that indicate the temperature settings for the black and white levels. A gray scale for temperature determination of the intermediate tones is also displayed. Other forms of image and data presentations include reversal of the black and white temperature range, electronic zoom up to $3\times$, three dimensional isometric display, single line scan, and various combinations of these. An accessory converter is available for producing thermograms in colour. This instrument is equipped with four objectives for magnifications of $3\times$, $10\times$, $40\times$, and $100\times$, but the objective $40\times$ is privileged. It offers a spatial resolution of 0.025 mm and temperature resolution of 0.1°C. The scanning spot size can be as small as 0.01 mm.

It is self-evident that in thermal microscopy objectives cannot be of the conventional type as optical glasses strongly absorb infrared radiation of a wavelength longer than 2 μm. The range of special materials that may be used in thermal imaging systems includes water-soluble salts such as cesium iodide as well as other infrared substances such as selenium, arsenic sulphide, thallium bromide-iodide,

TABLE 10.5

Basic parameters of the pyrometric thermomicroscope PMP-1 (Fig. 10.13)

Magnifying power/numerical aperture of the objectives	Field of view in the object plane [mm]	Resolving power (minimum size of the area being examined) [μm]	Range of the stage movement with automatic scanning [mm]*
$12\times /0.40$	0.66	100	16
$23\times /0.40$	0.39	50	8
$57\times /0.60$	0.16	25	4
$135\times /0.8**$	0.10**	12**	2

* Maximum speed of scanning 0.5 mm/s; range of the manual movement 50×25 mm.
** Only with $Hg_{1-x}Cd_xTe$ detector.

and especially polycrystalline germanium, silicon, and gallium arsenide. As an alternative, catoptric objectives are used in a thermal microscope developed by a group of Polish researchers from the Technical University of Gdańsk and the Technical University of Warsaw [10.35]. This instrument (Fig. 10.13 and Table 10.5) is capable of measuring the temperature of a given sample within the range of 0° to 360°C with an accuracy of 1%. Two detectors are employed: $Hg_{1-x}Cd_xTe$ and TGS. Their spectral ranges are, respectively, 3–8 µm and 2.5–20 µm. The first detector is cooled with liquid nitrogen, while the second functions at room temperature.

At present, the spatial resolution of thermal microscopes is relatively poor compared with that of conventional microscopes. However, this parameter is likely to be greatly improved in the near future.

For the past 25 years, *thermography* (the production of thermal maps by infrared scanning techniques) has provided scientists and engineers with valuable data about target temperature and surface or subsurface conditions of different solid-state objects which cannot be touched or disturbed by conventional thermal measuring techniques. Thanks to thermal microscopy, the benefits of thermography have recently been extended to such fields as microelectronics [10.26, 10.35, 10.36–10.38], micromechanics, biology, and other areas where objects or their internal defects are too small for normal macrothermal investigation. Thermal maps of the emitter sections of a voltage regulator chip, thermal microdefects of turbine blades, thermal microvariations across the surface of metallurgical materials are just some of the examples of very important applications of thermal microscopy in engineering technology.

*
* *

This chapter does not cover all aspects of ultraviolet and infrared microscopy since spectrophotometric problems are omitted here to be taken up again, however, in Chapter 14, which deals, among other things, with ultraviolet and infrared microspectrophotometry. However, it is worth noting here that IR microscopy has recently revealed its advantages for failure analysis of plastic encapsulated components (see J. Brown *et al.*, *J. Microscopy*, **148** (1987), 179–194).

11. Holographic Microscopy

Holography or *wavefront reconstruction*, invented by Denis Gabor [11.1–11.3], is an imaging process in which both the amplitude and phase variations of an object wavefront are photographically recorded as a diffraction-grating-like interference pattern, from which an object image is subsequently reconstructed. The object wavefront is recorded together with a reference wavefront. Both wavefronts are mutually coherent, can therefore interfere with each other and produce more or less complicated interference fringes that are exactly what is recorded. The reconstruction of the object image is typically carried out by using a light beam which is geometrically identical with or similar to that used as the reference wave during the recording process.

The amplitude and phase information contained in the object wavefront is materialized as the contrast and configuration of interference fringes of the recorded diffraction-grating-like interferogram. Photographic plates of high resolution (approaching 2000 or even 3000 lines per millimetre) are commonly used as a recording material. The plate on which the diffraction-grating-like interference pattern was recorded, next developed and fixed, constitutes the *hologram*. This term, based on Greek words meaning total recording,[1] here means a record of both amplitude and phase.

Unlike a hologram, a conventional photograph displays only information on the distribution of intensity or irradiance. Moreover, three-dimensional optical images are reconstructed from holograms, while photographs show only two-dimensional pictures. The depth of field or three-dimensionality is rendered thanks to the phase differences of waves reflected from or transmitted by different object regions. A photodetector cannot normally record the phase of light waves so that to make use of phase differences we must transform them via interference into differences of intensity distribution. Before the advent of holography, this procedure was applied by Zernike to phase contrast microscopy (see Chapter 5). Holography, however, is a technique quite different from the phase contrast method, although both have their origin in Abbe's two-step principle of image

---

[1] *Holos* = whole or total, *gràmma* = record.

formation in the microscope (see Subsection 3.3.1 in Volume 1). In holography, the first step corresponds to the hologram record and the second to the image reconstruction from the processed hologram. Here both steps are temporally separated from each other and carried out by means of a monochromatic and highly coherent light source such as a He–Ne laser, although there are some types of holograms from which images are also reconstructed in white light.

11.1. A bird's eye view of the history of holographic microscopy

Gabor's holographic method, which won him the 1971 Nobel Prize for Physics, has as its original aim an improvement in electron microscopy to enable examination of microobjects at molecular level with a resolution better than 0.5 nm. The idea was as follow: a hologram of a specimen is recorded by using a focused electron beam, and next an optical image is reconstructed from the processed hologram by using coherent light. All essential dimensions determining the shape of the reconstructing wave ought to be scaled upon the ratio of visible light wavelength λ_l to wavelength λ_e of matter (de Broglie) waves associated with electrons. This ratio determines image magnification which may be of the order 10^5 for electrons of about 50 keV energy.

In his earliest work, Gabor stated that the above idea was based on Sir Lawrence Bragg's X-ray microscope [11.4, 11.5], whose operating principle is similar to that of the holographic imaging process; an X-ray diffraction pattern of a cristalline lattice is fotographically recorded and next the lattice image is reconstructed from the developed diffractogram by using coherent light. However, this principle of X-ray microscopy was formulated and experimentally proved by a Polish physicist, Mieczysław Wolfke, 20 years before Bragg. In his work [11.6] Wolfke stated: "When normally illuminated by a parallel beam of monochromatic light, a diffraction pattern of a symmetric object without phase structure produces an image which is identical with the object picture",[2] and a little later: "this imaging method gives images magnified in proportion $(\lambda/\lambda') \times 10^5$ when the diffractogram is recorded by using X-rays (λ') and image is reconstructed in visible light (λ); if, moreover, a reconstructing optical system is suitably selected, total magnification 10^6–10^8 can be obtained, what is sufficient to visualize molecules in crystalline lattices".

[2] "Bei monochromatischer, paralleler, senkrechter Beleuchtung ist das Beugungsbild eines Beugungsbildes eines symmetrischen Objektes ohne Phasenstruktur identisch mit dem Abbild dieses Objektes".

It is surprising that Bragg, and before him Boersch [11.7], and next also
Gabor dit not cite Wolfke's work though it was published in a famous journal
with a large circulation. This is perhaps a misfortune typical of an invention
ahead of its time. I might perhaps be allowed to note here that on becoming
acquainted with Wolfke's work Gabor stated generously in a letter dated January
1968: "I have now read Wolfke's paper, and see that the priority for the double
Fourier trasformation must go to him, not to W. L. Bragg" [11.8]. Consequently,
Wolfke should be considered a precursor of holography in spite of an essential
difference between his method and Gabor's holographic process; namely, Wolfke
made use of a conventional diffraction pattern which contained information
on the intensities only, but not on the phases. Thus it does not readily appear
possible to obtain a complete image of the object configuration. The phase differences can be recorded if the diffraction pattern of the object is photographed
with coherent illumination and a coherent background or reference wave is added
to the object wavefront. This more general approach was first proposed by Gabor
and next modified by some other researchers [11.9]. With the present state of the
art, it is however impossible to produce electron or X-ray beams of the required
coherence. Moreover, difficulties exist in recording holograms of high resolution
at very short wavelengths. The very high magnifications mentioned above are
therefore theoretical values not at present obtainable in practice.

Before the advent of the laser, only a few contributions to holographic microscopy and related fields were reported [11.10–11.19]. At that time, the lack
of coherent light sources of high energy led to sophisticated but unpractical
designs for holographic systems. Moreover, the effective separation of the reconstructed twin images (real and virtual) which normally occur in the second step
of the holographic imaging process still posed a serious problem. Since lasers
have become commercially available, however, there has been a great amount
of work in this field. Although no practical results have been achieved to date
using Gabor's original idea, many variations of it have been successfully applied
in different areas of light microscopy.

Using their own off-axis holographic technique, Leith and Upatnieks [11.20]
devised a *lens-less holographic microscope* enabling enlarged images of good
quality to be obtained without the use of lenses. Off-axis holography enabled the
twin images to be effectively separated from each other. A very similar off-axis
system, but with multidirection illumination and moving scatterers, was used by
Stroke and Falconer [11.21] in their microholographic experiments with *lens-less
Fourier-transform holograms*, whose resolution may be better than that of conventional holograms.

A microholographic system with well-corrected microscope optics, located between the object under examination and holographic plate, was developed by van Ligten and Osterberg [11.22]. This system makes holograms of the magnified image and can be easily used as a *holographic microinterferometer*. Similar systems were also developed by other researchers [11.23, 11.24].

Another image holography system which does not require high quality microscope lenses was suggested by several authors and developed by Toth and Collins [11.25]. This system gives a reconstructed holographic image of magnification +1, which is then observed by means of a common microscope. Simultaneously, *in-line holography* was applied by Thompson, Ward, and Zinky[11.26] to the study of aerosols and other media of fine particles.

Kopylov devised a system using a magnifying hologram with properties similar to the Fresnel zone plate [11.27]. Holographic Fresnel zone microplates were next applied to soft X-ray microscopy [1.39, 1.40]. Stroke and Halioua [11.28–11.31], on the other hand, developed a method of a posteriori holographic image deblurring, making it possible to obtain a theoretical diffraction limit of resolution (0.1–0.2 nm) in high-resolution electron microscopy. An essential element of this method is a holographically made amplitude and phase filter.

A major problem that retards progress in more extensive applications of holographic microscopy in practice is the *granular* (*speckle*) appearance of the reconstructed image. This injurious effect is caused by the coherence of the laser light used in both the hologram record and image reconstruction, and produces serious difficulties in the observation of fine structures, especially of biological cells and tissues. Several systems have been developed to reduce *coherence noise* and *speckling* in holographic microscopy [11.32–11.39]; the most effective, in the present state of the art, is that developed in the Central Optical Laboratory, Warsaw [11.32, 11.35, 11.39].

This chapter covers all the more interesting and useful systems, methods, and techniques of optical holographic microscopy; electron and X-ray holographic microscopes are only occasionally mentioned as they are beyond the scope of this book.

11.2. Principles of holographic microscopy

The two phenomena basic to holography are interference and diffraction of light. Interference predominates at the moment of holographic recording, while diffraction predominates during the process of image reconstruction from the hologram.

11.2.1. Recording typical holograms of transparent objects

Let us consider a simple configuration involved in taking typical holograms of transparent objects. This configuration is schematically shown in Fig. 11.1a. A monomode light beam from a laser L is expanded by lenses L_1 and L_2 whose adjacent foci are coincident with each other and diaphragmed by a pinhole PH which improves homogeneity (coherence) of light. The expanded beam is split into two parts by a semitransparent mirror BS, known as a standard beam-splitter.

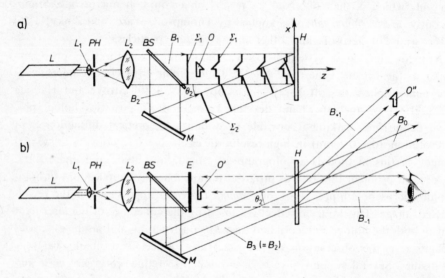

Fig. 11.1. Arrangement for recording typical holograms of transparent objects (a) and for holographic image reconstruction (b).

One of the split beams, called the *reference beam* (B_2), is allowed to fall on a plane mirror M and reflect towards a photographic (holographic) plate H. The other beam (B_1), referred to as the *object beam*, transilluminates the object O, whose hologram is to be recorded. Both beams interfere with each other, and the resulting interference pattern is recorded on the plate H. When this recorded pattern has been developed and fixed, the photographic plate forms the hologram from which images—real and virtual—of the object O can be reconstructed.

As was stated earlier, the hologram contains both the amplitude and phase information due to the object wavefront Σ_1. This fact can now be expressed in mathematical terms. For the sake of simplicity but without detriment to generality, let us assume that the object beam B_1 strikes the photographic plate H at right angles. The wavefront Σ_1 is plane before but deformed behind the object O which modifies both the phase ψ_1 and amplitude a_1 of the beam B_1. The reference beam B_2, on the other hand, passes alongside the object, so that its phase and amplitude are retained without any disturbances. However, as it falls obliquely on the photographic plate, its phase ψ_2 is not constant in the x–y plane but variable along the x-axis according to the following formula:

$$\psi_2 = \frac{2\pi}{\lambda}\Delta z_2 = \frac{2\pi}{\lambda} x \sin\theta_2 = kx\sin\theta_2, \tag{11.1}$$

where Δz_2 is the path difference between the x–y plane and the wavefront Σ_2, θ_2 represents the angle subtended by the propagation direction with the z-axis, and λ is the wavelength.

Now, let $\mathscr{A}_1(x, y)$ represent the complex amplitude[3] at the photographic plate H due to the object beam B_1 and $\mathscr{A}_2(x, y)$ the complex amplitude due to the reference beam B_2. In general, these quantities may be expressed as

$$\mathscr{A}_1(x, y) = a_1(x, y)\exp[-\psi_1(x, y)], \tag{11.2a}$$

$$\mathscr{A}_2(x, y) = a_2(x, y)\exp[-\psi_2(x, y)]. \tag{11.2b}$$

Consequently, the complex amplitude $\mathscr{A}$ due to the resultant wave in the x–y plane is given by $\mathscr{A}(x, y) = \mathscr{A}_1(x, y) + \mathscr{A}_2(x, y)$, and according to Eq. (1.19) the resultant intensity $I(x, y)$ can be expressed as

$$
\begin{aligned}
I(x, y) &= \mathscr{A}(x, y)\mathscr{A}^*(x, y) \\
&= a_1^2(x, y) + a_2^2(x, y) + \mathscr{A}_1(x, y)\mathscr{A}_2^*(x\ y) \\
&\quad + \mathscr{A}_1^*(x, y)\mathscr{A}_2(x, y) \\
&= a_1^2(x, y) + a_2^2(x, y) \\
&\quad + 2a_1(x, y)a_2(x, y)\cos[\psi_2(x, y) - \psi_1(x, y)],
\end{aligned} \tag{11.3}
$$

where $\psi_1(x, y)$ and $\psi_2(x, y)$ are the phases of the beams B_1 and B_2 at a given point in the x–y plane.

Let us now assume that the object O varies the wave phase along the x-axis only, i.e., $\psi_1(x, y) = \psi_1(x)$, and modifies the wave amplitude uniformly, i.e., $a_1(x, y) = \text{const} = a_1$. Then, just as the amplitude of the reference wave is

---

[3] See Subsection 1.2.1 for definition and notation using complex numbers.

assumed to be constant over the x–y plane and $\psi_2(x, y) = \psi_2(x)$ as supposed earlier (compare Eq. (11.1)), so

$$I(x) = a_1^2 + a_2^2 + 2a_1a_2 \cos\left[\frac{2\pi}{\lambda}x \sin\theta_2 - \psi_1(x)\right]. \qquad (11.4)$$

From this equation it follows that bright interference fringes or maxima of light intensity $I_{\max} = (a_1 + a_2)^2$ occur at those points of the photographic plate H, i.e., for those values x_m of the x-coordinate, which satisfy the following condition:

$$\frac{2\pi}{\lambda} x_m \sin\theta_2 - \psi_1(x) = m2\pi, \qquad (11.5)$$

where $m = 0, \pm1, \pm2, \ldots$ From the above equation it follows that

$$x_m = \frac{m2\pi + \psi_1(x)}{2\pi \sin\theta_2} \lambda. \qquad (11.6)$$

Similarly, dark interference fringes or minima of light intensity $I_{\min} = (a_1 - a_2)^2$ occur for those values x_m which satisfy the condition

$$\frac{2\pi}{\lambda} x_m \sin\theta_2 - \psi_1(x) = (2m+1)\pi, \qquad (11.7)$$

or

$$x_m = \frac{(2m+1)\pi + \psi_1(x)}{2\pi \sin\theta_2} \lambda. \qquad (11.8)$$

As can be seen from Eqs. (11.6) and (11.8), the position of bright and dark interference fringes depends on the phase $\psi_1(x)$ of the object wavefront, or—in other words—information on the phase of the object wavefront is coded by the position of the interference fringes recorded on the holographic plate. It is, however, important to note that there is a linear relation between x_m and ψ_1.

The question now arises of how we may code the amplitude of the object wavefront. This is simply the *visibility V* or *contrast* of interference fringes which is responsible for coding the wave amplitude; any variation of the amplitude of the object beam causes the interference fringe visibility V to be varied. By definition, $V = (I_{\max} - I_{\min})/(I_{\max} + I_{\min})$. As $I_{\max} = (a_1 + a_2)^2$ and $I_{\min} = (a_1 - a_2)^2$, we have

$$V = \frac{2a_1a_2}{a_1^2 + a_2^2}. \qquad (11.9)$$

This formula shows that a_1 is explicitly determined if a_2 is constant, as has been assumed. Moreover, if $a_2 \gg a_1$, the term a_1^2 is very small in comparison with a_2^2 and can be neglected, so that

$$V = \frac{2a_1}{a_2}, \tag{11.10}$$

which means that the visibility of interference fringes is a linear function of a_1. This property and the linear relation between x_m and ψ_1 are very important in holographic imaging.

The object considered above was a very simple specimen (uniformly transparent and varying the wave phase in a single direction, i.e., along the x-axis). The conclusions drawn are, however, valid for more and even highly complicated objects, which can affect the phase ψ_1 and amplitude a_1 of the object beam in an arbitrary manner, so that the recorded diffraction grating-like interference pattern (hologram) is also complicated. In illustration, Fig. 11.2 shows small magnified

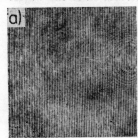 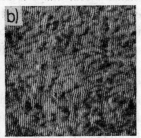

Fig. 11.2. Microscopic structure of holograms taken for non-disturbed (empty) object beam (a), and for a thin phase-amplitude object (b). The second pattern (b) represents an enlarged portion of the hologram shown in Fig. XLVIIIb. Print magnification 330×.

fragments of two holograms, one of which (a) was taken with a non-disturbed and the other (b) with a slightly disturbed object beam. In the first case, the object O, as shown in Fig. 11.1a, was excluded from the beam B_1, and in the second, this beam transilluminated a thin object, which only slightly modified both the phase and amplitude of incident light.

11.2.2. Holographic image reconstruction

Let us now consider the reconstruction of the object image from the recorded diffraction grating-like interference pattern such as shown in Fig. 11.2b. After exposure and development, the photographic plate H (Fig. 11.1a) constitutes a hologram (H, Fig. 11.1b), which is typically illuminated by a parallel beam B_3 of coherent light. Let us assume that this beam has the same inclination

θ_2 as the reference beam B_2 with which the hologram was taken. In this case, the result of the image reconstruction is without any magnification. The basic fact is that two images O' and O'' are reconstructed; one is referred to as *virtual* or *primary* (O'), and the second (O'') as *real* or *secondary*. The virtual image occupies the same position with respect to the hologram H as the object did with respect to the photographic plate H on which the hologram was recorded. This image can be conveniently observed through the hologram or photographed by means of a camera. The real image O'', on the other hand, occupies a position behind the hologram and is observed less conveniently than the virtual image O', but can be photographed directly on a photosensitive plate without any camera lens. The relation between the two images is the same as that between an object and its mirror image. If the observer looks at the virtual image, he sees it as a true copy of the object with all its essential features such as depth, perspective, and parallax. This is not the case when the real image is observed, since the object details nearer the hologram are seen further away and vice versa. This is not a stereoscopic but pseudoscopic image. For this reason, the virtual image is usually spoken of as the *true image*, while the other is referred to as the *conjugate image*.

The hologram can be considered as a sinusoidal diffraction grating (see Subsection 1.7.2 in Volume 1); consequently, it produces three light beams: the direct B_0 and two diffracted, B_{-1} and B_{+1}, of the first diffraction order, as shown in Fig. 11.1b. The direct beam is useless and only the diffracted beams are responsible for the reconstruction of the object images. This fact can be more clearly expressed by the mathematical formula described below known as the *basic equation of holography*.

Let the hologram H (Fig. 11.1b) be illuminated by a parallel beam B_3. The transmitted amplitude of this beam will depend on the amplitude transmittance of the hologram (τ_a) defined as the ratio of the amplitude of the emergent light to that of the incident light. This ratio depends on the exposure (E) defined as the product of the light intensity and the exposure time. In general, the relation between τ_a and E is nonlinear, as shown in Fig. 11.3a. Holographic recordings are typically restricted to the region AB, where the dependence τ_a on E is almost linear.

By way of digression, it is worth noting that τ_a–E curve differs essentially from the Hurter–Driffield curve (Fig. 11.3b), which is used for characterizing the photographic materials with which conventional photographs are taken. The H–D curve describes the dependence of the optical (photographic) density D on the decimal logarithm of the exposure.

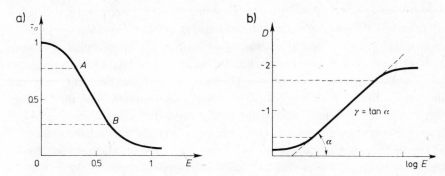

Fig. 11.3. Transmittance–exposure curve (a) and Hurter–Driffield curve (b).

Returning to Fig. 11.3a, we can assume that in the AB region the amplitude transmittance τ_a of the exposed and developed photographic plate is linearly related to the intensity of light during exposure, i.e., $\tau_a = KI$, where K is a constant of proportionality and I is defined by Eq. (11.3) relating to holographic records. If we now illuminate the hologram of amplitude transmittance τ_a with a parallel beam B_3 (Fig. 11.1b), which is identical to the reference beam B_2 (Fig. 11.1a), then the complex amplitude $\mathscr{A}'$ of the transmitted light close behind the hologram is simply equal to $\tau_a \mathscr{A}_2$ (here $\mathscr{A}_2$ is the complex amplitude of the incident beam B_3). We can therefore write the following equation:

$$\mathscr{A}'(x, y) = \tau_a(x, y)\mathscr{A}_2(x, y) = KI(x, y)\mathscr{A}_2(x, y)$$
$$= K[a_1^2(x, y) + a_2^2(x, y)]\mathscr{A}_2(x, y) + K\mathscr{A}_1(x, y)\mathscr{A}_2^*(x, y)\mathscr{A}_2(x, y)$$
$$+ K\mathscr{A}_1^*(x, y)\mathscr{A}_2(x, y)\mathscr{A}_2(x, y) = K_1(x, y)\mathscr{A}_2(x, y)$$
$$+ Ka_2^2(x, y)\mathscr{A}_1(x, y) + K\mathscr{A}_1^*(x, y)\mathscr{A}_2(x, y)\mathscr{A}_2(x, y)$$
$$= \underbrace{K_1(x, y)\mathscr{A}_2(x, y)}_{T1} + \underbrace{K_2(x, y)\mathscr{A}_1(x, y)}_{T2} + \underbrace{K\mathscr{A}_1^*(x, y)\mathscr{A}_2(x, y)\mathscr{A}_2(x, y)}_{T3},$$

$$(11.11)$$

where $I(x, y)$ is the recorded intensity defined by Eq. (11.3), while K_1 and K_2 are constant coefficients for given values of coordinates x, y in the plane of the hologram; here $K_1(x, y) = K[a_1^2(x, y) + a_2^2(x, y)]$ and $K_2(x, y) = Ka_2^2(x, y)$. These coefficients are not greater than unity.

The above equation is said to be the basic equation of holography valid for image reconstruction when using a light beam identical with the original reference beam with which the hologram was taken. Term $T1$ of Eq. (11.11) corresponds simply to a wave propagating in the original direction of the reconstructing beam with an amplitude reduction. Term $T2$ contains the object wave $\mathscr{A}_1(x, y)$ and so forms a wavefront reconstruction whose effect is identical with the effect of viewing the object itself. This term corresponds to the virtual image, which reproduces all the essential features of the object. The factor K_2 modifies (reduces) only the amplitude of the reconstructed wave in comparison with that of the original object wave, hence it only represents a brightness difference when the object and its reconstructed image are viewed directly. On the other hand, term $T3$ contains the complex conjugate of the object wave, $\mathscr{A}_1^*(x, y)$, therefore, forms a conjugate image which is real and lies on the opposite side of the hologram compared with the virtual image. This term, however, contains an additional phase factor resulting from the product $\mathscr{A}_2(x, y)\mathscr{A}_2(x, y)$, which not only tilts the reconstructed conjugate wave, but also introduces distortion in the image.

Let us now adapt Eq. (11.11) to the situation discussed previously and illustrated in Fig. 11.1, where we assumed $a_1(x, y) = \text{const} = a_1$, $a_2(x, y) = a_2$, and $\psi_2(x, y) = \psi_2(x) = kx\sin\theta_2$ (see Eq. (11.1)). Given these assumptions, $\mathscr{A}_1(x, y) = a_1\exp[-i\psi_1(x, y)]$, $\mathscr{A}_1^*(x, y) = a_1\exp[i\psi_1(x, y)]$, $\mathscr{A}_2(x, y) = a_2\exp[-ikx\sin\theta_2]$, $\mathscr{A}_2^*(x, y) = a_2\exp[ikx\sin\theta_2]$, $\mathscr{A}_2(x, y)\mathscr{A}_2(x, y) = a_2^2\exp[-i2kx\sin\theta_2]$, and Eq. (11.11) takes the form

$$\mathscr{A}'(x, y) = K_1 a_2 \exp[-ikx\sin\theta_2] + K_2 a_1 \exp[-i\psi_1(x, y)]$$
$$+ K_2 a_1 \exp[i\psi_1(x, y) - 2ikx\sin\theta_2]. \tag{11.12}$$

The first term in this equation corresponds to the direct beam B_0 in Fig. 11.1b the second to the diffracted beam B_{-1}, which produces the virtual image O', and the third to the diffracted beam B_{+1}, which produces the real image O''. As can be seen, the phase ψ_1 of the beam B_{+1} is modified by a factor $-2ikx\sin\theta_2$, compared with that of the beam B_{-1}, so that the real image is distorted. A distortion-free real image can, however, be obtained by changing the reconstructing beam so that its complex amplitude $\mathscr{A}_2(x, y)$ becomes complex conjugate $\mathscr{A}_2^*(x, y)$. This problem will be discussed later (see Subsection 11.4.2).

11.2.3. Holography of light-reflecting objects

Let us now consider holographic imaging of light-reflecting objects using an arrangement such as shown in Fig. 11.4. A single light beam from the laser L is divided into two parts by a semitransparent mirror BS. One part, B_1, falls on the object O and the other, B_2, constitutes the reference beam, which falls directly on the holographic plate H. Light reflected by the object towards H

interferes with the reference beam, and the result is the diffraction grating-like interference pattern recorded by the plate H which after exposure and development constitutes a hologram. Its grating structure is however, as a rule, more compli-

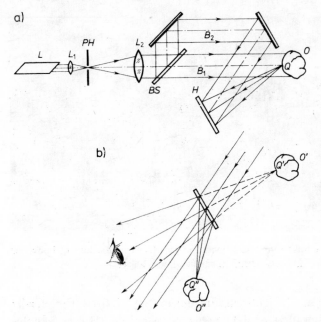

Fig. 11.4. Arrangement for recording typical holograms of light-reflecting objects (a) and that for holographic image reconstruction (b); L, L_1, PH and L_2 as in Fig. 11.1.

cated (Fig. 11.5) than that of holograms of transparent objects (Fig. 11.2b). Note that any point Q of the object O which reflects light diffusely is the source of a spherical wave. We therefore have the interference of many spherical wavefronts with a single plane wavefront due to the reference beam. Moreover, it is clear that each point of the photographic plate receives light scattered by all object points "visible" to the plate H. This fact causes the whole information on the object to be coded at a very small portion of the holographic plate, so that image reconstruction can be obtained using only a small fragment of the hologram without loss of the essential features of the image; only its brightness and resolution will be reduced in comparison with the image reconstructed from the whole hologram. In general, this property does not occur in holography of transparent objects unless a scatter plate or piece of frosted glass is used in the beam B_1 (Fig. 11.1a) before the object O.

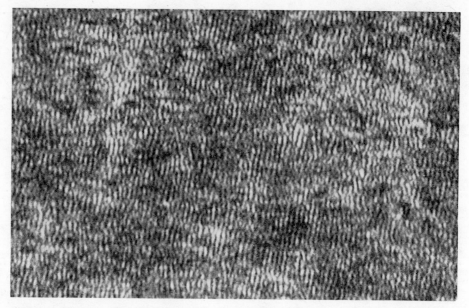

Fig. 11.5. Microstructure of a hologram of an opaque object (such as the human body, for instance), which reflects light diffusely. Print magnification $1300 \times$.

The image reconstruction of a light-reflecting object (Fig. 11.4b) is theoretically and practically the same as that of light-transmitting objects. However, holographic images of light-reflecting objects are more suitable for viewing by direct observation with the naked eye.

The holographic record is essentially composed of line and/or circular interference fringes of high spatial frequency. Any unwanted effect which superimposes an additional grating-like pattern of about the same spatial frequency will degrade the quality of the hologram. One such effect is the speckle pattern which degrades the resolution. If the optical path of the object beam changes with respect to that of the reference beam by about 0.5λ during the exposure time, the bright and dark interference fringes will be interchanged, and the diffraction grating-like interference pattern becomes less contrasted. In order to overcome this defect, it is important to maintain rigorous stability. In general, relative movements between all parts of the holographic set-up must be below $\lambda/10$ during hologram exposure. These and other technical and practical aspects of holography, and its numerous applications in many diverse areas are widely discussed in several books on optical holography [11.40–11.46]. One of these areas is microscopy.

It is, however, hoped that the contents of this section will prove sufficient to give the reader a suitable knowledge of the principles of holography, obviating the need to dip into more specialist tomes.

11.2.4. Classification of holograms

So far we have confined ourselves to *Leith-Upatnieks* [11.47] or *off-axis holograms* created by two interfering waves which arrive at the photographic material from different directions. If two interfering waves travel in the same direction, the recorded diffraction grating-like interference pattern is referred to as a *Gabor* or *in-line hologram* [11.1]. When two interfering waves are travelling in opposite directions, the recorded grating-like interference pattern constitutes a *Denisyuk* [11.48] or *reflection hologram*. This type is also called a *Lippmann* or *Lippmann–Bragg hologram* since it needs a Lippmann emulsion such as used in colour photography.

Off-axis and in-line holograms can be produced as either *amplitude* or *phase holograms* (compare adequate diffraction gratings discussed in Subsection 1.7.2, Volume 1). The former are generally recorded on silver halids materials (sometimes photochromic' media are also used), while the latter are produced by using thermoplastic sheets, some photopolymer materials, photoresists, elastomers, and dichromated gelatin [11.49]. The best known phase holograms are, however, produced by bleaching amplitude holograms recorded on silver halide emulsions (*bleached holograms*). In practice, there are no ideal amplitude or phase holograms. Silver halide amplitude holograms, for instance, show to some degree phase properties resulting from residual variations in thickness or the refractive index of the processed photographic film or plate. Bleached holograms, on the other hand, contain some residual variations in light absorption.

Both amplitude and phase holograms can be further classified by the kind of diffraction occurring in the space between the object and recording medium. The regions of diffraction were disscussed earlier in Volume 1 (see Subsection 1.7.1). According to the specification given there, lensless holograms recorded at distances $z = z_{Fre}$ to z_{Fra} are referred to as *Fresnel holograms*, whereas those recorded at distances $z = z_{Fra}$ to infinity are said to be *Fraunhofer holograms*. Any of these hologram types may be produced as either a thin or thick hologram. Theoretically, a thin or plane hologram is one for which the thickness of the recording material is smaller or only slightly greater than the period (interfringe spacing) of the diffraction grating-like interference pattern. Conversely, a thick or volume hologram is one whose recording medium is thicker than the interfringe spacing of the recorded diffraction grating-like interference pattern. In practice, the most popular silver halide holographic materials used for producing thin holograms have photosensitive emulsion from 6 to 8 μm thick (Agfa-Gevaert Holotest plates 8E75, 10E75, 8E56, and 10F56, for instance), while emulsions thicker than 10 μm are considerd as suitable for thick holograms (Kodak 649F plates).

If a lens (or lenses) are used between the object and its holographic record, we produce *image holograms*. These can be either defocused or focused. The latter are recorded in a plane which is coincident with the image plane of the lens; hence a focused hologram is also called an *image plane hologram*. If the recording material is coincident with the Fourier plane of a lens, a *Fourier transform hologram* is obtained (there are also lensless Fourier transform holograms).

Although holograms are classified in even more detail, the categories given above are likely to suffice for the purposes of our present discussion of holographic microscopy.

11.3. Holographic microscopy without objective lenses

A microholographic system in which no lens is used between the object and holographic plate is referred to as a lensless system. There are two basic types of this system, one after Gabor, known as an *in-line* (or *on-line*) *system*, and the other after Leith and Upatnikes, known as an *off-axis system*.

11.3.1. In-line microholographic system

The first holographic microscope stemmed from Gabor's early experiments [11.1, 11.2] aimed at improving electron microscopy. Although Gabor's original concept has not led to any practical results to date, many modifications based on it have been successfully applied in light microscopy, one of them being an in-line holographic system for the study of single microobjects. This system is shown in Fig. 11.6. A coherent laser beam is focused in the lens focus F', thus F' becomes the source of a divergent beam B with spherical wavefront Σ. A small object O is placed at a distance z_0 behind F'. This object diffracts light and becomes the source of a secondary beam B_1 with wavefront Σ_1, whose curvature differs from that of the original wavefront Σ (at a given distance from O). In this case, Σ_1 is the object wavefront, while Σ may be partially treated as the reference wavefront. Both wavefronts interfere with each other and a diffraction grating-like interference pattern arises, which is recorded on a photographic plate H. Interference occurs between light diffracted at the object and a portion of the same coherent beam which passes around the object. After development, the plate H constitutes a Gabor or in-line hologram from which image reconstruction is carried out by means of an arrangement shown in Fig. 1.6b.

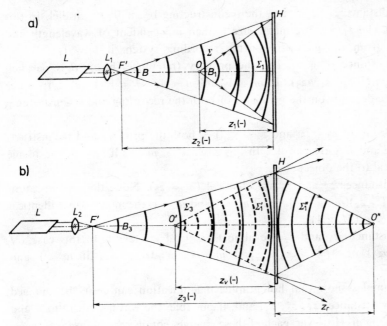

Fig. 11.6. Principle of in-line holographic microscopy: a) arrangement for hologram recording, b) arrangement for holographic image reconstruction. L—laser, L_1 and L_2—lenses, F'—their foci.

The reconstruction beam B_3 may be identical to the primary beam B uninfluenced by the object O or modified according to actual need; in particular, its wavelength and/or divergence may be changed. Transverse magnification M of reconstructed images, O' and O'', is given by the following formula [11.50]:

$$M = \frac{m}{1 \pm \dfrac{m^2 z_1}{\mu z_3} - \dfrac{z_1}{z_2}}, \tag{11.13}$$

where m denotes the lateral scaling up of the hologram by its possible magnified (or demagnified) reproduction using a photographic process, μ is the ratio

of wavelengths λ_2 and λ_1 used, respectively, for image reconstruction and hologram recording ($\mu = \lambda_2/\lambda_1$), and z_1, z_2, z_3 are the distances shown in Fig. 11.6.
The upper sign ($+$) in the denominator of Eq. (11.13) is for the virtual image
(O') and the lower sign ($-$) for the real image (O'').

With regard to Fig. 11.6, it will be interesting to consider some particular
magnifications resulting from Eq. (11.13):

(1) The distance $z_3 = \infty$, i.e., the reconstructing beam B_3 is parallel. In this
case, $M = m/(1 \pm z_1/z_2)$, the image size is then independent of wavelength and
results only from the geometry of the recording system if $m = 1$.

(2) The distance $z_2 = \infty$, i.e., the primary (reference) beam B is parallel.
Now, $M = m/(1 \pm m^2 z_1/\mu z_3)$. If $m = 1$ and $\mu = 1$, $M = 1/(1 \pm z_1/z_3)$; hence
the image size depends on the geometry of both the recording and reconstructing
systems.

(3) The distances $z_2 = \infty$ and $z_3 = \infty$, i.e., both the primary and reconstructing beams, B and B_3, are parallel. In this case, $M = m$ and if $m = 1$, the image
size is identical to the object size.

(4) The distances z_2 and z_3 are identical ($z_3 = z_2$). Now, the magnification
$M = m/[1 \pm (z_1/z_2)(m^2/\mu - 1)]$, but if $m = 1$ and $\mu = 1$, the image size is identical
to the object size ($M = 1$).

(5) The distances z_1 and z_2 are almost identical ($z_2 \approx z_1$). In this case, $M
= m/(m^2 z_1/\mu z_3)$ or $M = \lambda_2 z_3/\lambda_1 z_1$ if $m = 1$, or $M = z_3/z_1$ if $m = 1$ and
$\lambda_2 = \lambda_1$.

These examples show that holographic magnification can easily be changed
by varying the distances z_1, z_2, z_3, and, if possible, the wavelength ratio μ and
hologram scaling m. However, each of these changes entails certain disadvantages.
Magnification other than unity always suffer from aberrations which resemble
those of lenses (spherical aberration, coma, astigmatism, field curvature, and
distortion). In particular, any magnification due to an increase of the reconstructing wavelength λ_2 introduces aberrations which can be reduced by a corresponding scaling up of the hologram. This operation is valid if $\mu = \lambda_2/\lambda_1 = m$
and $z_3 = \pm m z_2$, hence $M = m$. Leith *et al.* [11.50] suggest that for $m = 1$
spherical aberration may be minimized if the distance z_2 is long and z_3
$\approx \pm (1/\mu)^{1/3} z_1$ or $z_1 \approx z_2$ and $z_3 = \infty$. These and other properties of holographic
imaging are discussed in more detail by Meier [11.51] and Smith [11.52].

Formula (11.13) does not contain the distances z_v and z_r due to the positions
of the virtual and real images (Fig. 11.6b). There is, however, a simple relation
between z_v or z_r and z_1, namely $z_v = M z_1$ and $z_r = -M z_1$. Moreover, it is
important to note that a hologram, like a lens, produces longitudinal or axial

magnification M_l, which is directly proportional to transverse magnification squared, i.e.,

$$M_l = \frac{1}{\mu}M^2, \tag{11.14}$$

which means that the axial dimensions of the image are considerably deformed in comparison with the lateral dimensions. Holography, however, enables this disproportion to be balanced if $M = \mu$; in this case, Eq. (11.14) shows that $M_l = M$. This possibility does not exist in conventional microscopy, where M_l is equal to M only for $M = 1$.

From Eq. (11.13) it follows that the transverse magnification M_l depends on five parameters: z_1, z_2, z_3, μ, and m. Manipulations with μ and m are troublesome, so in practice we use z_1, z_2, and z_3 rather than μ and m to produce different magnifications M. In any case, this procedure yields a maximum value of magnification equal to about 150 without special difficulties.

Broadly speaking, holographic image resolution is limited by the hologram area over which information on the spatial frequencies of the object has been recorded, always assuming that aberrations have been reduced so as to be virtually negligible. The larger this area, the better the lateral resolution (provided that the reconstructing beam completely covers the useful area of the hologram). A lensless holographic microscope, like a conventional microscope, does, of course, have a better resolving power when light of shorter wavelength is used. However, holographic microscopy requires highly coherent light which produces speckle patterns and so considerably reduces image resolution. In any case, present-day holographic microscopy does not offer images of better resolution than does conventional microscopy of equivalent magnification and comparable numerical aperture.

The holographic system shown in Fig. 11.6 is very simple and can be arranged by any professional or even amateur microscopist. Unfortunately, image reconstruction is inferior due to the fact that the virtual (O') and real (O'') images are localized along the same viewing axis. However, if an auxiliary lunette or low-power microscope is used, one of these two images may be observed as optimally focused against a background formed by the other image, which remains defocused. The advantages of this system are its mechanical stability and light coherence requirements, which are lower by comparison with other holographic systems.

The Gabor arrangement shown in Fig. 11.6 and its further modifications are valuable in the study of aerosols, bubble-chamber tracks, and submicron particles.

11.3.2. Off-axis microholographic system

It was mentioned above that the in-line holographic system suffers from axial overlapping of virtual and real images. This defect was overcome by Leith and Upatnieks [11.20], who introduced an improved microholographic system (Fig. 11.7a), This is an off-axis configuration considered at present as typical for optical holography.

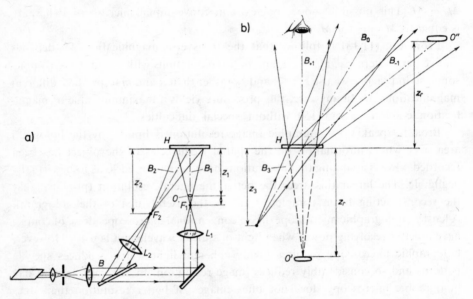

Fig. 11.7. Principle of off-axis holographic microscopy: a) arrangement for hologram recording, b) for holographic image reconstruction [11.20].

An expanded laser beam B is divided into two parts, which are then focused in focal points F_1 and F_2 of two lenses L_1 and L_2. Both foci are spatially separated from each other and function as point sources of two spherical waves. One of these waves transilluminates the object under study (O) while the other serves as the reference beam B_2. At a distance, this beam interferes with the object beam B_1 and a diffraction grating-like interference pattern arises, which is recorded on a holographic plate H. After development, the plate H (hologram) can be replaced in its original position and illuminated with the reference beam B_2, which reconstructs the virtual image axially separated from the real image. In this case, the virtual image occupies the some space does the object O, i.e., magnification $M = 1$ ($z_3 = z_2$, $m = 1$, $\mu = 1$, and Eq. (11.13) gives $M = 1$). In order to obtain a magnified holographic image, it is therefore necessary to place the hologram H at another (longer) distance z_3 from the light source of the reconstructing beam B_3 as shown in Fig. 11.7b.

Unlike Gabor's in-line configuration (Fig. 11.6), where the real image O'' occurs inconveniently in front of the virtual image O', the two images are now (Fig. 11.7b) satisfactorily separated off axis, and a part (B_0) of the reconstructing beam which passes directly through the hologram does not enter the observer's eye. Another advantage of the off-axis holographic system, compared with the in-line configuration, is the fact that it accepts not only small single objects but also extended specimens.

To illustrate this, Fig. XLVIIIa shows a biological object whose hologram is shown in Fig. XLVIIIb and holographic image in Fig. XLVIIIc. Notice the satisfactory resemblance between the object and its holographic image. Hologram recording and image reconstruction were performed by mean of a system such as shown in Fig. 11.7. A helium–neon laser was used as a source of coherent light. Coherent noise (large interference rings) seen in the hologram (Fig. XLVIIIb) originated in multiple beam interference between the glass plate surfaces during recording. This defect, in any case typical of most holograms, did not disturb the holographic image (Fig. XLVIIIc). The true hologram structure, which was responsible for the image reconstruction, is shown in Fig. 11.2b.

Upon reconstruction, a holographic image (real rather than virtual) of low magnification can additionally be studied with an auxiliary low-power microscope, in order to distinguish small details. The holographic image possesses all the qualities of three-dimensionality that the hologram was capable of recording. The recorded field-depth (d_r) depends only on the coherence length (l_c) of light used and on the optical path difference (Δ) between the reference and object beams. The latter quantity must always be smaller than l_c; otherwise, no interference fringes are formed on the holographic plate, hence no hologram is obtained. Roughly speaking, $d_r = l_c - (\Delta + \delta)$, where δ is the optical path difference introduced to the object beam by the object under study. Table 1.4 (see Volume 1) gives the coherence lengths of laser light sources used in holography.

11.3.3. Other lensless holographic systems for microscopy

An important advantage that holograms used in microscopy appeared to offer was their ability to record not only a great depth of field but also a large lateral area of the object under study. The latter property was effectively demonstrated by Stroke (see Ref. [11.9], pp. 221 and 246) in his project for a wide-field micro-holographic system based essentially on a Lippmann–Bragg hologram recorded close to the object under study where the reference and object beams had the same divergence but the opposite direction. A holographic recording of this kind enables the images to be reconstructed in white light and also makes possible multicolour holographic microscopy.

11.4. Holographic microscopy combined with objective lenses

The holographic systems described above do not incorporate any lens between the object and holographic plate. Another, rather more versatile class of micro-holographic systems is represented by configurations in which an objective is used to produce a premagnified image of the object under study; hence, image holograms are recorded.

11.4.1. Microholographic systems with premagnification and direct wavefront reconstruction

Let a microscope objective Ob (Fig. 11.8) be placed between an object AB and holographic plate H. The object AB occupies the object plane Π of the objective Ob and is transilluminated with a beam B_1. This beam is perturbed by the object and transformed by the objective into image forming beams B_1'. The geometrical image $A'B'$ of the object AB occupies the image plane Π' of the objective. A reference beam B_2 falls directly on the plate H and can interfere with the image forming beams B_1' because the object and reference beams originate, of course, from a single source of coherent light. The diffraction grating-like interference pattern is recorded and after photographic processing the plate constitutes a hologram. This is not, however, a hologram of the object AB but one of the image $A'B'$. If this image hologram is replaced and illuminated by the reference beam B_2 only (while the object beam is stopped), a holographic image is observed, which occupies the same space as does the geometrical image $A'B'$. The three-dimensional aspect of the holographic image is, however, reduced because of the limited depth of field of middle- and higher-power microscope objectives. Consequently, only a thin slice Δz of the object space can be recorded without spherical aberration. This defect may be unimportant for low-power objectives, say $5 \times /0.12$ or even $10 \times /0.25$, but is inadmissible for middle- and high-power objectives. In fact, for oil immersion objectives $(100 \times /1.25)$ the use of holography with premagnification is restricted to objects whose thickness does not exceed a few micrometres.

There are three possible locations for holographic plates in microholographic systems with premagnification: (1) behind the geometrical image $A'B'$ of the object AB (Fig. 11.8a), (2) before $A'B'$ (Fig. 11.8b), and (3) in the image plane Π'

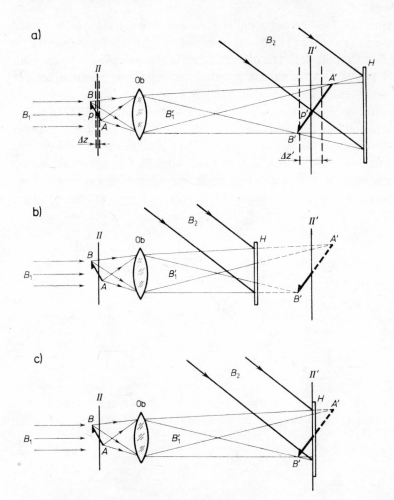

Fig. 11.8. Microholographic systems with premagnified object field: a) holographic plate H is located behind the image plane Π' of a microscope objective Ob, b) H is placed before Π', c) H is coincident with Π'.

of the objective (Fig. 11.8c). In the last instance, a focused (or image plane) hologram is recorded; otherwise, holograms are defocused. It is worth noting that the image plane hologram itself is similar in appearence to a photographic negative of the aerial image of the object under study. The only difference is that the negative aerial image is additionally coupled with a diffraction grating-like interference pattern formed by the reference beam and aerial-image-forming beams.

A holographic microscope using the first variant (Fig. 11.8a) was developed by van Ligten and Osterberg [11.22] and from time to time manufactured by the American Optical Corporation. The optical system of this microscope is schematically shown in Fig. 11.9. A laser beam B is typically divided by a splitter BS into two beams, one of which (the reference beam B_2) reaches the holographic plate H via the mirror M_2, auxiliary lens L, and beam-recombiner BR, whereas the other reaches H via the mirror M_1, condenser C, object under study O, microscope objective Ob (or objective and ocular), and beam-recombiner BR. Both beams interfere and produce a hologram. After photographic processing, the hologram is read out by means of an additional optical system which forms a coherent reconstructing beam of suitable divergence. In order to obtain a holo-

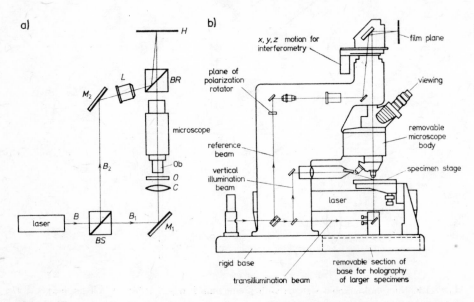

Fig. 11.9. Holographic microscope after van Ligten and Osterberg [11.22]: a) schematic diagram of the optical system, b) diagram of the design after American Optical Corporation.

gram of satisfactory quality, a good quality microscope optics is essential, since any residual aberrations inherent in the objective and stray light or diffraction patterns due to dust on the optical elements will be recorded on the hologram. When the object image is reconstructed, most of these defects will also appear in the holographic image to the detriment of its quality.

A disadvantage of the microholographic systems shown in Figs. 11.8a and 11.9 is the fact that the holographic image cannot be observed through a typical microscope ocular. This is not the case in the system shown in Fig. 11.8b [11.24], on which the first version [11.53] of the holographic microscope developed in the Central Optical Laboratory, Warsaw, was based. This version (Fig. 11.10) comprised typical parts of a standard microscope, i.e., condenser C, objective Ob and ocular Oc, and some special elements (polarizing interference beam-splitter IS, half-wave plates P_1–P_3, analyser A), which enabled a hologram of proper quality to be recorded. A linearly polarized light beam B emerges from a helium–neon laser with Brewster windows. This beam is expanded by means of the collimating lenses and divided by the polarizing interference beam-splitter IS into the object and reference beams, B_1 and B_2, polarized at right angles. The desired directions of light vibrations of these two beams are adjusted by rotating the half-wave plates P_2 and P_3. In general, these plates must be adjusted so as to obtain parallel vibrations of light beams which interfere with each other and produce the hologram H. This adjustment can additionally be optimized by using the polarizer (analyser) A. For obtaining holograms of high performance, it is necessary to maintain the optimum intensity ratio of the object and reference beams. This ratio depends on the optical properties of the object under study, and is adjusted by rotating the half-wave plate P_1. For image reconstruction, the hologram H is placed in the position where it was originally recorded. The holographic image, and also the direct (non-holographic) image of the object O is observed by means of the same ocular Oc. This property is very important for holographic microinterferometry (see Section 11.6 of this chapter).

The principle shown in Fig. 11.8b was then used by Russian researchers [11.54, 11.55] and other professional microscopists [11.56]. Figure 11.11 illustrates the holographic microscope MGI-1 developed by Ginzburg and her co-workers. An interesting feature of this instrument is the fact that the object beam B_1 is split into two parts, B_1' and B_1'', one of which (B_1') serves for side illumination of the object. The intensity ratio of the beam B_2' and the reference beam B_2 is adjusted by means of neutral filters of variable thickness (N_1 and N_2). This microscope can also be used as a holographic microinterferometer. Its basic technical data are given in Table 11.1.

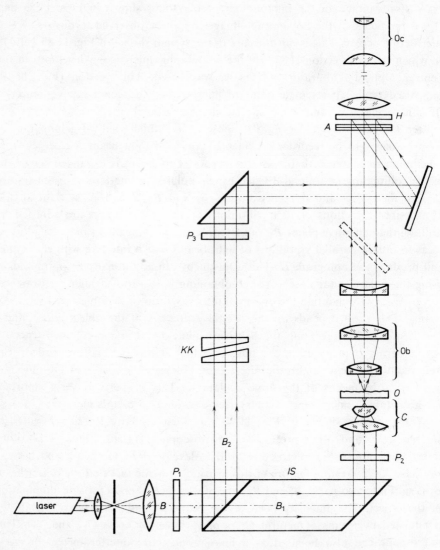

Fig. 11.10. Holographic microscope after Pawluczyk [11.53].

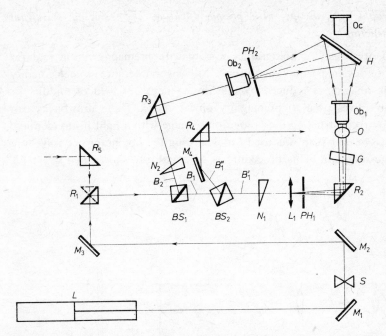

Fig. 11.11. MGI-1 holographic microscope [11.55]; *L*—laser, M_1 to M_4—mirrors, *S*—shutter, R_1 to R_5—reflecting prisms, BS_1 and BS_2—beam-splitters, N_1 and N_2—neutral filters, L_1—positive lens, PH_1 and PH_2—pinholes, *G*—glass plate, *O*—object under study, Ob_1 and Ob_2—microscope objectives, *H*—hologram, *Oc*—ocular.

TABLE 11.1

Technical data of the MGI-1 holographic microscope (according to Ref. [11.44])

Magnifying power	$40\times-1800\times$
Range of the variation of the optical path difference	$1-10~\mu m$
Accuracy of the optical path difference measurement	$\pm 0.16~\mu m$
Measuring accuracy for particle hologram analysis (size and distribution of particles)	$3~\mu m$
Free vibration frequency of the microscope	3 Hz
Hologram size	45×60 mm
Power of the He–Ne laser ($\lambda = 632.8$ nm)	8 mW
Dimensions of the microscope	$1\times0.7\times1.3$ m
Weight of the microscope	90 kg

11.4.2. Holographic microscopy with premagnification and reversed wavefront reconstruction

As mentioned previously, a holographic system with premagnification requires good quality optics free from aberrations. Moreover, the three-dimensionality of holographic imaging is reduced by the limited depth of field of middle- or high-power objectives used to premagnify object space. These drawbacks can be overcome by reconstructing the holographic image with a light beam identical to the reference beam which was used for recording the hologram but travelling in the opposite direction. This procedure is schematically shown in Fig. 11.12.

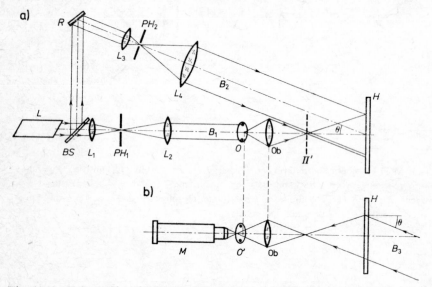

Fig. 11.12. Holographic microscopy with reversed wavefront reconstruction: a) hologram recording, b) image reconstruction; L—laser, BS—beam splitter, L_1 to L_4—collimating lenses, PH_1 and PH_2—pinholes, R—mirror (see text for further explanation).

The reference beam B_2 (Fig. 11.12a) is collimated (or divergent if necessery) and is incident on the holographic plate H at an angle θ. The object beam B_1 transilluminates the specimen under study (O), passes through the objective Ob and interferes with the reference beam B_2. The recorded hologram H may be placed behind or before the image plane Π' of the objective Ob, or even in the image plane itself. After photographic processing, the hologram must, however, be replaced exactly in the recording position in order to reconstruct correctly the object image O' (Fig. 11.12b). The reconstructing beam B_3 is identical to the reference beam B_2 and falls on the hologram H at the same angle θ, but from the opposite side. The reconstructed image O' is not virtual but real and represents an exact replica of the object O without any magnification and free from any aberrations which might have occurred during the process of hologram recording. If the original object wavefront was deformed by aberrations of the objective Ob, the reconstructed object wavefront passing back through the same objective is de-aberrated during the reconstruction process, i.e., aberrations are compensated by reversal wave reconstruction.

Let us now go back to the subject matter discussed in Subsection 11.2.2 and modify Eq. (11.11) in order to describe more precisely the phenomenon of reversed wave reconstruction. Without restricting in any way the general applicability of the results, we may consider a simpler situation such as that illustrated in Fig. 11.13, where the objective lens is omitted. The complex amplitude of the object

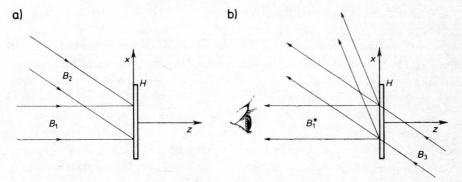

Fig. 11.13. Illustrating the principle and Eq. (11.15) of holography with reversed wavefront reconstruction: a) hologram recording, b) image reconstruction.

and reference beams, B_1 and B_2, are $\mathscr{A}_1(x, y)$ and $\mathscr{A}_2(x, y)$ at the surface of the hologram H. The reconstructing beam B_3 is identical to the reference beam B_2, but is reversed. This means that the complex amplitude $\mathscr{A}_3(x, y)$ of the beam B_3 is conjugate to $\mathscr{A}_2(x, y)$, i.e., $\mathscr{A}_3(x\,y) = \mathscr{A}_2^*(x, y)$. Consequently, Eq. (11.11) may be written as

$$\mathscr{A}'(x, y) = \tau_a(x, y)\mathscr{A}_2^*(x, y) = KI(x, y)\mathscr{A}_2^*(x, y)$$

$$= \underbrace{K_1(x, y)\mathscr{A}_2^*(x, y)}_{T1} + \underbrace{K\mathscr{A}_1(x, y)\mathscr{A}_2^*(x, y)\mathscr{A}_2^*(x, y)}_{T2} + \underbrace{K_2(x, y)\mathscr{A}_1^*(x, y)}_{T3}.$$

$$(11.15)$$

Term $T1$ of this equation corresponds to a wave propagating in the original direction of the reconstructing beam B_3 with an amplitude reduction. Term $T2$ contains the object wave $\mathscr{A}_1(x, y)$ and therefore forms a virtual image of the object except that now this image is deformed by a phase factor inherent in the product $\mathscr{A}_2^*(x, y)\mathscr{A}_2^*(x, y)$ which not only tilts, but also distorts the reconstructed image. Term $T3$, on the other hand, contains the conjugate object wave $\mathscr{A}_1^*(x, y)$ without any additional phase factor (coefficient K_2 modifies only the amplitude) and therefore forms the real image which is identical with the effect of viewing the object itself. This situation appears to be the opposite of that described by Eq. (11.11).

It is worth noting that for a non-parallel reference beam B_2, the reconstructing wave should be identical to B_2 but travelling in the opposite direction. It is self-evident therefore that a diverging reference beam would need a converging reconstruction beam.

As mentioned previously, the holographic image O' (Fig. 11.12b) is of unity magnification. It is therefore viewed through a microscope M which can be focused on any plane because of the aerial image O'. If, however, the hologram H (Fig. 11.12a) was recorded with an immersion objective, the space between the recording objective Ob and the objective of the viewing microscope must be filled with an appropriate immersion liquid.

The basic advantage of a holographic microscope with reversed wave reconstruction is its great depth of field. This property results directly from the compensation of aberrations described above: defocused, hence aberrated, portions of the recorded object space are reconstructed without defocusing aberration. On the other hand, the problem of coherent noise is more serious than in holographic microscopy with premagnification and unreversed wave reconstruction: there is an essentially greater concentration of speckles and other coherent artefacts over an image of unity magnification than over a largely magnified image.

Holographic microscopy with reversed wave reconstruction was first proposed by Toth and Collins [11.25] and next refined by McFee [11.57], Smith and Williams [11.58], and Cox and Vahala [11.59]. A paper by Smith [11.60] describes a microscope of this kind (Fig. 11.14) built in the Blackett Laboratory, Imperial College, London, which was found to give perfect aberration compensation for objects of thickness 200 μm using an immersion objective $100 \times /1.25$. This system of holographic microscopy has also been used by several other researchers for studying different microobjects and various microscopic processes. Some of these uses will be summarized in Section 11.8.

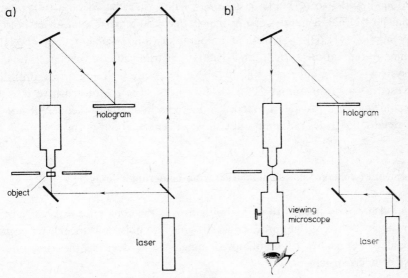

Fig. 11.14. Holographic microscopy with reversed wavefront reconstruction after Smith [11.60]: a) hologram recording, b) image reconstruction.

11.4.3. Other lens-assisted holographic systems for microscopy

The lens-assisted holographic systems described above are the most typical. There are, however, some other approaches to combined lens and hologram microscopy, most of which may be qualified as specialized arrangements for highly selective applications.

Brions et al. [11.61] describe an off-axis transmission holographic system in which a pair of lenses and a hologram are treated as a single rigid entity. This system was found to be capable of reconstructing a three-dimensional diffraction-limited image when reconstruction was carried out with a reference beam reversed back through the above-mentioned lens—hologram arrangement. The authors state that image reconstruction can be performed with wavelengths other than the recording wavelengths (provided achromatic lenses are used and the angle of incidence of the reconstructing beam is appropriately changed with respect to that of the reference beam). Two-micrometre resolution of the combustion of solid rocket propellants at high pressures has been achieved at a working distance of 6 cm. This arrangement differs from the more conventional holographic systems in its use of a pair of focusing lenses. The lenses provide the large numerical aperture needed for high resolution.

Mention should also be made of a holographic system based on a triangular interferometer [11.62], a microscope in which electrical superresolution of holograms is applied [11.63], a real-time holographic microscope with a special holographic illuminating system and rotary shearing interferometer [11.64], and a system of non-coherent microholography with a holographic optical element used as a beam-splitter [11.65]. This last system is not sensitive to vibrations and makes it possible to record holograms in spatially incoherent non-monochromatic light (desired temporal coherence length about 6 μm).

11.5. Problem of coherent noise elimination from holographic imaging

The essential property needed to make holograms is the coherence of laser light. However, highly coherent light produces easily coherent noise and speckle patterns which reduce the quality of holographic imaging and degrade the resolution of images of microstructures.

11.5.1. Coherent noise and speckle patterns

Coherent noise is mainly created by dust particles, scratches, finger prints, bubbles, striae, and other cosmetic or intrinsic defects present on and/or in the optical elements of a holographic system. All these defects scatter, diffract, reflect, and refract randomly coherent light, whose different rays interfere with each other and produce more or less chaotic diffraction–interference figures so that, in effect, a background results with irregular irradiance (Fig. 11.15).

Coherent light also causes diffraction fringes or rings to appear at the edges of objects under study. These diffraction artefacts can also be classed as coherent noise. They may be reduced by apodizing the exit pupil of the microscope objective, i.e., by introducing a filter whose transmittance varies gradually (exponentially, for instance) from 100% at the centre of the exit pupil to zero at its edge [11.66]. However, this method is not common practice.

Moreover, stray reflections of light from the surfaces of optical elements produce circular interference fringes (Newton rings), which are sometimes so distinct that they can be used for testing the quality of microscope objectives [11.67]. If the successive lenses of an objective are correctly aligned, the set of interference rings is coaxial with the optic axis of the objective; if not, the centres of ring interference patterns are more or less dispersed. This kind of coherent noise can only be eliminated by coating the lenses with effective antireflecting

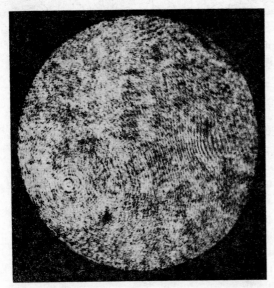

Fig. 11.15. Coherent noise observed in the field of view of a holographic microscope (courtesy of Dr. R. Pawluczyk).

Fig. 11.16. Photograph of a speckle pattern.

films correctly adjusted to the laser light wavelength (see Subsection 1.6.3 in Volume 1).

A specific type of coherent noise is due to the surface roughness of objects. It is a well known fact that as a rough or scattery surface is illuminated with laser light, a fine network of bright and dark patches is seen close by or on the surface (Fig. 11.16). The same speckle pattern is observed at an arbitrary distance from a ground glass transilluminated with a laser beam. This phenomenon is due to local interference effects at and close by the scattering surface and their random superposition at the far-field. The granular structure of the speckle pattern on the scattering surface corresponds to the resolving power of the microscope used to observe this surface. Coherent speckles therefore present a serious problem in holographic microscopy, especially as they obscure minute details in the object image. On the other hand, speckle patterns may be considered not only as injurious noise but also as carriers of useful information, a fact which has been exploited in a variety of practical applications [11.68].

It is interesting to note that the unwanted effects of coherent noise due to dust particles, scratches, stray reflections, or similar defects can be minimized by placing a ground glass before the transparent object at the time the hologram is recorded. The object is then diffusely transilluminated, which allows the useful size of the hologram and its degree of redundancy to be increased, and the signal-to-noise ratio (SNR) to be improved. Diffuse illumination, however, introduces a speckle pattern which is not desirable in holographic microscopy.

Any speckle pattern or other coherent noise disappears completely if the transilluminated ground glass is rapidly moved or rotated. In this case, the observer's eye or photographic plate combines incoherently a large number of differently distributed speckles and the overall pattern becomes uniform. Unfortunately, the rotatable ground glass cannot be used during the process of holographic recording because it would cause the diffraction grating-like interference pattern forming the hologram to move and blur. In any case, the ground glass or other diffuser should be immobile during the making of the hologram. On the other hand, rotatable ground glasses find a ready use in non-holographic optical systems with laser light sources.

11.5.2. Techniques for coherent noise reduction

There is no doubt that the problem of coherent noise held back progress in the practical application of holographic microscopy. Many techniques have been developed in order to overcome this limitation.

As was pointed out earlier, it is customary to introduce a diffuser between the transilluminated object and coherent light source in order to increase the redundancy of the hologram and thus reduce coherent noise due to dust particles, scratches, or other cosmetic defects present on optical elements. The diffuser, a ground glass for instance, causes a speckle pattern to occur in the holographic image, so that resolution is reduced unless the redundancy is properly regulated. Several ways of doing this have been proposed. These either depend on replacing the diffuser by a two-dimensional phase grating [11.69–11.72], or on using a multifrequency grating-diffuser-lens combination for object illumination during hologram recording [11.73]. This latter technique is similar to that described by Som and Budhiraja [11.74], who used a diffuser-lens-rotatable multisector mask in the object beam. All these methods are known as *speckle averaging techniques*; among them we may also include the Caulfield method [11.75], based on spatially multiplexed holograms. However, for diffusing (non-specularly reflecting) objects, speckle patterns cannot be reduced by these methods.

In holographic imagery, the reference beam produces coherent noise as well. Moreover, coherent noise and speckles arise during the second step of holographic imaging, that of image reconstruction, where the hologram itself is also a source of noise.

One of the earliest methods used to reduce coherent noise in holographic imaging systems consisted in taking a number of separate holograms of the same object and incoherently superposing the images reconstructed successively from each [11.76]. This method was used by Close [11.33] in his low-power microholographic system (high resolution portable holocamera) developed for the American moon-mission (Apollo Project).

Another multihologram method consists in making several exposures at the same place and on the same holographic plate, while the direction of the reference beam is changed and the diffuser moved in the object beam to a new position for each exposure [11.60]. A phase-shifter (a glass wedge, for instance) may also be used in place of the diffuser [11.77]. Image reconstruction is achieved by successively illuminating the multihologram with the particular reference beam (variously oriented) used to record each sub-hologram. If the successive reconstructions are generated quickly enough, the observer sees an overall image with reduced coherent noise. This method was used by Smith [11.60] in a holographic stereomicroscope. An interesting feature of this microscope is the fact that a number of collimated reference beams are generated by an array of holographic lenses.

A completely different method of reducing speckle noise in the holographic image makes use of a moving aperture in the exit pupil of the reconstructing objective [11.78]. The numerical aperture of the objective is reduced by the moving aperture whose diameter must be smaller than the available exit pupil area, so that there is a loss of light intensity and consequent reduction of resolution in the reconstructed image. This method has been slightly modified by van Ligten [11.36] and adapted to his holographic microscope (Fig. 11.9). The modification depends on using a multiplexed object beam. Multiplexing is generated by several mutually incoherent sources, which are, however, coherent with a single reference beam. Hence, the hologram contains several sub-holograms originating from each light source of the object beam and the reference beam. During reconstruction, the moving aperture permits the holographic images to be reconstructed from the individual sub-holograms. If succeeding reconstructions are quickly generated, the observer sees an overall image with reduced noise. If there are m sub-holograms, RMS intensity fluctuation is reduced by a factor $\sqrt{m}$.

Interesting possibilities for coherent noise reduction are offered by image plane holograms (see Fig. 11.8c). In practice, any hologram recorded not only in the image plane Π' of an objective but also close to this plane can be classified as an image plane hologram [11.79, 11.80]. The basic feature of these holograms is that they allow image reconstruction using a large illuminating source with little spatial coherence [11.81–11.84] or even white light from a slide projector [11.59]. Since the mean speckle irradiance is directly related to the degree of spatial coherence of light, reducing spatial coherence during image reconstruction can also reduce speckle noise in the reconstructed holographic image. This method of speckle reduction is very simple and was studied by Cox and Vahala [11.59], who concluded that speckle noise was indeed reduced in holographic microscopy with reversed wave reconstruction, but that a careful choice of recording parameters was required in order to retain the resolution needed to study microscopic objects.

A drawback of this method is that speckle noise varies with the position of the image plane under observation. If the conjugate image plane Π' is $\Delta z'$ from the hologram (Fig. 11.8a), the speckle power spectrum is unaltered (not reduced) over this plane when $\Delta z' = 0$. In other words, all speckles whose source is in the object plane Π will be observed in the image plane Π' if the hologram is recorded exactly at position $\Delta z' = 0$ (Fig. 11.8c).

A common feature of the methods described above is that noise reduction is primarily achieved during either the reconstruction or recording process. For effective holographic microscopy, this would not appear to be entirely satisfactory. In the experience of the author and his co-workers, coherence noise and speckle pattern should be reduced during both the recording and reconstruction processes

for holographic microscopy to be effective. A method which satisfies this requirement was developed in the Central Optical Laboratory, Warsaw [11.32, 11.35,
11.85, 11.86] and is described below.

11.5.3. *Holographic microscopy with unidirectional suppression of coherence noise*

In papers published in the early 1970s, the authors from the above-mentioned
laboratory stated that coherence noise could be effectively reduced in laser interference systems if the laser light source oscillates parallel to the interference
fringes. This suggested that a similar approach could be used in holographic
systems which were, in fact, classified as two-beam interference systems. After
some successful initial experiments [11.35], a holographic microscope with unidirectional suppression of coherence noise was built and presented at an international conference [11.87]. Finally, Pawluczyk formulated a complete theory for
holographic microscopy of this kind [11.39].

The method described here depends essentially on unilaterally deflecting the
laser beam and changing its angle of incidence on the object plane. This operation
is performed by means of a very simple device shown in Fig. 11.17, where *B*

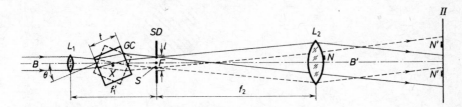

Fig. 11.17. Device for unidirectional suppresion of coherent noise in holographic microscopy.

is the laser beam, *GC* is a glass cube rotated about an axis *X* normal to the incident
beam, L_1 and L_2 are two positive lenses arranged as a telescopic system (their
focal lengths are f_1' and f_2), and *SD* is a diaphragm with a slit *S* of variable length.
The slit diaphragm is situated in a plane where the common point of the adjacent
foci of both lenses is to be found. This focal point is a secondary light source *F*.
As the glass cube rotates, the source *F* suffers variable lateral displacement (*p*)
due to the rotation of the cube. A simple application of Snell's law of refraction
and elementary trigonometric identities show that *p* is defined by Eq. 2.67a,
in which θ is a changing angle of incidence and θ' the respective angle of refraction
in the glass cube of thickness *t* and refractive index *n*.

The displacement of the source F causes the parallel laser beam B' to be emerged from the lens L_2 at a variable angle. Consequently, an arbitrary object plane II will be scanned unilaterally in a direction normal to the axis of rotation of the glass cube. The amplitude of scanning is controlled by the variable length (l) of the slit S. It is self-evident that the degree of spatial coherence of light is degraded along the length of the slit S and so along the scanning direction in the object plane II. If there is a dust particle or other defect N on the surface of the lens L_2 for instance, the diffraction pattern (noise) N' created by the defect N changes its position according to the oscillatory deflection of the laser beam B'. Since this deflection is rapid, the observer's eye or a photographic plate integrates incoherently the changes of the noise light intensity in the plane of observation, thus averaging the detrimental noise effects.

Let us now consider a simple holographic system as shown in Fig. 11.18. Apart from the scanning device described above, this system contains only a beam-splitter BS and two mirrors M_1 and M_2. The laser beam B is split into the object beam B_1 and the reference beam B_2. Both split beams produce interference fringes, which can be recorded on a holographic plate H. In general, the fringes

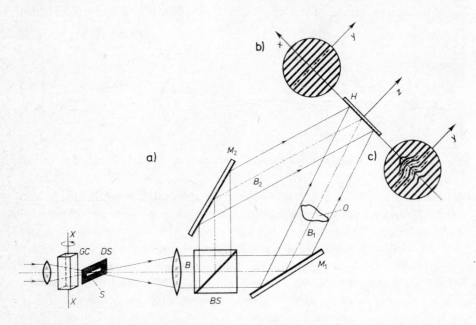

Fig. 11.18. Principle of the unidirectional suppression of coherent noise during the process of hologram recording, by using the device shown in Fig. 11.17.

are hyperbolic (compare Fig. 1.59 in Volume 1), but over a limited area they can be considered as rectlinear (Fig. 11.18b) when viewed along the z-direction. If the glass cube GC is immobile, the interference fringes do not move and are disturbed by coherence noise (Fig. 11.19a). On the other hand, if the glass cube is rotated about the X-axis, the interference fringes are periodically displaced in a direction parallel to the fringes. The interference pattern is practically the same as previously when the glass cube was immobile except that now coherence noise is reduced (averaged), and the interference fringes becomes more distinct and free from local irregularities or diffraction rings as shown in Fig. 11.19b.

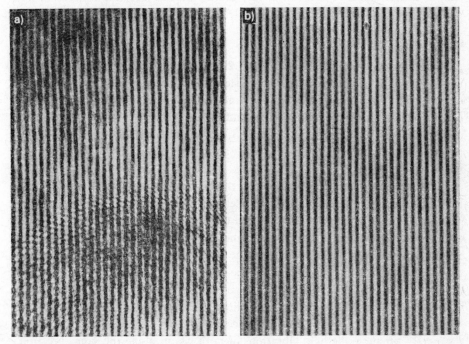

Fig. 11.19. Small fragments of aerial empty holograms ("empty" means that the object is removed from the object beam): a) without and b) with suppression of coherent noise, by using the technique shown in Fig. 11.18. Print magnification 1500 ×.

So far we have confined ourselves to the holographic system alone, in isolation from the object to be studied. Now, let this object (O) be placed in the path of the beam B_1 (Fig. 11.18a); the wavefront of this beam becomes deformed and the interference fringes will no longer be rectilinear in the x–y plane, but irregularly bent (Fig. 11.18c). What happens to the interference pattern if the glass cube

GC is rotated? Someone not familiar with this system of coherent noise reduction would expect the fringes disappear since, in general, they are no longer parallel to the scanning direction or, in other words, perpendicular to the rotation axis (X) of the glass cube. This is not the case, however. Fortunately, the interference fringes retain their contrast and configuration provided the possible movement of the empty interference pattern (Fig. 11.18b) in the x-direction is not longer than 1/10 of the interfringe spacing. The rotation of the glass cube GC causes both the undeformed and deformed fringes (Figs. 11.18b and c) to move along their current path as shown by the arrows. However, both the deflection amplitude and the deflection phase of the beams B_1 and B_2 must be identical in the recording plate H. Consequently, the optical path lengths of both beams must be identical (or nearly identical).

The application of this method to the reduction of coherent noise in holographic microscopy is shown in Fig. 11.20. Besides the scanning device shown in Fig. 11.17, the microholographic system contains a polarizing interference beam-splitter BS, two condensers C_1 and C_2, two objectives Ob_1 and Ob_2, ocular Oc, three half-wave plates P, P_1, and P_2, a polarizer (analyser) A, a number of mirrors (M_1–M_9), a phase shifter PS, a compensating plate CP, a viewing tube (lens L_3 + mirror M_9 + ocular Oc), and some other optical elements not shown in Fig. 11.20.

As can be seen, the optical system of the microscope consists of two identical branches similar to those of the Mach–Zehnder interferometer. The object to be studied is placed in the object plane Π_1 of the objective Ob_1, while the glass plate GP_2 should be located in the object plane Π_2 of the objective Ob_2 in order to compensate the optical path difference due to the object slide GP_1. The hologram H is recorded in the image plane Π' of the objective Ob_1; this plane being coincident (or nearly coincident) with the image plane of the objective Ob_2. The half-wave plate P, together with the beam-splitter BS, make it possible to control the intensity ratio of the object and reference beams. Two other half-wave plates, P_1 and P_2, orientate the light vibrations of both beams along the same direction. This operation can additionally be optimized by means of the polarizer A.

It is important that the angular magnification of the optical system should be the same in both branches of the microscope. This requirement is fulfilled if the objectives Ob_1 and Ob_2, condensers C_1 and C_2, and other elements making up each of the two branches of the microscope are identical. At the same time, the objectives and condensers should be adjusted at equal distances from the holographic plate H.

The ocular Oc enables both the aerial image of the object O and the recon-

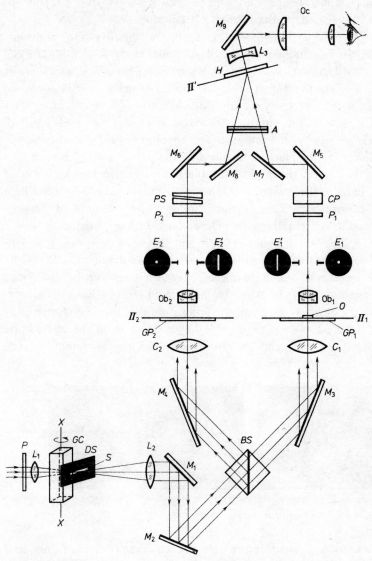

Fig. 11.20. Schematic diagram of a holographic microscope with unidirectional suppression of coherent noise (a simplified version of Pawluczyk's design).

structed holographic image to be observed alternately or simultaneously. To observe the holographic image only, the object beam is stopped and the hologram H is illuminated by the reference beam alone.

It is interesting to look into the exit pupils of the objectives Ob_1 and Ob_2. If the glass cube GC is immobile and the object under study is set aside, we can see these pupils as dark circles E_1 and E_2 with small light spots, which are in fact Airy patterns of the point light source. If, however, the glass cube is rotated about its X-axis, these light spots are transformed into bright lines against dark circles E_1' and E_2'. When the optical system is correctly aligned and adjusted, these bright lines are parallel to each other and equal in length, while their direction is perpendicular to the plane determined by the optical axes of both microscope branches. The longer the bright lines, the more effective the reduction of coherent noise, but also the greater the reduction of the spatial coherence of light and hologram performance. Thus, a compromise must be arrived at between these two contradictory requirements. Depending on the magnifying power and numerical aperture of the objectives Ob_1 and Ob_2, this compromise is optimized by a suitably adjusting the length of the slit S of the diaphragm DS. Figure 11.21 shows the overal exit pupil of the microscope when the glass cube is immobile (Fig. 11.21a) or rotated (Fig. 11.21b). One light spot or bright line comes from the exit pupil of the objective Ob_1 and the other from that of the objective Ob_2. The separation of these two spots or bright lines depends on the intersection angle of the object and reference beams incident on the hologram H. The angle may be changed, but typically is equal to about $20°$.

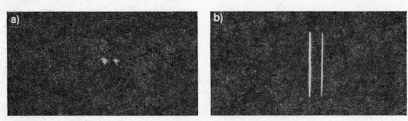

Fig. 11.21. Photographs of the exit pupil of the microholographic system shown in Fig. 11.20: a) the glass cube GC is immobile, b) GC is rotated.

The glass cube GC is rotated during both the recording of the hologram and image reconstruction; coherence noise is therefore reduced in these two steps of holographic imaging. This is an essential advantage of this system compared with other methods of the coherent noise reduction. Since we record an image plane hologram, coherent noise due to a defect near to the object plane Π_1 (and/or

Π_2) or the image plane Π' is reduced less effectively than the noise caused by a defect of the same size and contrast located further from these planes. This property holds good for any plane optically conjugate to the image plane Π' (Fig. 11.20). Moreover, the effectiveness of noise reduction depends on the size and contrast of the sources creating coherent noise. Large and very contrasty noise sources cause the reduction of noise to be performed incompletely.

Figure 11.22 represents a photograph of the holographic microscope constructed in the Central Optical Laboratory, Warsaw, according to the principle shown in Fig. 11.20; its technical data are given in Table 11.2. This microscope functions in both transmitted and reflected light; its optical system is thus more complicated than that shown in Fig. 11.20 and adapted for holographic microinterferometry. Two sequences of photomicrographs shown in Figs. XLIX and L illustrate the performance of this microscope from the viewpoint of the effectiveness of coherent noise reduction.

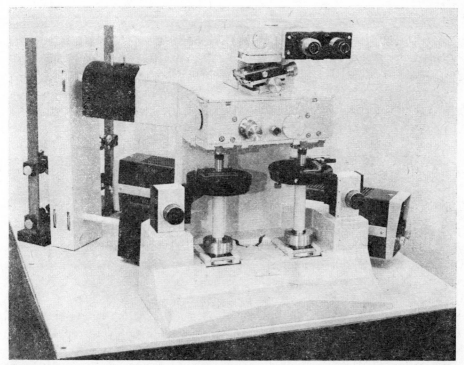

Fig. 11.22. Holographic microscope constructed in the Central Optical Laboratory, Warsaw, according to Pawluczyk's design.

TABLE 11.2

Technical data of the holographic microscope shown in Fig. 11.22

Objectives (magnifying power/numerical aperture)	$10 \times /0.24$; $20 \times /0.40$
	$40 \times /0.65$
	$100 \times /1.25$ O.I.
Numerical apertures of condensers	0.15 and 0.35
Resolving power	better than 1 μm
Visual magnification	$50 \times -1500 \times$
Efficiency of the coherent noise reduction	higher than 99%
Size of the holographic plate	45×60 mm
Diameter of the useful area of the hologram	20 mm
Range of the variation of the optical path difference	from 0 to ± 5 mm
Accuracy of the optical path difference measurement	± 0.06 μm
Dimensions of the microscope (without table)	$90 \times 90 \times 55$ cm
Weight of the microscope (without table)	ca 100 kg
Power of the He–Ne laser ($\lambda = 632.8$ nm)	5 mW

The photomicrographs show that noise reduction during either the recording process or image reconstruction alone does not yield a holographic image of acceptable quality (Figs. XLIXb and c). A satisfactory visual appearance of reconstructed images is only achieved if coherent noise is reduced during both these processes (Figs. XLIXd and Lb). By comparing Figs. XLIXb and c we can see that noise reduction during the recording process is, however, more important than during image reconstruction.

When compared with a, photomicrographs b and d of Fig. XLIX (as well as b of Fig. L) show that the coherent noise reduction method described in this subsection is highly effective; it helps to improve considerably the overall quality and resolution of microholographic images. The visual appearance of reconstructed images is quite good, particularly where the recognition of details near the resolution limit is concerned. It has been stated [11.39] that this limit is generally equal to λ/A, where A is the numerical aperture of the microscope objective and λ is the wavelength of light used (for the He–Ne laser, $\lambda = 632.8$ nm). The resolution value mentioned above is therefore of the same order as in conventional microscopy with the condenser numerical aperture A_c reduced to zero (compare Subsection 3.6.5 in Volume 1). Hence it is clear that resolution in holographic microscopy (or in any other coherent light microscopy) cannot be better than in conventional microscopy with incoherent or partially coherent illumination.

Photomicrograph b of Fig. L also shows that the resolution limit of this microholographic system is somewhat smaller than λ/A in the directions along which the coherent noise is suppressed, i.e., along directions parallel to the focal lines shown in Fig. 11.21b. This result is self-evident since the rotating glass cube GC (Fig. 11.20) causes the numerical aperture of the object beam to be enlarged in a plane perpendicular to the rotation axis of the glass cube.

Thanks to its simplicity, the method of coherent noise reduction described in this section is universal and can easily be adapted to any holographic system. It is therefore worth noting that this method was also used by Russian researchers in their microholographic system (Fig. 11.23) for the study of cryogenic properties of human erythrocytes ([11.88], pp. 218 and 219). They undoubtedly borrowed the idea from Pawluczyk's papers (compare Ref. [11.44], pp. 106 and 228, and refs. 54 and 58 cited there).

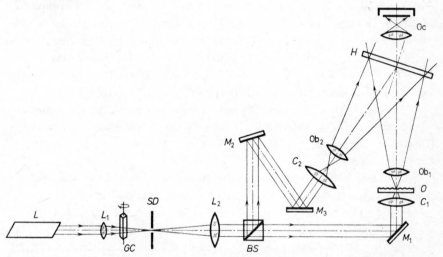

Fig. 11.23. Schematic diagram of a Russian holographic microscope with coherent noise suppression based on the principle originally developed in the Central Optical Laboratory, Warsaw; L—laser, L_1 and L_2—collimating lenses, GC—rotatable glass cube, SD—slit diaphragm, BS—beam splitter, M_1 to M_3—mirrors, C_1 and C_2—condensers, Ob_1 and Ob_2—microscope objectives, O—object under study, H—hologram, Oc—ocular.

11.6. Holographic interference microscopy

Whenever observations and/or measurements of phase microobjects are required, interferometric techniques may be used to great advantage. In conventional microinterferometric systems, the optical elements must be made with great precision. The mirrors in the Linnik microinterferometer, for instance, must be well polished. Similary, wavefronts in the Mach–Zehnder type interference microscope manufactured by E. Leitz Wetzlar can be shaped by two exactly identical branches of a complicated and very expensive optical system. The advent of holographic interferometry has completely changed all this, since a hologram can record and reconstruct a wavefront of quality sufficiently high to replace, say, the comparison wavefront[4] in the above-mentioned microinterferometer.

11.6.1. Advantages of holographic interference microscopy

Several variations of interference microscopy are known to date. The most popular techniques are those based on the lateral wavefront shear (see Chapter 7), and the Mach–Zehnder or Michelson interferometers with a comparison wavefront. Each technique needs a separate instrument. Thus the Mach–Zehnder microinterferometer available from E. Leitz Wetzlar, for example, cannot be used for wavefront shearing interferometry, and vice versa, the Interphako system available from Carl Zeiss Jena—a wavefront shearing interferometer—cannot be employed for interferometry with a comparison wavefront.

This drawback does not apply to holography, since here various interfero-metric techniques can be exploited by means of a single holographic arrangement. This feature is simply the hologram's ability to record both the amplitude and phase of any wave.

Let us consider a transparent specimen consisting of an object O of refractive index n and variable thickness t, surrounded by a medium of refractive index $n_1 < n$ (Fig. 11.24a). After passing through the specimen, the plane wavefront Σ becomes deformed according to variations in optical path difference δ introduced by the object O. The deformed object wavefront Σ_1 can be compared interfer-entially with the comparison plane wavefront Σ', and δ may be determined by suitably adjusting and measuring the optical path difference Δ between both wavefronts. These wavefronts can be either parallel (Fig. 11.24b) or inclined (Fig. 11.24c) to each other. In the first case, we have uniform, and in the second, fringe field interference. The comparison wavefront Σ' may be formed by a holo-gram and superposed on the object wavefront Σ_1 actually formed by the object

---

[4] In classical interferometry, this wavefront is usually called the reference wavefront, but its function is different from that of the reference wave used for recording the hologram. To avoid misunderstanding, the reference wavefront of an interferometer will be referred to as a comparison wavefront throughout this chapter.

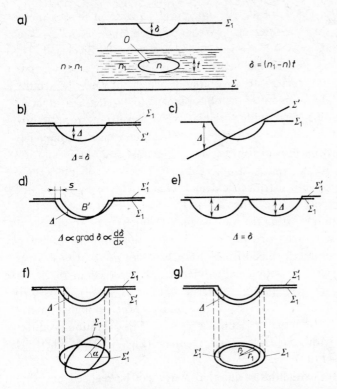

Fig. 11.24. Illustrating the discussion of holographic interference microscopy (see text for explanation). Primed sigmas indicate the wavefronts reconstructed from the hologram.

under study. The converse is also possible, i.e., Σ_1 may be reconstructed from a hologram and compared interferentially with a plane wavefront Σ' not deformed by the object. Moreover, it is possible to record holographically two wavefronts, Σ_1 and Σ', on a single hologram and reconstruct them together so as to obtain an interference image of the object from the superposition of both reconstructed wavefronts. This resembles the situation obtaining in common-path interferometers.

Let us now suppose that a hologram is taken of the object wavefront Σ_1, but that the object O is left in its original position. If the hologram is replaced exactly in its recording location after photographic processing, it will produce an object wavefront Σ'_1 exactly coinciding with the original object wavefront Σ_1. If there is a slight displacement (s) between them, as in Fig. 11.24d, a differential interference image will occur, generally referred to as the differential interference contrast (see Chapter 7). The optical path difference Δ observed in this image does not express the optical path difference δ in the specimen, but a *lateral gradient* of δ in the direction x along which the wavefronts are mutually displaced. The wavefronts Σ_1 and Σ_1' can also be inclined to each other; in this case, we have fringe field differential interference. Moreover, the displacement s may be arbitrarily changed by the lateral movement of the object under study. Hence, a situation typical of shearing interferometry with variable lateral wavefront shear is produced. For a particular value of s, the reconstructed object image is completely separated from the object itself as in Fig. 11.24e. In this case, we obtain uniform field interference with totally duplicated object image. This kind of interference can easily be transformed into a fringe field interference pattern by tilting one wavefront (Σ_1') with respect to the other (Σ_1). This is achieved by the lateral movement of the hologram, starting from its original position.

If the hologram is left in its original recording location and the object is rotated by an angle α, an interference image with *azimuthal wavefront shearing* is obtained (Fig. 11.24f). Here the optical path difference Δ observed in the interference image expresses an *azimuthal gradient* of δ (if α is small). It is also possible to reconstruct the wavefront Σ_1' with slightly lower or higher lateral magnification than that of the actual object wavefront Σ_1. This operation produces an interference image with *radial wavefront shearing* (Fig. 11.24g) where Δ expresses a *radial gradient* of δ.

All these and other potentialities of interferometry can be realized by using a single microholographic system, such as shown in Fig. 11.20. This is an obvious advantage of holographic microinterferometry against which has to be set the serious drawback of coherent noise. If the latter is not reduced, holographic interference microscopy cannot be an effective tool in the study of the majority of problems occurring in both biomedicine and the materials sciences.

If one wavefront, say Σ_1' (Fig. 11.24), is reconstructed from the hologram and the other (Σ_1) is the actual object wavefront, we have an interference situation which is commonly called *real-time* or *live-fringe interferometry*. As was mentioned earlier, it is also possible to record two wavefronts, say Σ_1 and Σ_1, separately on a single holographic plate and reconstruct them together by means of the reference light beam, which was earlier used for the hologram recording, the original object beam being stoped during the reconstruction process. This double-exposure holographic recording forms a permanent record of two fixed states of the object and is therefore referred to as *lapsed-time* or *frozen-fringe microinterferometry*. These two approaches are basic procedures for holographic microinterferometry and will be discussed in more detail in the following subsections.

11.6.2. Real-time holographic interference methods

The procedure for operating a real-time holographic microinterferometer simply consists in viewing the object under study through the processed hologram, with illumination identical to that used during hologram recording. The reference beam used for making the hologram now reconstructs the wavefront which was previously recorded, while the object beam contains an actual object wavefront derived directly from the object under study. Both wavefronts are mutually coherent, can therefore interfere with each other and produce an interference image of the object. Two typical variants are possible. One of them is illustrated by Figs. 11.24b and c, where the reconstructed wavefront Σ' is an "empty" wavefront, which contains no information on the object to be studied and can therefore be treated as a comparison wavefront for the evaluation of deformations of the actual object wavefront Σ_1. In the other variant, the reconstructed wavefront is also perturbed by the object under study (Figs. 11.24d–g). These two variants will be discussed further below.

Real-time interference method with reconstructed empty wavefront. Let us consider a parallel object beam B_1 whose plane wavefront Σ is not perturbed by the object to be studied (Fig. 11.25a). This wavefront is accompanied by a reference beam B_2 whose wavefront Σ_2 is also plane. Both beams illuminate a holographic plate H, and a hologram of the wavefront Σ is recorded. The plate H is removed from its holder for photochemical processing and then returned exactly to its previous location. Now, a transparent object O is inserted into the beam B_1 (Fig. 11.25b). The object introduces changes in wave phase and amplitude and the incident wavefront Σ becomes a deformed wavefront Σ_1. The beam B_2 previously used as the recording reference beam now reconstructs the wavefront Σ. The reconstructed wavefront is denoted as Σ'. This is coherent with the direct object wavefront Σ_1 since both are derived from a single laser beam B and can therefore interfere with each other. If the relocation of the hologram H is exact, a single infinitely enlarged interference fringe covers the field of view (Fig. 11.25c). This fringe is dark since the phase of the reconstructed wavefront Σ' is reversed by π from the actual object wavefront Σ_1 owing to the negative development of the hologram. Narrow interference fringes can, however, be observed in the region where the deformed part of the object wavefront Σ_1 comes away from the plane wavefront Σ'. The fringes trace the lines of equal optical path difference due to the object O. If next some slight lateral displacement of the hologram H occurs, a series of interference fringes will be brought into the field of view (Fig. 11.25d).

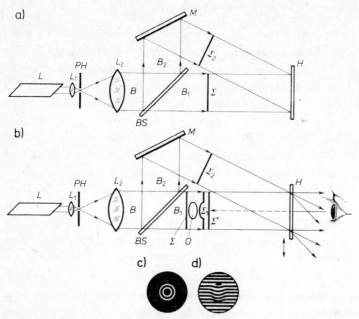

Fig. 11.25. Principle of real-time microinterferometry with the comparison wavefront (Σ') reconstructed from the hologram H; L—laser, L_1 and L_2—collimating lenses, PH—pinhole, BS—beam splitter, M—mirror, O—object under study (see the text for further explanation).

This dispacement (shown in Fig. 11.25b by a double arrow) causes the wavefront Σ' to be inclined with respect to Σ_1 as illustrated in Fig. 11.24c. The larger the hologram displacement, the greater the wavefront inclination and thus the greater the number of interference fringes obtained in the field of view (compare Fig. LV). As the optical properties of the object change, the interference pattern varies correspondingly and we can thus observe variations of the object as a function of time. For this reason, this technique is also called live-fringe microinterferometry because the interference fringes can actually (in real-time) be used to monitor variable processes and transient events in the object or specimen under study.

The situation shown in Fig. 11.25c illustrates uniform field interference and that in Fig. 11.25d fringe field interference (see photomicrographs of Fig. LV for more practical illustrations of these two variants of interference images).

Real-time shearing interference method. This technique is similar to the previous one, except that now a holographic recording is made not of the empty wavefront

Σ but of the wavefront Σ_1 perturbed by the object under study (Fig. 11.26a). After photochemical processing, the hologram H is returned to its exact original position and illuminated as during recording (Fig. 11.26b). The beam B_2 previously used as the reference beam reconstructs the object wavefront Σ_1 from the hologram. The reconstructed wavefront Σ_1' exactly covers the actual object wavefront Σ_1. Both wavefronts are mutually coherent and can interfere with each other. If the relocation of the hologram is exact and no additional perturbations have been introduced to the object wavefront during the period between recording (Fig. 11.26a) and reconstruction (Fig. 11.26b), a single dark interference fringe

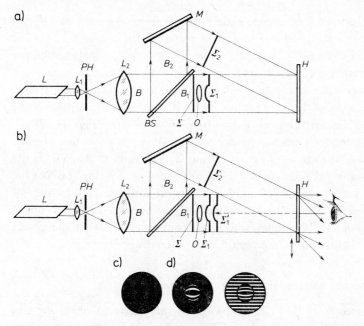

Fig. 11.26. Principle of holographic shearing microinterferometry in real time (see text and Fig. 11.25 for explanation).

covers the object O and the entire field of view. If it is completely transparent, the object under study is not perceived at all or only faintly visible (Fig. 11.26c). When the object O is next slightly laterally displaced, differential interference contrast occurs (Figs. 11.26d and 11.24d). Continuing to displace, the object leads via partially duplicated forms of the object image to two totally sheared images (Fig. 11.24e). This technique is referred to as shearing interference microscopy with uniform background and a continuously variable amount of wavefront shear. Uniform field interference (Figs. 11.26c and d) may be transformed into fringe field interference (Fig. 11.26e) if the hologram H (Fig. 11.26b) is slightly displaced in a transverse direction (shown by a double arrow) from its zero locations to a new position. When applying this procedure, the density of fringes may be suitably controlled to maintain the optimum visibility of the object interference image.

It should be noted, however, that in practice it may not be possible to obtain the ideal zero condition (Fig. 11.26c) when one infinitely spread dark fringe covers the field of view, owing to hologram emulsion shrinkage at the photographic processing stage.

An adjustable form of holographic plate holder with micrometers has several advantages for the correct positioning and repositioning of the hologram. The interpretation of interference fringe configurations and optical path difference measurements are often simplified if some additional optical elements (compensators or phase shifters, for example) are introduced into the holographic micro-interferometer. Facilities of this kind are available with the holographic microscope constructed in the Central Optical Laboratory, Warsaw (see Fig. 11.22).

So far we have confined ourselves to an invariable object as a function of time. If this is not the case, the actual object wavefront Σ_1 will differ in shape from the reconstructed wavefront Σ_1' and under the zero condition too (Fig. 11.26c) the region occupied by the object under study will appear to be covered by narrow interference fringes corresponding to changes of the wavefront Σ_1. If object variation is continuous, the fringes will move continuously across the object image; this, too, is live-fringe holographic interference microscopy.

11.6.3. Double-exposure holographic interference microscopy

Interference microscopy of this kind is based on the ability of photosensitive material to record two or more superposed holograms, and, as in the case of the previous real-time technique, we can distinguish two procedures. One of them uses an empty wavefront and object wavefront, and the other two wavefronts perturbed by the object under study.

Double-exposure interference microscopy with an empty wavefront. Let us return to the arrangement of Fig. 11.25a and make a holographic record of a plane wavefront Σ. The holographic plate H (Fig. 11.27a) is not developed, however, but left in position, and an object O is inserted into the beam B_1 (Fig. 11.27b). The wavefront Σ is now perturbed in its phase and/or amplitude by the object and becomes the deformed wavefront Σ_1. We now make a second exposure of the holographic plate H. This doubly exposed plate is photochemically processed, then replaced in its original position and illuminated only with the beam B_2, which was previously used as the reference beam (Fig. 11.27c). Now, two wavefronts, Σ' and Σ'_1, are reconstructed simultaneously by this beam, while the object beam B_1 is stopped by an opaque screen S. The wavefronts Σ' and Σ'_1 interfere with each other and produce an interference image of the object O. This time, in contrast to the technique shown in Fig. 11.25, the object under study is not visible directly, but its previous presence manifests itself in the form of an interference pattern produced by the reconstructed wavefronts Σ' and Σ'_1.

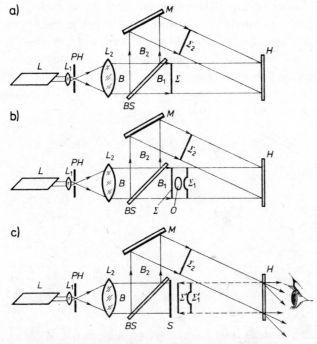

Fig. 11.27. Principle of double exposure holographic microinterferometry (see text and Fig. 11.25 for explanation).

At the reconstruction stage this technique does not require the exact replacement of the processed hologram in its recording position as in the real-time method, and is especially suitable for recording transient events using a pulsed laser for the double-exposure of a single holographic plate.

Double-exposure shearing interference microscopy. Two object wavefronts, Σ_1 and $\Sigma_{1'}$, emitted by the same object O under different conditions, are recorded successively on a single holographic plate H (Fig. 11.28a). If this doubly exposed hologram is processed and then replaced in its previous position and illuminated with the single beam previously used as the reference beam B_2, these two object wavefronts are simultaneously reconstructed. They are marked by Σ_1' and $\Sigma_{1'}'$ in Fig. 11.28b. Their mutual interference supplies information about possible object variation during the period between the first and second exposure. The two different conditions mentioned above can be created by a lateral displacement of the object, for instance; this operation produces a shearing interference image. Another possibility is to maintain the object in position and record two different states due to certain processes or transient events. In this case, the reconstructed wavefronts Σ_1' and $\Sigma_{1'}'$ produce an interference pattern which displays the object change between the first and second exposures. Many other states can also be considered as different for the purposes of double-exposure; the second object wavefront $\Sigma_{1'}$ can be rotated or demagnified as shown in Figs. 11.24f and g.

The double-exposure hologram forms a permanent record of two fixed states

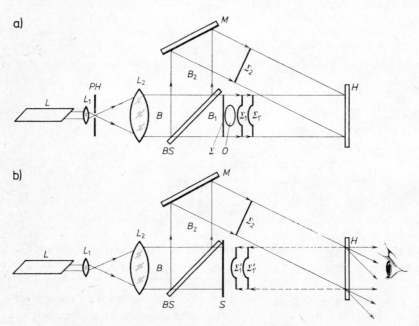

Fig. 11.28. Principle of double exposure shearing holographic microinterferometry (see the text and Figs. 11.25 and 11.27 for explanation).

of an object and therefore holographic interference microscopy of this kind is also called *lapsed-time* or *frozen-fringe microinterferometry*. This technique has advantages over real-time holographic microinterferometry for measurement purposes, although it cannot differentiate between positive and negative changes in optical path difference without additional data. Such changes are, on the other hand, easily distinguished in the real-time procedure since the interference fringes move in opposite directions for advancing and retarding phase differences. Another disadvantage of the double-exposure method is that the transformation of uniform into fringe-field interference is not so easily achieved as in the real-time method. Its main advantage, on the other hand, is the cancellation of unwanted emulsion-shrinkage effects. This appraisal of the pros and cons of both techniques shows that real-time and double-exposure interference are in fact complementary.

So far we have considered microinterferometric systems for transparent objects. However, all techniques and procedures described above are also valid for opaque objects examined in reflected light.

11.6.4. *Is it possible to accept holographic interference microscopy without coherent noise reduction?*

In contrast to holographic interferometry of macroscopic objects, holographic microinterferometry is almost useless for the study of microscopic objects unless coherent noise and speckling are reduced. This problem is more serious in microinterferometry than in common holographic microscopy (see Section 11.5).

In some papers it has been stated that "excelent" or "high quality" holographic interference images were obtained although no method of coherent noise reduction was used. Such over-enthusiastic statements are only likely to be made by someone who is not familiar with conventional interference microscopes which in fact produce excellent images such as no present-day holographic systems are yet capable of rivalling.

Holographic interference microscopy is mainly intended for measurement purposes. In each microinterferometric measurement, the order of interference fringes must be defined correctly and the fringe centres fixed precisely. Coherent noise and speckle patterns badly deform those fringes, frequently deplacing their centres or even changing their traces, and in general blur the appearence of the holographic interference image. In short, it is clear that holographic microinterferometry without coherent noise reduction can in no way be considered an effective technique for the quantitative study of microobjects. It is for this reason that efforts were undertaken in the present writer's laboratory to develop an effective method for coherent noise reduction. This method was described in Subsection 11.5.3. Its basic advantage is the fact that it functions in both the hologram recording and image reconstruction processes and can equally well be used in common holographic microscopy and holographic microinterferometry. Photomicrographs of Figs. LI–LIII illustrate the effectiveness of this method and the performance of the holographic microscope constructed in the Central Optical Laboratory, Warsaw (Fig. 11.22). This instrument is suitable for both common holographic microscopy and microinterferometry. The photomicrographs of Figs. LI–LIII were taken using the real-time and double-exposure techniques described in Subsections 11.6.2. and 11.6.3.

11.7. Holographic phase contrast microscopy

The ability of a hologram to record the phase of any wave should not be taken to mean that ideal phase objects can be observed by using a common holographic microscope. The fact that the hologram contains information on the phase distribution of the object wavefront permits only this phase distribution to be reconstructed. The reconstructed phase distribution or wavefront, like the phase object itself, remains invisible because the human eye and other light receptors known to date are insensitive to phase states and changes in relative phase shifts of light waves. This situation resembles that in common light microscopy, where special phase contrast devices (see Chapter 5) or interference techniques must be used for the observation of phase objects.

Designs for phase contrast holographic microscopes have been suggested by several authors [11.3, 11.89–11.91]. In particular, phase contrast holographic imaging makes use of a phase plate to modulate the phase and/or amplitude of the zero-order diffraction term, while the hologram is recorded and/or the image is reconstructed. This procedure was suggested by Gabor [11.3] and first realized in practice by Ellis [11.89]. Basically, this depends on the suppression of the zero-order diffraction light at the time the hologram is recorded. A suitable spatial filter is used, and the object beam which transilluminates the phase specimen is focused on the holographic plate by means of a well-corrected objective. The spatial filter is placed in front of this plate and blocks the zero-order diffraction light. After development, the hologram is placed in its original position, while the spatial filter as well as the specimen are removed from the apparatus. Reconstruction of the phase contrast image is accomplished by using two coherent beams: one—the so-called *restoration beam*—is identical with the beam that earlier transilluminated the specimen, while the other—the *reconstruction beam*—is identical with the reference beam previously used for making the hologram. Introducing a continuous phase compensator and/or variable attenuator in the restoration beam permits variable phase contrast to be produced during of the reconstruction of the holographic image.

Another holographic phase contrast system has been described by Tsuruta and Itoh [11.9]. A hologram of unit magnification is made using technique similar to Anderson's procedure [11.90], except that the phase specimen is transilluminated with a diffused beam and the spatial filter is omitted. After passing through the specimen, the object beam is focused on a holographic plate and interferes with a reference plane wave falling obliquely upon the holographic plate. After development, the hologram is placed exactly in its original position, the specimen together with the reference beam is removed, and the hologram is illuminated with a convergent beam, identical with the formerly used to transilluminate the specimen. In this situation, a bright point image appears at the focus of an additional objective placed coaxially with the previous reference beam behind the hologram. Placing a phase plate across this point image enables phase variations in the specimen to be clearly observed by the eye located close behind the phase plate. With this system, Tsuruta and Itoh have examined phase phenomena (*striations*) in liquids. Although these were macroscopic objects, there seem to be no essential objections to employing this procedure in microscopy.

In general, phase contrast holographic techniques are less convenient than holographic interference methods for the study of phase objects. Some examples of holographic interference images of biological organisms, which belong to the class of phase objects, are shown in Figs. LIV and LV. The photomicrographs of these figures were taken with the holographic microscope shown in Fig. 11.22. Coherent noise has, of course, been reduced at both the hologram recording and image reconstruction stages. Transformation of a dark-field interference (Fig. LIVa) into a bright-field interference (Fig. LIVb) has been achieved by using a phase-shifter and thus obtaining negative and positive image contrast.

11.8. Application of holographic microscopy

There are many papers describing different applications of holographic microscopy and microinterferometry for both biological research and materials sciences microscopy. As it is impossible to discuss here all applications proposed to date, only some of the most typical or of outstanding interest will be summarized here.

11.8.1. Examples of biological applications

The first biomedical applications of holographic microscopy were announced in the last 1960s and early 1970s and related to recordings of dynamic microscopic subjects such as marine plankton [11.92, 11.93], flow, stream, and motion of plasmodium particles [11.94], cinemicroscopy of small animal microcirculation [11.95], superresolved quantitative reconstruction of red-blood cells [11.96], and some other problems. These earliest applications were reviewed by Cox [11.97, 11.98] and Sokolov [11.99].

Several interesting papers next appeared in the second half of the 1970s. Glycerination of Amoebe proteus was examined by Opas [11.100], who used the holographic microscope shown in Fig. 11.22. By using real-time and double-exposure interference techniques, he was able to show that the process of glycerination led to a redistribution of cell material, and was accompanied by reversible deformation to the cell cortex. The role of the cortex in the maintenance of cell shape was found to be predominant.

Next, Baranowski [11.101] used the same holographic microscope as Opas for the three-dimensional analysis of movement in Physarum polycephalum plasmodia. The real-time interference images were recorded on cine film (Fig.

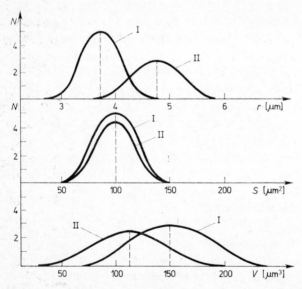

Fig. 11.29. Normalized distributions of erythrocytes as a function of radius (r), surface area (S) and volume (V) of red-blood cells; graphs I—for adults, II—for children [11.88].

LVI), which allowed simultaneous recording of changes of both the protoplasm flow direction and thickness of the protoplasmic channel. The results obtained by Baranowski indicate that the protoplasm flow direction changes and channel thickness variations show a similar rhythm.

A large research project was undertaken by Ginzburg and her co-workers ([11.88], pp. 215–221), who examined red-blood cells using the MGI-1 holographic microscope (see Figs. 11.11 and 11.23). Real-time holographic interferograms were statistically processed in order to determine the basic parameters of erythrocytes. The red-blood cells examined suffered from different diseases or were maintained in various (e.g., cryogenic) conditions. Comparisons were also carried out between the red-blood cells of children and those of adults (Fig. 11.29).

The microholographic studies of marine plankton initiated by Knox were then continued by Heflinger et al. [11.102] as well as by Carder [11.103].

Mention should also be made of holographic fundoscopy [11.104], studies of the human eye [11.15] and contractile activity in muscles [11.106].

11.8.2. Applications in the materials sciences

Applications of holography in materials sciences microscopy were initiated simultaneously by several researchers in the late 1960s. Bedaria and Pontiggi [11.107] employed a holographic arrangement with pre-magnification for crystal studies in metals. Holograms were recorded in reflected light and image reconstruction was carried out by means of an additional magnifying lens. This first holographic study of crystal objects was then perfected by McFee [11.57], who examined birefringent crystals growth from the melt, using a holographic system with reversed wavefront reconstruction and real-time microinterferometry. Microholographic studies of crystals growth and related phenomena were also carried out by several other researchers [11.108, 11.109].

Some stimulating examples of the application of holography to interference microscopy of solids were given by Snow and Vandewarker [11.23], and a general discussion of lens-assisted holographic microscopy was provided by Magill and Wilson [11.110]. Their work was mainly concerned with the application of the holographic interference microscope to studies of a small vibrating element known as a *resonistor*. Both real-time and double-exposure holographic microinterferometry was applied to the study of this element. Van Ligten and Lawton [11.111] discussed the potential use of holographic microscopy and microinterferometry for inspecting integrated circuits, localizing their defects and testing their overall quality. Pierattini [11.112] used real-time and double-exposure microholographic interferometry for observing the dynamics of phase variations, whereas Presby [11.113] applied time-resolved differential holographic microscopy to the study of liquid distribution in capillary tubes. Rhodes and Cournoyer [11.114] used holographic interferometric microscopy to the study of diffusion mechanisms in polymeric substances. Surface roughness measurements by holographic interferometry were described by Snow and Vandewarker [11.23], van Ligten and Lawton [11.111], and Ribbens [11.115]. It is worth noting here that holographic microinterferometry is also suitable for examining surfaces which reflect light diffusely. Conventional microinterferometry, on the other hand, can only be used for studying smooth, mirror-like surfaces.

Russian researchers (see Ref. [11.44]) employed the MGI-1 holographic microscope (Fig. 11.11) for testing lenticulated discs. These optical elements consist of a matrix of small thin lenses with long focal lengths. Their shape was examined by using double-exposure uniform-field holographic microinterferometry. The same microscope was also used for refractive index profile characterization of optical fibres.

Microscopic holography of small parts is also the subject of a work by Karger and Holeman [11.116], while Attwood *et al.* [11.117] used holographic microinterferometry for diagnostic purposes in laser fusion experiments to study the laser produced plasmas with frequency tripled probe pulses. Similar studies of plasmas produced by 1-ns CO_2 laser pulses were also carried out by Fedosejev and Richardson [11.118]. An ultraviolet holographic microinterferometer for plasma probing was then designed by Pierce [11.119].

Quite recently, Smith [11.60] developed a holographic stereomicroscope for an improved technique of classifying fossil ostracods. These objects are small aquatic bivalve crustaceans whose fossils are used in the oil prospecting industry for the dating of sedimentary rock samples. The new technique allows the simultaneous presentation in the field of view of the stereoscopic microscope of holographically created images of known ostracods alongside the image of the unidentified specimen.

11.8.3. *Holographic imaging and analysis of three-dimensional distribution of particles*

One of the basic applications of holography is concerned with the study of microscopic particles (fog, mist, sprays, rocket and engine exhausts, pollution, condensation, etc). Holographic recording allows a dynamic three-dimensional distribution of particles to be stored in a hologram so that a frozen (stationary) image can be reconstructed for their detailed study in different planes of the recorded object space. The depth of field is considerably larger than in conventional photomicrography and evaluated to be equal to $50d^2/\lambda - 100d^2/\lambda$ (or even more), where d is the diameter of particles and λ the wavelength of incident light [11.120–11.122]. The most popular technique involves forming in-line Fraunhofer holograms, although off-axis holography is also in use. The basic principles of these techniques, their theory, advantages and limitations are discussed in many papers [11.123–11.131].

Holographic techniques for particle size analysis were initiated by Thompson [11.26]. Since those beginnings, a great deal of scientific research and development work has been carried out. Work of the first decade is reviewed by Thompson [11.123], and that of the second is cited in twelve papers devoted to particle sizing and spray analysis published in *Optical Engineering*, vol. 23 (1984), pp. 554–640.

A detailed description of holographic particle analysis is beyond the scope of this book. The present discussion merely serves to give a bird's eye view of a field that is not strictly speaking holographic microscopy, but the holography of dynamic large objects consisting of microscopic particles.

Pulsed holographic microscopy. The most widely used light sources in holography are *continuous wave* (*cw*) *gas lasers*, especially He–Ne and argon lasers, which are relatively inexpensive and simple in use, but pose some limitations. These limitations have their origin in the low-power outputs of cw lasers and in the low sensitivity of the high-resolution photographic materials needed for hologram recording. This means that long exposure times are required as well as extreme stability in the holographic arrangement and of the object under study. Many of these limitations can, however, be completely overcome by using *pulsed lasers*. Among these the ruby laser has a privileged position. The radiation of this laser is at 694.3 nm wavelength, but can still be recorded by photographic emulsions (Agfa-Gevaert Holotest 8E75 or 10E75 plates, for instance).

Conventional ruby lasers are, however, not suitable for holography because of insufficient coherency of radiation (see Table 1.4 in Volume 1). Great efforts were made to increase the coherence length of these lasers. Today no problem is involved in building ruby lasers with a coherence length of as much as 1 m and a pulse width equal to 20 ns. This is achieved by adding some special components (Q-switch, selectors of transverse and axial modes) to a laser head. For holography, the ruby laser is usually equipped with a Pockels cell, which is capable of producing two light pulses whose temporal separation varies from 100 μs to 1 ms, for instance. Pulse energy can be from 50 mJ to 10 J, depending on the size and number of amplifiers added to the main laser head. Two pulses are necessary for two-exposure holographic interferometry and for the study of trajectories and velocity of particles.

The light energy required to take a properly exposed hologram depends on the size and optical properties of the object to be holographed and its distance from the holographic plate. If an object needs energy of 1 J for the proper exposure of the holographic plate, a 10-second exposure is necessary when using a cw laser of 100 mW, while a pulsed laser can deliver this energy in 20 ns. This means that during the exposure the tolerable speed of the object in the direction of the light beam is about 3 m/s. If the object is illuminated from the side, its speed can be much higher [11.132].

A holographic apparatus with a pulsed laser usually incorporates a cw helium-neon laser for aligning the pulsed recording system and for image reconstruction. The latter process, as well as the final stage of image magnification, is frequently done by a closed-circuit television (CCTV) readout system. The CCTV system also eliminates the danger of looking directly at the laser beam with the naked eye when an image is reconstructed from an axial hologram.

Holographic particle analysing system. The most obvious application of holography in the study of the three-dimensional distribution of particles depends on the detailed high resolution analysis of a volume occupied by a number of dynamic particles. Many lens-less and lens-assisted systems for both in-line and off-axis holography of particles are described in the literature [11.120, 11.123, 11.131, 11.133–11.135]. In-line holograms of a particle field are formed by passing a laser pulse through a volume of particles. Part of the light pulse is diffracted (scattered) by the particles and the remainder passes directly through the volume. The direct (unscattered) light acts as a reference wavefront which interferes with the diffracted (object) light. The resulting interference pattern is recorded on a photographic plate. Off-axis holograms of particles are formed by introducing a reference light pulse which is separated from the pulse passing through the volume of particles under recording. A single original laser pulse is, of course, used and split in two portions. One portion constitutes the object beam (this illuminates the particle field), while the other serves as the reference beam. The object beam illuminates the volume of particles directly, but it is sometimes convenient to use illumination via a frosted glass or similar diffusing medium.

The images of particles are reconstructed by illuminating the developed hologram with a cw laser beam whose geometry is similar to that of the reference beam, which has been used during the recording process. There are formulae which predict the position of particles as a function of the recording and reconstruction parameters. Obviously, the real image is in this case more convenient for the quantitative analysis of the three-dimensional distribution of particles. This image is observed directly by an auxiliary microscope or on the screen of the CCTV system. Commonly a magnifying lens or a microscope must be inserted between the reconstructed holographic image and the TV camera in order to obtain sufficiently magnified final images of individual particles.

Figure 11.30 shows a typical arrangement for the off-axis holographic study of aerosols emerging from a nozzle. The reconstructed image of the spray forms an exact three-dimensional model of the original distribution of particles at the time of recording. Individual particle images can be focused and displayed on a TV monitor. It is possible to translate the hologram H so that the whole

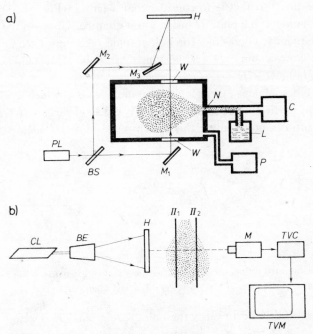

Fig. 11.30. Typical arrangement for the off-axis holographic analysis of aerosols emerging from a nozzle: a) hologram recording, b) image reconstruction; PL—pulsed laser, BS—beam splitter, M_1 to M_3—mirrors, W—windows of test chamber, H—hologram, N—nozzle, L—liquid, C—compressor, P—pump, CL—continuous wave laser, BE—beam expander, M—microscope, TVC—television camera, TVM—television monitor.

volume of particles can be studied. As the hologram is axially moved, the various image planes II_1, II_2, ... come into sharp focus at their proper position in space.

The resolving power of a hologram is directly dependent upon the solid angle of diffracted and/or scattered light which can be effectively recorded. Therefore, as the distance (z) between the particle volume (Fig. 11.30a) and the holographic plate H increases, the required effective size of the hologram must be increased to maintain the same image resolution. Otherwise, the overall quality of the reconstructed image decreases as z and $1/d$ increase (d is the diameter of particles). This rule may involve more than simply enlarging the size of the holographic plate since the useful size of the hologram is limited by the potentiality of the photographic material to record effectively information for later image reconstruction. Moreover, the contrast of interference fringes which form the hologram drops as z increases. Under these circumstances, a lens-assisted holographic set-up is sometimes more useful than a lens-less system for the study of a given particle volume. A way of interpreting this fact is to consider that a lens L (Fig. 11.31) forms an image H' of the holographic plate H inside a test chamber closer to the particle volume O. Consequently, conditions for holographic recording become improved since we have now a small $O-H'$ distance (or equivalent $O'-H$ distance) instead of a large one $O-H$.

There are two main types of lens-assisted holographic systems, which are widely used for the study of particle objects. One of them is the single lens system shown in Fig. 11.31. For off-axis holography, this system must be supplemented

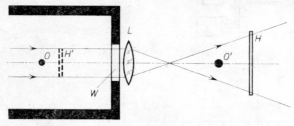

Fig. 11.31. Advantages of a lens-assisted holographic system for the study of a particle volume O; H'—equivalent position of the hologram H, W—window of a test chamber.

by an oblique reference beam directed towards the holographic plate H. During the image reconstruction, the same lens L is returned into the reconstructing light beam, which is conjugate to the original reference beam. The second type is the collimating transfer system shown in Fig. 11.32. Two positive lenses, L_1 and L_2, are separated so that their adjacent foci are coincident with each other. Thus, a parallel beam incident on the object O is maintained as a collimated reference beam at the holographic plate H (in-line holography). Any magnification M of the re-imaged object field is possible by suitable selection of focal lengths, f_1 and f_2, since $M = f_2/f_1$. The transferred collimated beam is easily reproduced during the image reconstruction. This system is free from significant spherical aberration and distortion, and is especially convenient for in-line holography.

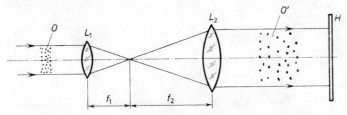

Fig. 11.32. Collimating transfer system for holography of particles.

A defect of lens-assisted holographic systems is the coherent noise generated by light scattering from dust and optical irregularities that may exist on or within lenses. Therefore, the smaller the number of lenses, the better the system is for noise-free holographic recording.

An in-line system is the simplest type of holography for the study of particle objects. There are, however, many situations in which off-axis holography is alone applicable. In particular, in-line holographic systems are useless when the density of particles is high or their total cross section is large; in this case, off-axis holography must be used.

There is an additional unfortunate complication in the holographic study of particle fields since the small particles give rise to large diffraction rings in coherent light so that particle images are confused. However, Thompson and his co-workers [11.123, 11.136] used Fraunhofer diffraction rings to predict particle positions and analyse particle sizing.

It has been stated [11.137] that it is rarely possible to obtain sufficiently distinct images of particles whose diameter is smaller than 5 μm if in-line holography is employed. The same problem occurs in off-axis holography when a parallel or divergent object beam illuminates a volume of very small particles. To deal with this problem, an off-axis holographic system has been devised (Fig. 11.33) in which the object beam B_1 is convergent in such a manner that its focal point is before the holographic plate H. At this point there is a small opaque screen S, which stops the direct (undiffracted) portion of light of the object beam. Consequently, the plate H is only illuminated by the diffracted portion of the

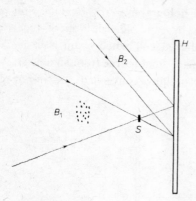

Fig. 11.33. Off-axis holographic system with convergent object beam for the study of small particles [11.137].

object beam and the reference beam B_2. This modification of the holographic system makes it possible to count and size particles down to 1 μm diameter [11.138].

Analysis of natural fog droplets. The holographic study of natural fog droplets was initiated by Thompson and his co-workers (see a review article by Thompson [11.123]). In their first experiments, the in-line holograms were recorded with the use of lens-less systems. However, in the fog measuring devices the droplet volume was magnified by a lens of 5× factor. The device was capable of recording holograms at a rate of 30 per minute on Kodak SO-243 film. The ruby pulsed laser operating at $\lambda = 694.3$ nm had a peak power of 10 kW and a pulse duration of 50 ns. The fog volume recorded on each hologram was 7 cm³, and the system was capable of recording droplets down to 5 μm diameter. A cw helium–neon laser with a collimating beam expander provided a parallel beam for image reconstruction. The film on which the holograms were recorded was moved through x, y, and z coordinates by a motor-driven carriage. The reconstructed images of fog droplets were displayed on a CCTV monitor at several hundred (ca 300) times magnification. The droplets were sized visually by using a reticle on the monitor screen. Data collected during 1964 and 1965 were satisfactory and correlated well with the total water content measured at the same time.

This technique was then used by several other researchers and meteorologists [11.139–11.141]. In one work [11.141], the Thompson procedure was applied but without the use of a magnifying lens in the hologram recording process and using the divergent image-reconstructing beam instead of the parallel beam.

It was decided to analyse a volume of natural fog droplets of 22.5 cm³ from a single hologram in order to find if the spatial distribution of fog droplets followed to Poisson statistics. The result obtained was ambiguous. Double exposure holograms of the droplet field were also recorded with a suitable time interval between the exposing light pulses (Fig. 11.34). The double pulsed hologram enables two separated images of a single droplet to be reconstructed. The velocity vector of the droplet is determined by knowing the time interval between laser pulses and measuring the vector distance between the droplet images.

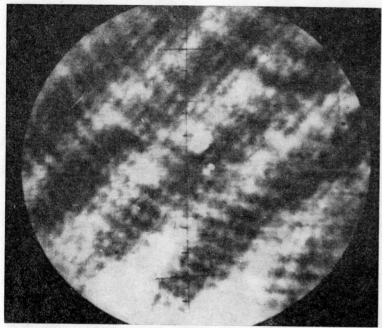

Fig. 11.34. Two separated images of a single droplet of fog reconstructed from a double exposure hologram. Laser pulse interval 0.3 ms, pulse width 20 ns, droplet diameter 15 μm, droplet velocity ca 4 cm/s [11.141].

Analysis of oil-mist. Recent efforts to reduce the consumption of technical oil have led to a more economical method of lubrication by using oil-mist. Optimization of this method, taking into account oil-mist generation, transportation, and deposition on lubricated surfaces, requires information about the dimensions of the oil droplets. Obviously, these are produced in the size range 0.1–30 μm, but droplets close to 2 μm in diameter are required for the most effective lubrication.

The oil-mist was holographically studied by Zachara [11.138], who used a set-up for pulsed holography built in the Central Optical Laboratory, Warsaw (Figs. 11.35 and 11.36). The ruby laser operating at $\lambda = 694.3$ nm had a pulse duration of 30 ns. The oil-mist stream leaving the reclassifier nozzle was recorded at a distance of 5 mm from the nozzle outlet, which is the typical distance of lubricated surfaces in industrial applications. Real images of oil-mist droplets were reconstructed by using a cw helium–neon laser and observed directly through a microscope or on the screen of a TV monitor.

Fig. 11.35. Layout for pulsed and double-pulsed holography built in the Central Optical Laboratory, Warsaw.

Preliminary experiments showed that the oil droplets as small as 1–5 μm in diameter were hardly visible in the reconstructed object fields if a parallel or divergent object beam was used. This limitation was overcome by using a convergent object beam (Fig. 11.33). Such a configuration resembles central dark-ground illumination (see Subsection 6.2.4).

Figures 11.35 and 11.36 show that pulsed holographic microscopy occupies a good deal of laboratory space, requires many optical elements and electronic

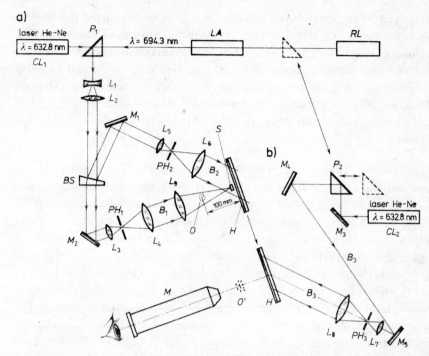

Fig. 11.36. Set-up for pulsed holography used by Zachara [11.138] for the study of oil-mist: a) hologram recording, b) image reconstruction; *RL*—main ruby laser head, *LA*—ruby laser amplifier, P_1 and P_2—reflecting prisms, CL_1—continuous wave laser (He–Ne laser) for adjusting purposes, CL_2—continuous wave laser (He–Ne laser) for holographic image reconstruction, L_1 to L_8—collimating lenses, *BS*—beam splitter, L_9—convergent lens, M_1 to M_5—mirrors, PH_1 to PH_3—pinholes, *O*—oil mist under study, *S*—small opaque screen for stopping the direct (undiffracted) light of the object beam B_1, B_2—reference beam, *H*—hologram, B_3—image reconstructing beam, O'—holographic oil-mist image, *M*—viewing microscope.

devices, and has to use some automation for coinciding the laser pulses with a given phase of the event under study. This technique is therefore very expensive and may only be used in larger laboratories.

Energetic aerosol analysis. Continual efforts to reduce emissions from Diesel engines have led to the development of holographic techniques to measure particle size and density of fuel sprays, to analyse combustion phenomena and study rocket injection problems [11.123, 11.142–11.144]. In these applications, pulsed holography (with 10 to 20 ns pulse lengths) offers new research possibilities since under appropriate conditions the entire volume of reacting and non-reacting fuel sprays can be instantaneously recorded and then reconstructed to show the size, position, shape, and three-dimensional distribution of particles or droplets. The technique is particularly useful where the energetic aerosol is contained in a fast atomized liquid, gaseous flow, or generated explosively and therefore extremely dynamic. Although dynamic and explosive sprays create some extremely specialized technical problems, several successful devices have been developed and today pulsed holography offers a powerful research tool for energetic and rocket engineering.

Other applications. There are many other applications of holographic particle analysing techniques [11.123]. In Ref. [11.93], for instance, a repetitively pulsed argon ion gas laser is described, operating at $\lambda = 514.5$ nm and with 50 μm pulse lengths, which was used for the cinemicroholographic study of living marine plankton, whose components varied in size from about 1 mm to 10 μm. Thompson and Zinky [11.145] used holographic particle sizing technique for detecting submicron particles in liquids; latex beads of 0.365 μm diameter were detected in water, for instance. A very attractive application of holography at microscopic level is used in nuclear physics for bubble chamber recording. This application was first described by Ward and Thompson [11.136] and independently by several other researchers [11.83]. In 1981, Royer [11.146] used in-line holography in a bubble chamber at the CERN (Geneva). He recorded holograms with the help of a single-mode pulse laser. Bubble tracks of 25 μm in diameter were reconstructed with a resolution of 2 μm.

A valuable contribution to the holographic study of trajectories of particles in bubble chambers was made by Budziak from the Institute of Physics, Jagiellonian University, Cracow, and his co-workers in the Joint Institute of Nuclear Research, Dubna, USSR. In a paper [11.147], he reports electron track registration results from helium and hydrogen streamer chambers at 5 atm pressure of filling gas. The track images were obtained by holographic recording with the aid of a dye impulse laser. The use of pulse holography made it possible to localize the trajectories of ionization particles, define more accurately the location of interactions, and carry out ionization measurements.

A serious drawback that retards progress in the large-scale engineering use of pulse holography for particle diagnostics is the time-consuming process of analysing particle holograms. Much effort is therefore being put into the development of automated systems (Fig. 11.37) for fast analysis of particle holograms using processors and computers [11.148–11.153]. An interesting approach to this problem was described by Tschudi *et al.* [11.148], who used holographic

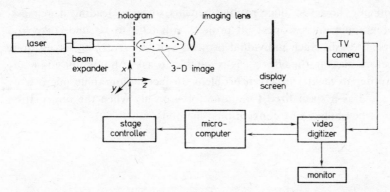

Fig. 11.37. Typical set-up for automatic analysis of particle hologram [11.149].

filters in an automatic particle size analyser. Then Bexon *et al.* [11.133] adapted a Quantimet (Imanco) image analyser for this purpose. The adaptation is not on the whole satisfactory, however, and suffers from several limitations [11.141]. Quite recently, Stanton *et al.* [11.149] have developed an automatic system based on the analysis of patterns in two or more non-image planes, which allows both the particle size and location to be recovered with an accuracy greater than that obtained by the plane-by-plane search for sharpest focus. However, this system can only be used with spherical particles and not particles of arbitrary shape. Nevertheless, it represents a significant step forward in automated systems for quantitative holographic particle analysis.

The method developed by Stanton *et al.* has many parallels in the paper by Vikram [11.152] on far-field holography of non-image planes for size analysis of small particles. This recommends the use of in-line Fraunhofer holography when reconstruction is analysed at non-image planes where the size of the reconstructed pattern is found to be much larger than the image size in the image plane. This situation should generally be very helpful for the accurate size analysis of smaller particles of known shapes.

In any case, the problem of automatic holographic particle sizing has clearly not yet been resolved satisfactorily and further work is required before this technique becomes suitable for routine use.

*

*　　　*

As has been indicated in this chapter, holographic microscopy has several basic advantages compared with conventional light microscopy. At present, these advantages do not, however, offer routine benefits since no leading firm mass-produces holographic microscopes; all projects still tend to be undertaken on a "do-it-yourself" basis. Such individual projects can frequently be set up quite easily, but before making the effort it is resonable to ask if holography is really necessary in order to tackle the given problem. In fact, holographic microscopy should be treated as a specialized technique for use only when the information required cannot be obtained by conventional methods.

12. Laser Projection, Scanning, and Other New Microscope Systems

Commercially available laser have indeed stimulated progress in light microscopy. In holographic microscopy, which has been discussed in the previous chapter, the use of lasers was motivated by the coherency of their radiation. However, they have other properties, such as monochromacy (directly connected with coherency), collimation, and high power density of radiation, which have made it possible to improve some of the old projects or construct quite new microscope systems. Some of them will be discussed in this chapter.

12.1. Laser projection microscope with brightness amplifier

In general, for projection microscopy passive image displays, such as diffuse screens, are used. However, these systems are frequently useless in biology because of the need of strong illumination. Highly intense light may damage or even destroy the object under examination, especially when living cells or tissues are investigated. One of the ways to overcome these drawbacks is the use of a laser amplification system. The idea was given by Rabinovitch and Chimenti in 1970 [12.1], and then developed by Russian researchers [12.2–12.7].

The simplest system of a projection microscope with laser amplification is shown in Fig. 12.1. Its basic component is an active (lasing) medium AM, located between a microscope objective Ob and the primary image plane Π', which is reimaged, via a concave mirror M, into the secondary image plane Π'', where a screen is placed. The object under study (O) is located in the object plane Π of the objective Ob and illuminated with laser light through Ob. Part of incident light is reflected or scattered by the object O, goes back to the objective, and in passing along the active medium AM becomes considerably amplified, and thus can produce a highly magnified object image on a large screen S. The lasing medium AM permits us to lower significantly the intensity of epi-illumination, and thus protect the object from overheating or other distortions. On the other hand, the image-forming light reflected from the object may be amplified by lasing action, so that the image brightness increases to any reasonable values limited by the saturation level of the active medium AM. However, it is also possible to use this medium in such a manner as to obtain a focused light beam as powerful as necessary, which is capable of microsurgery or some other destructive actions (stimulation, photochemical reaction, etc.) on cells and other biological objects.

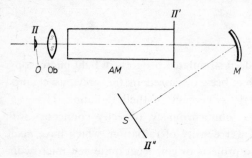

Fig. 12.1. Scheme of the laser projection microscope with amplification of image brightness [12.2] (see the text for explanation).

As a lasing active medium, copper vapour can be employed. The copper atoms are excited (at an operating temperature of 1600°C) by the current pulses produced by a generator based on a hydrogen thyratron with repetition frequency equal to 12 kHz.

The authors of this system obtained superradiance at $\lambda = 510.6$ nm, and then at $\lambda = 578.2$ nm. A light background was separated from these wavelengths by means of interference filters. Highly bright images were obtained, with resolution/magnification equal to 5.2 µm/825×, 3.7 µm/1250×, 2.6 µm/2250×, 1.3 µm/3250×, and 0.6 µm/6500× for objectives 1× /0.05, 2× /0.07, 4× /0.1, 8× /0.2, and 20× /0.4, respectively [12.4]. Effective amplification of radiation (P_{out}/P_{in}) from 1500 to 6100 was achieved for the input power P_{in} ranging from 1.6 mW to 0.4 mW, respectively. It is interesting to note that the lower input power the higher the amplification ratio.

A more advanced version of this system is schematically shown in Fig. 12.2. This type of the *laser projection microscope* (LPM) differs from the previous one mainly in having a concave spherical mirror M_0, which reflects the light passing through the object under study. Consequently, transmitted light contributes to the amplification process and the object O can also be studied in transmitted light. This version is therefore preferable to the first prototype (Fig. 12.1), especially when object of low contrast are examined. The improved type, like the

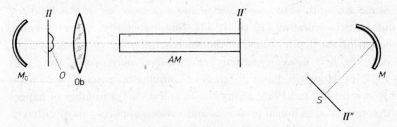

Fig. 12.2. Schematic diagram of an improved version of the laser projection microscope [12.5]; M_0 and M—concave mirrors, Π—object plane, O—object under study, AM—active (lasing) medium, Π'—image plane, Π''—secondary image plane, S—projection screen.

previous one, contains copper vapour as a lasing amplifier medium AM, which operates with pulse frequencies up to 10 kHz and pulse duration ranging from 10 to 15 ns. Laser pulses of these frequencies are not perceived by the observer's eye, and a continuous and stable object image is observed. On the other hand, each individual pulse of light is sufficiently strong to record the image on a conventional photoemulsion. Thus, a combination of the LPM and an appropriately triggered high-speed camera permits recording image sequences on a film. Consequently, movements and structure changes of transparent objects can be recorded. With the use of this technique, different isolated cells of animals and plants, protozoa, unicellulars, and other biological objects were examined [12.5]. The obtained results show that such a combination of the LPM and high-speed photography opens new possibilities for studying structure changes in living microorganisms and cells occurring in the millisecond and the submillisecond ranges. Thus, for instance, it may be possible to study the correlation between such short-time structure changes and the electric activity of nervous cells.

It is hoped that further progress in laser amplifier microscopy will soon be reported by Russian researchers.

12.2. Scanning optical microscopy

Scanning optical microscopy had its beginnings in the early 1950s when Roberts and Young described a *flying spot microscope* for the study of biological specimens, especially by using UV light [12.8]. This microscope made it possible to form an image on the screen of a display tube (or CCTV monitor) without exposing the cells and tissues to high, and hence destructive, doses of UV radiation [12.9], as in the case of conventional UV microscopy, which has been discussed in

Chapter 10. Some attempts were also made to use a microscope of this kind for metrology and particle counting [12.10, 12.11]. Simultaneously, some scanning systems for increasing the depth of field of the microscope were proposed [12.12–12.14], and next the first *tandem-scanning reflected-light microscopes* were described [12.15, 12.16]. However, basic progress in scanning microscopy was made in the period between 1972 and 1985, giving rise to an enormous number of papers dealing with theoretical or technical problems and various applications of different scanning microscopes [12.17–12.65]. Recently, the first commercially available *laser-scanning microscope* developed by a leading firm (C. Zeiss Oberkochen) has appeared on the market.

It is impossible to discuss here all the concepts of scanning optical microscopy, and only a few systems will be presented. For more intensive information, the reader is referred to the book [12.32] edited by Ash.

12.2.1. Confocal scanning microscopy

Scanning microscopy of this kind has been developed mainly by C. J. Sheppard and T. Wilson. It uses a laser light source and has some of the advantages of the well-known scanning electron microscope. The structure of the object on which the optical system of the microscope directs the laser beam changes the beam so that the sum of all the transmitted and/or reflected rays produces an exact image of the object. The laser beam manifests itself on the object as a small, monochromatic, coherent light spot of diameter defined by an Airy disc (see Eq. (3.4) in Volume 1). Obviously, optical systems of high numerical aperture are used, so that the light spot diameter is smaller than 1 μm. The light beams coming from the object are picked up by a photodetector and put together electronically to form an image on a monitor screen. As this image is obtained electronically, its contrast may vary within wide limits, and small differences in the contrast, frequently not perceived directly by the human eye, can be displayed on the monitor.

Figure 12.3 shows the general arrangement of a scanning optical microscope adapted to the study of specimens in reflected light. As can be seen, the optical system is relatively simple, but the electronic system is highly advanced. The specimen is mechanically scanned in a TV-like fashion relative to the focused spot of laser light. The rays reflected from the specimen are collected by a photodetector (for instance, a photodiode or photomultiplier), and the resulting photoelectric signal, appropriately amplified, is used to modulate the intensity of an oscilloscope (cathode ray tube) spot, which is deflected synchronously with the

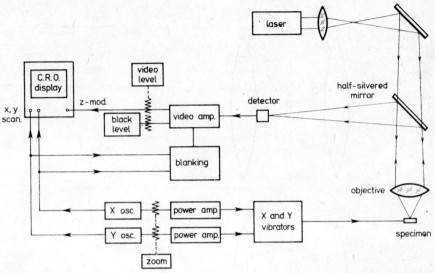

Fig. 12.3. Schematic diagram of the scanning optical microscope—a version for reflected light [12.33]. Courtesy of T. Wilson.

specimen scan. In this way, an image of the object is formed and displayed on the oscilloscope screen. The magnification of the object image can be varied electronically by changing the scanning speed. The specimen is scanned by means of a motorized, electronically controlled, high-precision scanning stage. This is the most important unit of a microscope of this kind.

There are three basic configurations of the optical systems of scanning microscopes. The first is described above and shown in Fig. 12.4a. It involves the use of a focused or point source of light and an incoherent photodetector. The scanning process is through object plane Π (Fig. 12.4a), involving the use of either a scanning stage or a movable point light source S. The second configuration (Fig. 12.4b) is based on a conventional microscope, in which it is not a focused spot but a patch of light from an extended source S that illuminates the object O through a condenser C. The object is then imaged by an objective Ob in the objective image plane Π', where a point photodetector PD is located. In particular, such a detector consists of a photomultiplier PM, whose window is masked by a pinhole diaphragm PH. The scanning process can be carried out either in the object plane Π by means of a motorized scanning stage or through the image plane Π' by using a movable (scanning) point photodetector. The object image is built up point by point as in the previous configuration. This principle (Fig. 12.4b) is mainly used in classical *scanning microscope photometers* (see Section 14.1 in Volume 3).

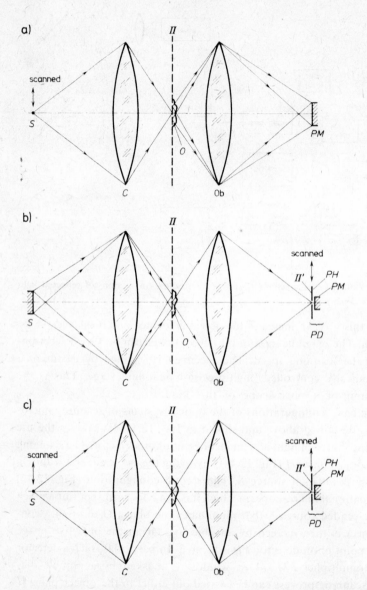

Fig. 12.4. Types of scanning optical microscopes [12.33]: a) point light source and extended photodetector, b) extended light source and point photodetector, c) point light source and point photodetector; S—light source, C—condenser, Ob—objective (these two optical components are usually identical), Π—object plane, O—object, PM—extended photodetector (photomultiplier), PH—pinhole, PD—point photodetector (PD = PM + PH). Courtesy of T. Wilson.

It can be shown that both configurations have the same imaging properties. The arrangement of Fig. 12.4a incorporates a point source of light and an *extended* (*incoherent*) *photodetector*, whereas that of Fig. 12.4b has, conversely, an extended light source and a *point* (*coherent*) *photodetector*.

The third configuration (Fig. 12.4c) is referred to as a *confocal* arrangement. It incorporates both a point light source S and a point photodetector PD. The term "confocal" means that the condenser C and the objective Ob are focused onto the object Ob. If these two optical components are identical and the condenser focusing is so perfect that a diffraction limited spot of light is formed on the object, then the objective Ob accepts the light from the same object point that the condenser illuminates; here the two optical systems play equal parts and imaging is generally improved [12.23, 12.25]. The required movement of the object with respect to the focused spot of light is achieved either by a scanning stage, as shown in Fig. 12.3, with a stationary light spot, or by deflecting the illuminating beam across a stationary object; both procedures have their advantages and disadvantages [12.33]. The second is at present easily realized by means of two scanning mirrors oscillating about perpendicular axes. In this case, however, the point photodetector PD (Fig. 12.4c) must follow synchronically the image-forming light beams emerging from the successive points of the object which is being scanned. This requirement is difficult to satisfy in practice for a pure confocal system. Therefore, in most confocal scanning microscopes high-precision scanning stages are used.

As mentioned above, the image quality achieved by a confocal scanning microscope is higher than that produced by other configurations. This advantage is illustrated in Fig. 12.5, which shows the point spread functions for the confocal and the conventional scanning microscope. As can be seen, the diameter of the Airy disc produced by the former is smaller than that formed by the latter. Consequently, the confocal configuration offers a somewhat better resolution and a higher image contrast (Fig. LVII). A further improvement of these parameters can be achieved by apodizing the central part of the exit pupil of the objective (Ob, Fig. 12.4c) so as to form an annular pupil. Objectives with a pupil of this kind are not advantageous in conventional bright-field microscopy since their point spread functions have high secondary maxima (compare Figs. 3.4 and 3.5 in Volume 1), which tend to damage the image quality of extended objects; this does not apply to confocal scanning microscopy [12.39]. However, the depth of field in this system is smaller than that in classical configurations; on the other hand, optical sectioning is more effective.

Figure 12.6 shows the optical part of a high resolution confocal scanning

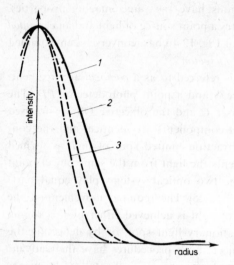

Fig. 12.5. Point spread function (intensity distribution in the Airy pattern) produced by various microscope configurations [12.33]; Curve *1*—conventional microscope or scanning configurations shown in Figs. 12.4a and b, curve *2*—confocal scanning microscope with circular objective aperture (Fig. 12.4c), curve *3*—confocal scanning microscope with annular objective aperture. Curves *2* and *3* are narrower than curve *1* by a factor 1.4 and 1.75. Courtesy of T. Wilson.

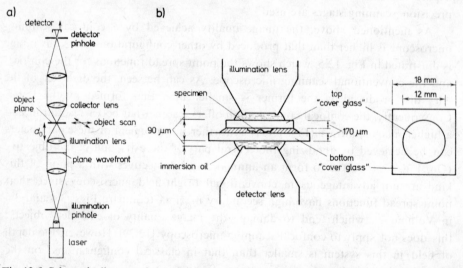

Fig. 12.6. Schematic diagram of a confocal scanning light microscope with high aperture immersion lenses [12.29].

microscope developed by Brakenhoff and his co-workers [12.29], in which high power planapochromatic immersion objectives (100 × /1.3) have been used as both the illumination (spot-forming) and the image-forming systems. The specimen is located between two cover slips. To avoid gravity influences on the immersion oil, the two confocal optical systems are arranged vertically above each other.

Present-day scanning microscopes use laser light sources, whose collimated and monochromatic radiation can easily be focused to obtain a fine light spot of the required intensity. However, the high coherency of laser light may give rise to speckle noise (see Section 11.5). Fortunately, this effect, especially in trans-mitted-light scanning systems, is here less injurious than in other microscopes with laser illumination, and it is possible to obtain noise-free imagery with the depth discriminations property [12.52].

In principle, all specialized techniques of conventional microscopy are suitable for scanning microscopes if a proper mode of detection is used [12.53]. As mentioned above, there are two basic detection modes: *incoherent* and *coherent*. In the first, the detector signal represents the sum of local intensities of the object wave far from the object, while in the second it represents the vector sum of the complex amplitude distributions of light in the object. The detection configurations may be different, and some of them are sketched in Fig. 12.7. The first two sketches

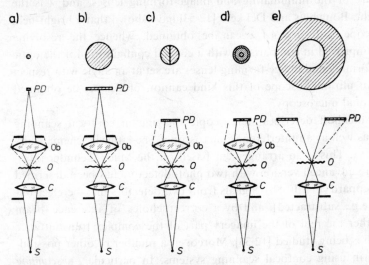

Fig. 12.7. Detection systems in scanning microscopy [12.53]: a) coherent, b) incoherent with extended photodetector, c) incoherent with split-detector, d) incoherent with multiple ring detector, e) incoherent with annular detector; *S*—point light source, *C*—condenser, *O*—object, *Ob*—objective, *PD*—photodetectors.

have already been discussed. They incorporate coherent (a) and incoherent (b) detectors, such as shown in Figs. 12.4c and a. The remaining diagrams of Fig. 12.7 represent other incoherent detections with detectors consisting of two semi-circular sections (c), concentric rings (d), and a single annulus (e). When the signals from the two semicircular sections of the *split-detector technique* (Fig. 12.7c) are properly combined, an imaging method, referred to as the *differential phase contrast*, is obtained as a counterpart of the Nomarski DIC technique discussed in Chapter 7. Hamilton and Sheppard [12.47] have stated that this counterpart has even a number of important advantages over the usual Nomarski DIC system.

Both the phase and the amplitude information can also be obtained by using the detector shown in Fig. 12.7d when the signals from concentric rings are properly combined. However, if we use an annular detector (Fig. 12.7e), which records not direct but only diffracted light, *dark-field scanning optical microscopy* may be obtained by analogy to dark-field electron microscopy.

In scanning microscope systems described above, the optic axes of the spot-forming and image-forming lenses are coaxial. The limit spatial frequency of these systems is given by $\mu = 1/\varrho = 2A/\lambda$, where ϱ is the resolving power, A is the numerical aperture of the illuminating and image-forming lenses, and λ is the wavelength of light. Bouwhuis and Dekkers [12.53] have shown that a bright-field scanning microscope with $\mu = 4A/\lambda$ can be obtained, whence the resolving power ϱ may be improved in comparison with a coaxial configuration if the optic axes of the spot-forming and image-forming lenses are set at an angle with respect to each other. An ultramicroscope of this kind cannot, of course, be obtained by using conventional microscopy.

In [12.30], Brakenhoff describes stereoscopic imaging in confocal scanning light microscopy as well as an interference microscope. The latter is schematically shown in Fig. 12.8a. This is an arrangement based on the Mach–Zehnder inter-ferometer. Its more advanced version with two photodetectors has been described by Wilson and Sheppard (Fig. 12.8b). Signals from two detectors may be electron-ically combined (e.g., substracted), and by a correct choice of reference beam one can image either the real of the imagery part of the complex transmittance of the object which is being studied [12.33]. Moreover, a number of other possibil-ities are available in using confocal scanning systems. In particular, a *scanning fluorescence microscope* can be very useful, especially if specimens of extremely weak fluorescence must be examined and highly concentrated (focused) excita-tion is necessary. Here, the laser scanning microscope offers the additional advan-

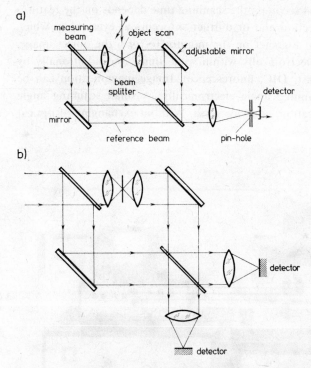

Fig. 12.8. Schematic diagrams of optical scanning interference microscopes: a) system with a single photodetector [12.30], b) configuration with two photodetectors [12.33].

tage of exciting each object point for a very short period only, which eliminates the fading of fluorescence.

The field of scanning light microscopy and its capabilities has now grown to vast proportion, and for a fuller introduction to this specialized and attractive field the reader is referred to Wilson and Sheppard [12.54].

12.2.2. Laser-scan microscope developed by C. Zeiss Oberkochen

This microscope (Fig. 12.9) functions on a principle similar to that of the well-known scanning electron microscope, but instead of an electron beam a laser beam is used in it. The laser beam is focused on the object as a narrow, coherent, monochromatic light spot of a diameter below the micrometre limit. The movement of this spot across the object is realized by high-precision mirrors and a corresponding lens system. A standard laser-scan microscope (LSM) produces

512×512 spots within 2 to 64 seconds; the scanning time depends on the features and size of the object. Reflection and/or diffraction occurs at every spot, which is picked up, stored, and put together on the monitor to form the object image. The image contrast varies electronically within wide limits, and additionally by optical means (phase contrast, DIC, fluorescence). Image magnification can be changed by varying the scanning angle electronically. A small scanning angle corresponds to high magnification, and vice versa. Without exchanging of optical

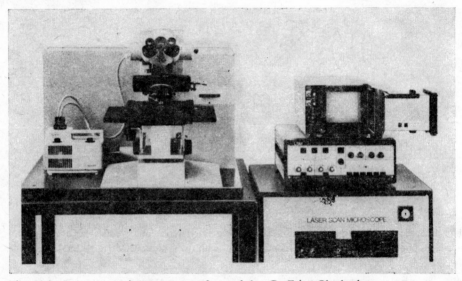

Fig. 12.9. Laser-scan microscope manufactured by C. Zeiss Oberkochen.

parts, the object image can be elecronically varied in size in the ratio 1:8. The LSM model is set up on the basis of a normal research microscope, and permits the use of both the conventional methods of light microscopy and those of laser scanning microscopy. The following techniques can be used: in the scanning mode of operation—transmitted light, incident light, fluorescence, phase contrast/incident-light DIC; in the normal microscopy mode—transmitted light, incident light, phase contrast/incident-light DIC.

The LSM model includes several scanning modes: (1) frame—scanning of a square object field and display of the scanned object area on the monitor (CCIR standard, 625 lines); (2) Y-scan—scanning of a square object field and display of the intensity distribution of the image (pseudo-three-dimensional image); (3) line—scanning with the X-scanner alone, while the Y-scanner remains stationary and can be positioned at will (display of the intensity of the scanned line); (4) monitor line—an adjusting aid in the line mode, i.e., the object field is scanned once, and the position of the Y-scanner is marked by a white line; (5) spot—both scanners are stationary and their positions are selectable. Since in this case the laser spot can be kept at a point for any desired length of time, the laser-scan microscope can also be used as a micromanipulator even at high magnification; new fields of application (in gene technology, for instance) are therefore possible.

The following laser types can be used with this microscope: Ar laser, 10 mW at $\lambda = 488$ nm or 5 mW at $\lambda = 514.5$ nm; He–Ne laser, 5 mW at $\lambda = 632.8$ nm, and He–Cd laser, 5 mW at $\lambda = 441.6$ nm.[1]

12.2.3. Tandem-scanning reflected-light microscope

This microscope was contrived by M. Petran in 1964. He wanted an instrument that could be used to study live neurons in live, unfixed, unstained brain [12.15, 12.16, 12.55]. The first model was built in Czechoslovakia in 1965 by Petran and Hadravsky. Recently, this microscope has been improved considerably and accepted as a valuable instrument for studying internal structural features in the surface layers of bulk soft tissues with no need to prepare sections, muscles, nerves, mineralized tissues (including fossils), bones, teeth, especially asteocyte lucunae in bone, enamel prisms in teeth, and the transparent filling materials placed in anterior teeth [12.56–12.59]. Just as the scanning microscopes discussed previously, it has small depth of field but considerable depth of penetration into intact specimens.

The tandem-scanning reflected-light microscope (TSRLM) was in fact the first configuration of the confocal mode, but its construction differs considerably (Fig. 12.10) from that of the *object-scanning microscopes* of Wilson and Sheppard, and of Brakenhoff which has been described in Subsection 12.2.1. The basic difference is in the scanning mode, which is achieved by means of a simple and reliable device whose idea was given by Nipkow in 1884. It is a rotatable opaque screen with holes, known as the *Nipkow disc*. However, the original Nipkow disc had one row of holes arranged along one scroll of the Archimedean spiral at the outer zone of the disc. The Petran and Hadravsky device, on the other hand, incorporates many Archimedean spirals with holes properly arranged in a quasi-hexagonal array (Fig. 12.11). The Nipkow disc occupies the image plane Π'

---

[1] All the data listed above are taken from *Zeiss Information*, No. 1/1985, CM-H-II/85 PUoo.

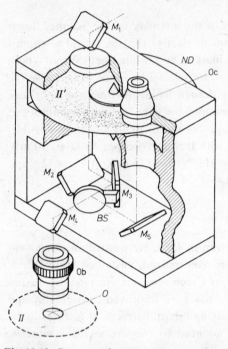

Fig. 12.10. Cut-away view of the head of the tandem scanning reflected light microscope [12.55] (see the text for explanation).

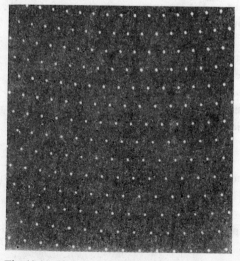

Fig. 12.11. Photograph of a fragment of the Nipkow disc used in the tandem scannig microscope shown in Fig. 12.10 [12.55].

(Fig. 12.10) of a standard microscope objective (160 mm tubelength). The object under study is illuminated through one side of the Nipkow disc *ND* and objective Ob (via mirrors M_1–M_3, beam-splitter *BS*, and mirror M_4). The objective Ob produces a sharply focused image of the disc holes in the object plane *II* occupied by the specimen *O*. The latter is illuminated across the entire field of view since a large number of the holes are simultaneously projected onto the specimen. However, the illumination is most intense in those parts which lie in the focused-on layer coincident with the object plane *II*. The light reflected from that layer is collected by the same objective Ob and, via mirror M_4, beam-splitter *BS*, and mirror M_5, focused onto the diametrically opposite side of the disc *ND*. The image-forming light can pass through the holes, and the object image is observed by means of an ocular Oc or projected into a photo-camera or a TV-camera. Each hole on the observation side of the disc *ND* has a conjugate hole on the illuminating side. This is a very important feature of the Nipkow disc used in the microscope in question.

The above principle is more clearly illustrated in Fig. 12.12. Here H_1 is a hole in the Nipkow disc *ND* on its illuminating side, Ob is a microscope objective for both illumination and image formation, and H_2 is the conjugate hole on the image-forming side of the disc *ND*. The hole H_1 is focused by Ob as a light patch

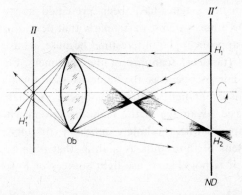

Fig. 12.12. Schematic ray diagram for the tandem scanning microscope shown in Fig. 12.10.

H_1' on the plane *II*. Rays reflected from this plane are focused on the image plane *II'*, which is coincident with *ND*, pass through the hole H_2 conjugate to H_1, and enter an ocular of the Ramsden type, whose focal plane is coincident with the disc *ND*; these rays are only responsible for the object image. The light reflected from planes below *II* is focused in front of the disc *ND* and therefore is largely intercepted by the opaque parts of this disc, while the light reflected from planes above II is focused behind the disc ND and is also largely cut out. Thus, the specimen layer with reflective features coincident with the focused object plane II will appear with the highest brightness and also high contrast. In any case, the property is common to all confocal systems. It should be noted, however, that many holes act simultaneously on the illumination and the image-forming parts of the TSRLM. Since the Nipkow disc ND is rotatable, its structure is not visible by the ocular Oc; the object image is solely observed in the same manner as through the eyepiece of a conventional microscope. Moreover, it is important to note that any high-power light source can be employed in this instrument. In the latest models, Petran and Hadravsky have used a tungsten ribbon lamp and a highest pressure mercury arc lamp (the latter for working with poorly reflecting specimens).

A wide review of the areas of application which are envisaged for the TSRLM is given by A. Boyde *et al.* [12.56–12.58].

12.2.4. Other scanning light microscopes

As mentioned earlier, a large variety of scanning light microscopes are described in the literature. The most characteristic systems have been presented above. However, it is worthwhile to note some other systems, particularly that developed by Svishchyov [12.60, 12.61], which appears to be interesting because it uses two slit diaphragms for synchronous (tandem) scanning of the specimen (Fig. 12.13). The goal in developing this microscope was the same as that for the TSRLM, i.e., the study of thick transparent light scattering objects, especially bulk tissues, in reflected light. The basic advantage of the scanning system with slits lies in the economy of light compared with the hole diaphragm; the image brightness achieved by using slits is tens of thousands as high as that obtained by using holes [12.60] Consequently, it is possible to use light sources of lower power for specimen illumination.

It is also worthwhile to mention a scanning optical microscope developed by Bogdankevich *et al.* [12.62] as an instrument intended for measuring purposes and photolitography in semiconductor technology. Its characteristic feature is a semiconductor laser used as a light source to scan a given object.

More than ten years ago, Sawatari described a laser scanning microscope in which a heterodyne technique was applied [12.20]. This technique depends on the use of an optical system similar to the configuration of the Mach–Zehnder

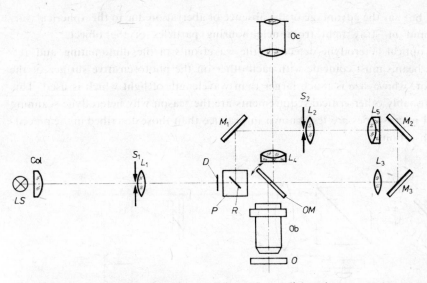

Fig. 12.13. Optical system of tandem scanning reflected light microscope with two slit field diaphragms [12.60]; LS—light source, Col—collector, S_1 and S_2—slit diaphragms, L_1 to L_3—collecting lenses, D—opaque strip (stop), P—glass cube with a reflecting strip R, OM—oscillating (scanning) two-sided mirror, Ob—objective, O—object, L_4 and L_5—transfer lenses, M_1 to M_3—mirrors, Oc—ocular. Slits S_1 and S_2, opaque strip D, reflecting strip R, and axis of oscillation of the scanning mirror OM are at right angles to the plane of drawing.

interferometer (Fig. 12.14). A single laser beam is split into two beams. One of them illuminates the object to be scanned, while the other serves as a reference beam, whose frequency[2] ω is shifted by an acoustooptical deflector. The illumination beam scans the object synchronously with the scan of the reference beam. Part of the light scattered from the object interferes with the reference beam to produce the beat frequency. The amplitude or the phase of the beat is displayed on a cathode ray tube (CRT). Imaging characteristics of this system are similar to those of the laser scanning microscope described in Subsection 12.2.2. Heterodyning, however, has additional advantages in that the object illumination can be at least one order of magnitude less intense and that ambient light does not affect the image of the object. Besides these characteristics, the heterodyne system is capable of giving images of pure phase objects and light polarizing objects. Fujii *et al.* [12.63] showed that the resolving power of the heterodyne laser microscope is essentially the same as that of the conventional light micro-

[2] See Subsection 1.2.1 in Volume 1 for definition.

scope, but has the advantage of the absence of aberration due to the spherical surfaces and of stray light from neighbouring particles of the object.

In optical heterodyne detection, the wavefronts of the illuminating and reference beams must coincide with each other on the photosensitive surface of the
detector, whose size is much larger than wavelength of light which is used. This
and probably other critical requirements are the reason why heterodyne scanning
optical microscopes are less known in practice than those described in the preceding subsections.

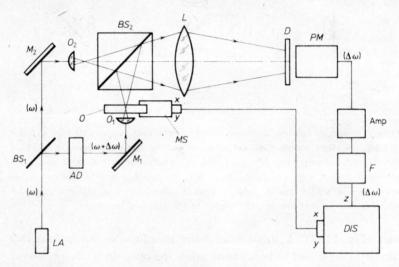

Fig. 12.14. Schematic diagram of an optical heterodyne scanning microscope [12.20]; *LA*—laser,
BS₁ and *BS₂*—beam splitters, *AD*—acoustic shifter of frequency, M_1 and M_2—mirrors, O_1
and O_2—objectives, *O*—object, *MS*—mechanical scanner, *L*—lens, *D*—diffusor, *PM*—photodetector (photomultiplier), Amp—amplifier, *F*—electronic filter, *DIS*—display.

It is also worthwhile to mention the phase sensitive scanning optical microscope described by Jungerman *et al.* [12.48]. It measures amplitude and phase
of light and is insensitive to mechanical vibrations. The recorded phase information
makes it possible to measure variations in surface height with an accuracy of about
10 nm.

A large group of specific scanning optical microscopes is formed by instruments
of the Quantimet-type based on TV scanning system. These are well known in
practice and referred to as automatic image analysers, which are discussed in
Chapter 18.

12.3. Nonlinear microscopy with second harmonic generation

There are many transparent materials whose structural details cannot be seen in either conventional or specialized light microscopes, though the size of the details is larger than the wavelength of light. No structure is normally visible in a transparent material if there are no variations (or there are extremely small variations) in the refractive index accompanying the structural inhomogeneities. Materials of this kind include, for example, certain polycrystalline cubic substances. However, structural inhomogeneities mentioned above are accompanied by large spatial variations in the nonlinear refractive indices. This fact has been used by Hellwarth and Christensen [12.66, 12.67] to render the structure visible with nonlinear optical effects. Consequently, 15 years ago, a new kind of microscopy, referred to as *harmonic optical microscopy*, was initiated by the researchers mentioned above. The idea of this microscopy is shown in Fig. 12.15. A mono-

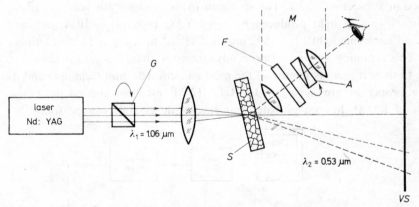

Fig. 12.15. Second harmonic optical microscopy [12.67] (see the text for explanation).

chromatic light beam from a high-power Nd:YAG laser is focused on a sample (S) of nonlinear material which is able to generate the second harmonic wavelength, i.e., $\lambda_2 = \lambda_1/2 = 532$ nm ($\lambda_1 = 1064$ nm is the wavelength of infrared light emitted by Nd:YAG lasers). The incident laser beam is polarized linearly by a Glan polarizer (G), and a viewing microscope (M) is tilted at an angle to this beam. The microscope is fitted with an analyser (A) and a barrier filter (F), which suppresses infrared rays scattered by the sample S. We can also use a far-field viewing screen (VS) to see the angular distribution of second harmonic light. Under these conditions, the magnified image of a nonlinear material is viewed through the microscope M in green second harmonic light generated in the sample S by the incident infrared beam.

Objects suitable for nonlinear microscopy of this kind are, for example, polycrystalline ZnSe plates, which show no structure under an ordinary polarizing microscope because their grains are optically isotropic. Hellwarth and Christensen have discovered embedded among some larger grains a densely packed array of randomly-oriented single-crystal platelets about 60 μm in diameter and 0.5 μm thick, with their (1, 1, 1) crystallographic axes near-normal to the platelet faces. No optical inhomogeneity of this kind was visible in an ordinary polarizing microscope. Some other inhomogeneities or defects have also been observed in single-crystal GaAS plates; slabs of CdTe and CdS films deposited on a glass substrate have also been observed by using the system shown in Fig. 12.15 [12.67].

A more sophisticated instrument—the *scanning harmonic optical microscope* (Fig. 12.16)—has been developed by Gannaway, Wilson, and Sheppard [12.68–12.70]. This microscope is a version of the confocal scanning optical microscope described in Subsection 12.2.1. The specimen (S) is mechanically scanned across a stationary spot of light produced by a Nd:YAG laser ($\lambda_1 = 1064$ nm), and the second harmonic light ($\lambda_2 = 532$ nm) is detected by using a photomultiplier (PM). This scanning system is more advantageous than the set-up shown in Fig. 12.15 since it permits the use of the most efficient detection technique and its imaging properties are superior. In Ref. [12.70] excellent second harmonic pictures of KD^{x}P, lithium niobate, and zinc oxide crystals can be found.

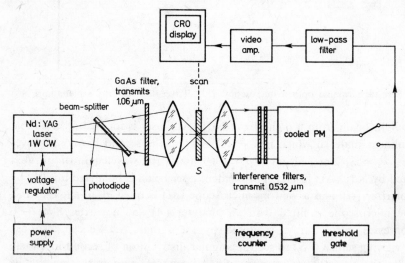

Fig. 12.16. Scanning second harmonic optical microscopy [12.70].

12.4. Raman microscopy

Raman scattering, like the generation of second harmonic light, is classified among nonlinear optical phenomena. However, in harmonic optical microscopy a contrasty image is obtained from the variations in the generation of the second harmonic wavelength (λ_2) in the specimen by incident radiation of fundamental wavelength (λ_1); the image obtained displays essentially spatial variations in the nonlinear polarization of the specimen which is transparent to both fundamental and to second harmonic frequencies, ω_1 and $\omega_2 = 2\omega_1$. On the other hand, the purpose of Raman microscopy is to obtain information about molecular vibrational properties of a given object.

Raman microscopy has been developed mainly by Delhaye, Dhamelincourt, and Bisson [12.71, 12.72]. Its basic principle is shown in Fig. 12.17. A laser beam is focused on the surface of a solid sample. Incident photons can be partially transmitted, reflected, and elastically (resonantly) scattered while retaining their original (fundamental) frequency ω_1. Such interactions of photons with matter

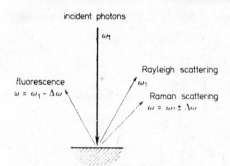

Fig. 12.17. Underlying principle of Raman microscopy.

are used to obtain optical images in conventional or laser-scan microscopes. However, if the light emerging from the specimen contains other frequencies than ω_1, it transfers information relating not to the elements (atoms) but to the molecules, ions, or crystals that are present in the specimen. In particular, fluorescence emission and Raman scattering belong to radiation of this category. The frequency of fluorescence light is always smaller than that of incident (excitation) light, while the frequency of Raman light is higher or lower than ω_1.

As mentioned above, the Raman effect permits the detection and identification of the molecular composition and structure of the examined object. Unfortunately,

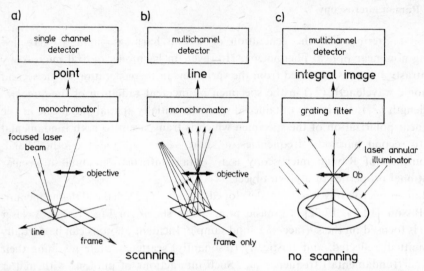

Fig. 12.18. Possible systems of Raman microscopy [12.71]: a) scanning technique along x, y-coordinates with focused laser beam, b) scanning technique with straight line of light along y-coordinate, c) no-scanning technique using annular illumination. Systems a and b are also qualified as Raman microprobes.

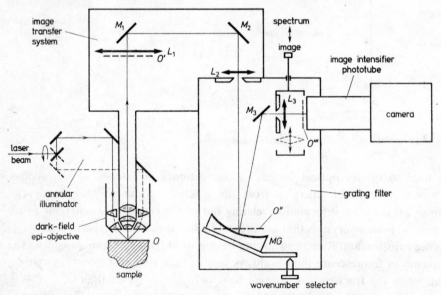

Fig. 12.19. Schematic diagram of Raman microscope with annular dark-field illumination and holographic diffraction grating [12.71].

the effect is usually weak and high power light sources are required. Moreover, Raman scattering intensity is obviously suppressed by Rayleigh scattering and/or fluorescence light, by the overheating of the sample, and by other disturbing factors. To overcome these limitations, different systems of Raman microscopy have been devised and tested (Fig. 12.18). The most attractive solution is shown in Fig. 12.19. It uses the principle illustrated in Fig. 12.18c and produces the image of the whole object area illuminated uniformly by laser radiation. A conventional microscope with dark-field epi-illuminator is used together with a stigmatic concave diffraction grating MG (2000 lines/mm, one metre focal length). The latter selects the appropriate spectral line of Raman light. The primary image O' of the object O is projected onto the photocathode of an image intensifier camera by means of auxiliary lenses $L_1 - L_3$ and mirrors $M_1 - M_3$.

According to Delhaye and Dhamelincourt [12.71], the Raman microscope can be regarded as a valuable analytical tool for the study of various materials (rocks, plastics, crystals) and chemical reactions in microsamples. For instance, Delhaye and Dhamelincourt examined the surface of a potassium nitrate crystal and a mixture of chromate and ferrocyanide microcrystals with dust particles, while Sombert et al. [12.73] studied phases of molibdenium oxide coated by aluminae.

12.5. Optoacoustic and photothermal microscopy

When a focused laser beam with periodically modulated intensity strikes a specimen of a solid, the latter is subjected to periodic local heating, due to absorption of light, and a *thermal wave* is generated at the specimen surface. The term "thermal wave" is a notion that results from the solution of the equation of periodic thermal diffusion. This process manifests itself as a highly damped wave whose length is determined by the thermal diffusion length. The thermal wave causes periodic expansion and contraction of the specimen structure, and an *acoustic wave* occurs. Both waves are of the same periodicity and each of them is able to reveal almost the same variations in thermal and/or geometrical features of the specimen. These features may be imaged by either acoustic or thermal detection. Consequently, in the first case we speak of *optoacoustic* (or *photoacoustic*) *imaging* and in the second—of *photothermal imaging*. If the position of the focused laser beam is slowly changed via a scanning system, an image (map) of the local thermal properties of the specimen may be obtained.

The earliest method of optoacoustic imaging consists in using a gas-microphone or a photoacoustic cell. The thermal wave causes the gas pressure to be modulated, and the modulation can be recorded or even heard as sound. A more modern method is based on piezoceramic detection. It was developed mainly by Busse and Rosencweig [12.74, 12.75]. An arrangement of an optoacoustic microscope with piezoceramic detection is shown in Fig. 12.20. An intense beam of light

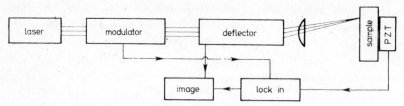

Fig. 12.20. Block-diagram of an optoacoustic microscope [12.74].

(from an argon-ion laser, $\lambda = 514.5$ nm, power 200 mW) is intensity modulated at 185 kHz by means of an acousto-optic modulator. This modulation frequency permits us to obtain thermal waves of ca 7 μm length. The laser beam is then deflected in such a way that its focused spot of ca 5 μm diameter scans across the sample glued to a piezoceramic detector (PZT). The alternating current output voltage of this detector is analysed by a lock-in amplifier and then displayed as a function of the sample coordinates. The thermal waves generated in the sample by periodic heating (with frequency 185 kHz) scatter off defects and thermal irregularities below the surface sample, which causes periodical changes in the photoacoustic signal. When the laser beam scans across the sample, any change in the photoacoustic signal indicates the presence of a defect at the point of the sample upon which laser beam is just falling; thus, the photoacoustic signal provides true thermal-wave imaging of the sample.

Another arrangement for photoacoustic microscopy has been devised by Quimby [12.76]. It (Fig. 12.21) differs from the previous system (Fig. 12.20) in that it is the beam position that is modulated rather than the beam intensity, so that an entire line scan photoacoustic image is displayed directly on an oscilloscope screen. The result is quite similar to that obtained in the previous system, but acquired in a time measured in milliseconds rather than minutes (as mentioned earlier, the scanning process in the system based on the modulation of intensity of a laser beam is slow). Consequently, Quimby's system is referred to as real-time photoacoustic microscopy.

The optoacoustic systems shown in Figs. 12.20 and 12.21 require a physical contact of the sample with the detector of acoustic waves. This is, of course, a de-

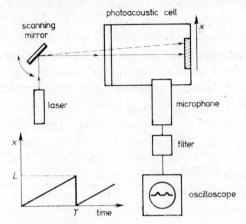

Fig. 12.21. Real-time photoacoustic microscopy [12.76]. By contrast to the system shown in Fig. 12.20, the position of the focused laser is a ramp modulated as shown in the inset, while the beam intensity is constant (T—temporal period, L—spatial period).

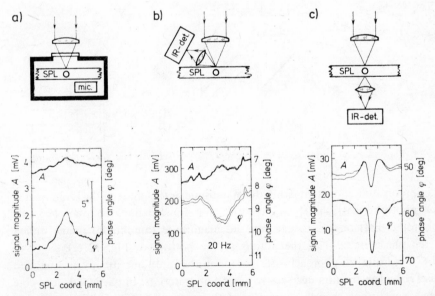

Fig. 12.22. Modes of imagery of thermal waves [12.74]: a) photoacoustic detection, b) photothermal detection in reflection, c) photothermal detection in transmission. The laser beam scans across 1 mm aluminium sample (SPL) provided with a 0.4 mm subsurface hole.

fect which does not occur if photothermal detection is applied. Photothermal waves modulate the temperature of the sample, and this modulation is recorded by an infrared detector. The detection procedure is the same as that described in Section 10.4, where only thermal microscopy was considered, but the photothermal imaging discussed here differs from thermal imaging. As shown in Fig. 12.22, photothermal imaging is possible in both reflection and transmission configurations. The latter is illustrated in more detail in Fig. 12.23 (according to Busse [12.77]). An intense beam of light from an argon-ion laser (*L*) is intensity modulated by an acousto-optic modulator (*M*), then deflected by a mirror scanner (*SX*, *SY*), and focused on the object (*O*) by a microscope objective (Ob). The focused ligh spot which scans the object is observed by a telescope (*T*) associated with a beam-splitter (*BS*) and an appropriate filter (*F*). Infrared radiation is recorded by a detector (*D*), which cannot be translated synchronously with the movement of the focused light spot if the scanned sample region is small.

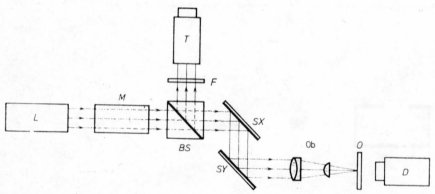

Fig. 12.23. Photothermal transmission microscopy [12.77] (see the text for explanation).

It is important to note that either the magnitude or the phase of the photoacoustic (or photothermal) signal can be used for constructing the photoacoustic (or photothermal) image. In general, the magnitude (amplitude) image differs from the phase image (see the graphs at the bottom of Fig. 12.22).

The technique of photoacoustic (or photothermal) microscopy is qualified as a useful tool in the non-destructive testing of materials. In particular, it is able to detect subsurface as well as surface defects of solids. This ability makes it possible to locate voids and cracks in metals, ceramic, semiconductors, and other materials [12.76, 12.78].

It is worth noting that microscopy of this kind was recently utilized to investigate the possibility of characterization of coal by its thermal properties [12.79]. Coal surfaces were examined through the detection of iron pyrite band, macerals, SiO_2, and sites of pyrolisis by argon-ion and CO_2 lasers. It was stated that the absence of sulfur and other material alterations produced variations in the photoacoustic data of the pyrolisis site. A band of iron pyrite was clearly observed photoacoustically as were CO_2 and Ar laser pyrolized areas. The change in photoacoustic data due to laser pyrolisis was confirmed by a scanning electron microscope with X-ray fluorescence capability.

For the interested reader, more details regarding the photoacoustic microscopy may be found in Chapter 10 of the book *Laser Optoacoustic Spectroscopy* by V. P. Zharov and V. S. Letokhov [12.80].

*

*   *

All the microscope systems described above use laser light sources, can therefore be qualified as instruments belonging to the area of light microscopy referred to as "laser microscopy". This area is now so large that a separate book is necessary to give full insight into in [12.81]. It is worth noting, however, that the instruments presented in this chapter are not yet commercially available except of the Zeiss laser-scan microscope and similar instruments produced by a few other manufacturers.

In any event, this new field of light microscopy is currently in the stage of the fascinating development. Quite recently, for instance, the coherent anti-Stokes Raman scattering microscope has been described [12.82]. This instrument can image spatial distribution of biological compounds with resolution better than 1 μm, and excellent molecular discrimination, using chemically similar molecular species, can be obtained by means of digital image processing techniques.

Finally, it is also worth mentioning that, in any case, photoacoustic microscopes should not be confused with acoustic microscopes like the ELSAM manufactured by E. Leitz Wetzlar [12.83]. Acoustic microscopy is quite a different domain, which is beyond the scope of this monograph.

Epilogue to Volume 2

The specialized techniques described in this volume are continuously being improved, and so it is impossible to give an up-to-date picture of them in a book whose period of production is relatively long. To compensate, this epilogue is added. It was written when the final proofs of the volume had been completed, and covers quite recent events related to the subject of the book.

One of the most important developments has been that of Nanovid microscopy, which enables submicroscopic colloidal gold particles to be visualized in cells and tissue sections using the video enhanced contrasts (VEC) technique [E2.1]. The immuno-gold microscopy mentioned in Subsection 9.8.1 has therefore become a more competitive alternative to immunofluorescence microscopy based on fluorochromes.

The use of colloidal metallic particles as a label in immunoassays was introduced by Leuvering and his co-workers in 1980 (see Ref. [E2.2] and papers cited there). Today, many metals or insoluble metal compounds (e.g., gold, silver, silver iodide, barium sulfate) can be used as labels in immunoassay techniques. Gold is one of the most advantageous reagents since it is not present in body fluids, its optical properties are specific, and it is suitable for the preparation of colloidal particles. Immunochemically reactive gold particles are obtained by adsorption, under specific conditions, of antibody molecules onto the surface of the gold particles, with a diameter of around 20–50 nm. Normally, such small particles are not suitable for observation using a common light microscope, but they can easily be observed when using the VEC method and AVEC-DIC technique (see Section 7.3 and Epilogue to Volume 1), with a Hamamatsu C-1966 video camera and microscope objectives of maximal numerical aperture. The image contrast of gold particles in a bright-field microscope is enhanced by applying an offset to the video signal and increasing the gain. Optical mottle or background noise (empty microscope field of view) is stored in a frame memory by an electronic processor and continuously subtracted from the incoming actual image of the specimen labelled with colloidal gold particles. Additionally, electronic noise is reduced by averaging the images over a selected number of frames and the image contrast can be further enhanced by digital gain expansion. The gold particles can be localized accurately within or on cells and tissue sections in relation to other cellular structures or organelles by using the AVEC-DIC technique. Moreover, epi-illumination permits one to distinguish the gold particles from cellular organelles by light reflection and lack of phase effects.

Nanovid microscopy permits us to follow dynamic processes in living cells and the behaviour of individual cell components labelled with discrete markers of a sufficiently small size (colloidal gold particles of 20–50 nm in diameter) to disturb the physiology of cells as little as possible. Microscopy of this kind is well suited to study the dynamic behaviour of cell surface components and receptors, to observe their internalization and to follow their route through different intercellular regions [E2.3]. Discrete cytoplasmic components can be localized and their dynamic translation may be observed using microinjected gold particles. Due to the sharp image contrast generated by the Nanovid system, automatic quantification of numbers and movements of gold particles is made possible by a computer coupled interactively to the Hamamatsu image processor. Using their Nanovid system, the authors of Ref. [E2.1] investigated the microtubule associated motility, cell membrane dynamics and receptor mediated endocytosis. When Nanovid microscopy uses the AVEC-DIC technique, its potentiality approaches that of an electron microscope [E2.4].

Fluorescence microscopy, discussed in Chapter 9, is a technique which frequently suffers from a lack of sufficiently intense fluorescent light. Recently, an ultra-high-sensitivity video camera (the C-2400 VIM model; VIM = Video Intensified Microscopy) has been developed for imaging at extremely low light levels, typically those encountered in luminescence and fluorescence applications [E2.5]. The camera uses a high-performance two stage MCP image intensifier. Its noise is extremely low and its dynamic range extends from visible intensity levels to single photons ($1-10^9$ photons/mm$^2 \cdot$ s), allowing images of fluorescent particles to be recorded at very low light levels (down to single photon events).

The VIM camera (C-2400 or C-1966-20C) is especially suitable for the analysis of organelle motility in nerves [E2.6]. The movement of organelles is made visible by having fluorescent probes bound to their surface. For instance, monoclonal antibodies labelled with fluorochromes can be used for some probes.

This camera is available for a spectral range of either 280–650 nm or 280–850 nm. A choice of image pickup tubes allows the camera to be optimally configured for conventional analog or photon counting imaging. Moreover, the C-2400 video camera is characterized by great geometrical image stability due to its use of a crystal oscillator.

In the last few years, time-lapse recording and monitoring of living cells cultured in vitro has changed enormously. In the opinion of A. Glauert, as expressed in Ref. [E2.4], "the microscope itself appears almost lost among monitors, video recorders, frame stores and associated computers". In this category of equipment, we can also include the time-lapse video system with its animation unit, described recently by T. D. Allen [E2.7].

Three-dimensional imaging of microscopic objects is also a topic which has roused considerable interest recently, mainly stimulated by confocal scanning light microscopy[1] [E2.8–E2.12] and by the development of computer-based methods for the automated three-dimensional analysis of microscopic objects [E2.13].

There are three main types of scanning system: scanning the specimen stage, scanning a laser beam over the stationary specimen stage, and scanning confocal pinholes. (A hybrid scanning system combining the two first types is also used). Microscopes with all these scanning systems are described in Section 12.2. However, only one of them, the Zeiss (Oberkochen) laser-scan microscope is qualified there as a commercially available instrument. Today, a confocal laser scanning microscope, developed mainly by T. Wilson and C. Sheppard, is also commercially available, from Bio-Rad Lasersharp Ltd., Abingdon, England [E2.14, E2.15]. It uses the first of the scanning systems mentioned above.

The confocal scanning microscope enables thin ($< 1\,\mu$m) sections of a specimen to be recorded without the need for physical slicing. If a series of focused consecutive sections is recorded digitally, then it is possible, with the help of a computer, to display three-dimensional images of the specimen. Stereoscopic images can also be produced by the tandem scanning reflected-light microscope described in Subsection 12.2.3 [E2.16]. It is also worth mentioning the ability of holography to produce true 3-D images of microscopic specimens from a series of two-dimensional sections recorded consecutively on a single holographic plate [E2.17].

The developments summarized above were presented and discussed for the first time at two specialized meetings ("3-D Microscopy" and "Gold Label") and at the Micro-76 Conference ("Industrial and Biomedical"), organized by the Royal Microscopical Society in London, respectively, in April, June and July 1987.

So far, contributions to scanning microscopy have come mainly from four

---

[1] The confocal scanning light microscope (CSLM) is also termed confocal laser scanning microscope (CLSM) or confocal scanning optical microscope (CSOM). The last term is rather unfortunate.

leading European laboratories. Recently, however, a beam-deflection type laser scanning microscope for differential phase contrast images using the split-detector technique has been developed by a group of researchers from Olympus Optical Co., Japan [E2.18]. Real-time scanning (30 images/s) has been realized with acousto-optic deflectors [E2.19].

Today, laser scanning microscopy and VEC microscopy are indeed the "star attractions" of optical (light) microscopy, and are developing into powerful tools both in research and in practice.

References

Chapter 5

[5.1] Zernike F., Das Phasenkontrastverfahren bei der mikroskopischen Beobachtung, *Z. Techn. Phys.*, **16** (1935), 454–457; *Phys. Z.*, **36** (1935), 848–851.

[5.2] Köhler A. and Loos W., Das Phasenkontrastverfahren und seine Anwendung in der Mikroskopie, *Naturwiss.*, **29** (1941), 49–61.

[5.3] Zernike F., Phase contrast, a new method for the microscope observation of transparent objects; in: Bouwers A. (ed.), *Achievements in Optics*, Elsevier Publishing Company, New York and Amsterdam 1946, pp. 116–135.

[5.4] Barer R., A vector theory of phase contrast and interference contrast, *J. Roy. Micr. Soc.* **72** (1952), 10–38, 81–98; **73** (1953), 30–39; **74** (1954), 206–215; **75** (1955), 23–37.

[5.5] Goldstein D. J., A simple quantitative analysis of phase contrast microscopy, not restricted to objects of very low retardation, *J. Microscopy*, **128** (1982), 33–47.

[5.6] Richter R., Eine einfache Erklärung des Phasenkontrastmikroskops, *Optik*, **2** (1947), 342–345.

[5.7] Nebe W., Anwendung verallgeineinerter Phasenkontrastverfahren im Makro- und Microbereich, *Feingerätetechnik*, **30** (1981), 117–119.

[5.8] Wilson T. and Sheppard C. J. R., The halo effect of image processing by spatial frequency filtering, *Optik*, **59** (1981), 19–23.

[5.9] Françon M., *Le microscope à contraste de phase et le microscope interférentiel*, CNRS, Paris 1954.

[5.10] Beyer H., Untersuchungen über den Einfluss der Gestalt der Aperturblende auf die mikroskopische Abbildung beim Phasenkontrastverfahren, *Jenaer Jahrbuch 1953*, Kommissionsverlag VEB Gustav Fischer, Jena 1953, pp. 162–209.

[5.11] Pluta M., Non-standard method of phase contrast microscopy; in: Barer R. and Cosslett V. E. (eds.), *Advances in Optical and Electron Microscopy*, Vol. 6, Academic Press, London and New York 1975, pp. 49–133.

[5.12] Yamamoto K. and Taira A., Some improvements on the phase contrast microscope, *J. Microscopy*, **128** (1982).

[5.13] Gerlovin B. Yu., Dependence between imaging performance of phase contrast devices and the size of their phase ring, *Optiko-mekh. Prom.*, no. 8/1984, 21–24 (in Russian).

[5.14] Bennett A. H., Jupnik H., Osterberg H., and Richards O. W., *Phase Microscopy*, John Wiley and Sons, Inc., New York 1951.

[5.15] Szyjer H. and Pluta M., Simple and accurate method for measuring the sensitivity of phase contrast devices; in: *Optics in Science and Practice, Abstracts of the 1983 European Optical Conference*, Rydzyna, Poland, pp. 128–129.

[5.16] Wilska A., Observations with the anoptral microscope, *Mikroskopie*, **9** (1954), 1–80 (see also: Clark C. L. (ed.), *The Encyclopedia of Microscopy*, Reinhold Publishing Co., New York 1961).

[5.17] Wilska A., A new method of light microscopy, *Nature*, **171** (1953), 353 and 697–698.

[5.18] Pluta M., Optical properties and applications of thin soot layers in phase-contrast and amplitude-contrast microscopy; in: *Proceedings of the Second Colloquium on Thin Films*, Akadémiai Kiadó, Budapest 1967, pp. 257–266.

[5.19] Zernike F.; in: Strong J., *Concepts of Classical Optics*, W. H. Freeman and Co., San Francisco 1958, pp. 525–536.

[5.20] Pluta M., Phase contrast microscope with non-reflecting phase rings, *Kosmos, Seria A: Biologia*, **7** (1958), 587–593 (in Polish).

[5.21] Pluta M., Phase contrast equipment with soot phase rings for the microscopical observation of transparent objects, *Pomiary. Automatyka. Kontrola*, **4** (1958), 459–463 (in Polish).

[5.22] Pluta M., A highly sensitive phase-contrast device, *The Microscope*, **17** (1969), 235–248.

[5.23] Pluta M., Dispositif à contraste de phase ayant des anneaux de phase positifs préparée à partir du noir de fumée et de la substance diélectrique, *Mikroskopie*, **22** (1967), 326–336.

[5.24] Veselý P., Tumour cell surface specialization in the uptake of nutrients evidenced by cinemicrography as a phenotypic condition for density independent growth, *Folia Biologica* (Prague), **18** (1972), 395–401.

[5.25] Veselý P., Malý M., Čumpelík J., Pluta M., and Tuma V., Improved spatial and temporal resolution in an apparatus for time-lapse phase contrast ciné light micrography of cells in vitro, *J. Microscopy*, **125** (1981), 67–76.

[5.26] Ross K. F. A., *Phase Contrast and Interference Microscopy for Cell Biologists*, Edward Arnold, London 1967.

[5.27] Beyer H., *Theorie und Praxis des Phasenkontrastverfahrens*, Akademische Verlagsgesellschaft Geest Portig. K.-G., Leipzig 1965.

[5.28] Pluta M., A phase contrast device with positive and negative image contrast, *J. Microscopy*, **89** (1969), 205–216.

[5.29] Pluta M., Variable phase contrast device, *Pomiary Automatyka Kontrola*, **11** (1965), 33–35 (in Polish).

[5.30] Göke G., Veränderlicher Phasenkontrast, *Mikrokosmos*, **70** (1981), 374–378.

[5.31] Osterberg H., The polanret microscope, *J. Opt. Soc. Am.*, **37** (1947), 726–730.

[5.32] Osterberg H., The theory of measuring unresolvable particles with the phase microscope, *J. Opt. Soc. Am.*, **37** (1947), 523–524.

[5.33] Osterberg H. and Pride G. E., The measurement of unresolved, single particles of uniform thickness by means of variable phase microscopy, *J. Opt. Soc. Am.*, **40** (1950), 64–73.

[5.34] Osterberg H., The measurement of unresolved opaque particles by means of variable phase microscopy; in: Françon M. (ed.), *Contraste de phase et contraste par interférence*, Éditions de la Revue d'Optique, Paris 1952, pp. 227–234.

[5.35] Osterberg H., Phase and interference microscopy; in: Oster G. and Pollister A. W. (eds.) *Physical Techniques in Biological Research*, Vol. 1, Academic Press, New York 1955, pp. 378–437.

[5.36] Richards O. W., The Polanret variable densiphase microscope, *J. Microscopy*, **98** (1973), 67–77.

[5.37] Nomarski G., A variable achromatic phase-contrast microscope. *J. Opt. Soc. Am.*, **58** (1968), 1568 (see also French Pat. 1591113).

[5.38] Pluta M., Single-ring Polanret phase-contrast system, *J. Microscopy*, **148** (1987), 11–19.

[5.39] Pluta M., Simplified Polanret system for microscopy, *Appl. Opt.*, **27** (1988), in press.

[5.40] Beyer H. and Schöppe G., Interferenzeinrichtung für Durchlichtmikroskopie, *Jenear Rundschau*, **10** (1965), 99–105.

[5.41] Beyer H., Auflicht-Interferenzmikroskop Epival-interphako, *Jenaer Rundschau*, **16** (1971), 82–88.

[5.42] Pluta M., A new polarization interference microscope, *The Microscope*, **18** (1970), 113–122.

[5.43] Pluta M., Variable phase-contrast and interference microscopy, *Optik*, **39** (1973), 126–133.

[5.44] Pluta M., Stereoscopic phase contrast microscope, *The Microscope*, **16** (1968), 32–36.

[5.45] Veselý P. and Pluta M., Indication of the surface location of refractive motile spots wreathing the central area of some tissue cells, *Folia Biologica* (Prague), **18** (1972), 374–375.

[5.46] Gruzdev A. D. and Kiknadze I. I., On the connection of polytene chromosomes with the nuclear membrane, *Tsitologiya*, **12** (1970), 919–921 (in Russian).

[5.47] Benford J. R. and Seidenberg R. L., Phase contrast microscopy for opaque specimens, *J. Opt. Soc. Am.*, **40** (1950), 314–316.

[5.48] Popielas M., A phase and amplitude contrast investigation of reflecting materials, *The Microscope*, **20** (1972), 101–110.

[5.49] Ruthmann A., *Methods in Cell Research*, G. Bell and Sons, Ltd., London 1970.

[5.50] Barer R., Refractometry and interferometry of living cells, *J. Opt. Soc. Am.*, **47** (1957), 545–556.

[5.51] Ross F. K. A., The immersion refractometry of living cells by phase contrast and interference microscopy; in: Danielli J. F. (ed.), *General Cytochemical Methods*, Vol. 2, Academic Press, New York 1961, pp. 1–60.

[5.52] Tatekura K., Index profile determination of single-mode fiber by the phase contrast method: a proposed technique, *Appl. Opt.*, **21** (1982), 4260–4263.

[5.53] Bożyk M., Application of phase contrast to the optical fibre refractive profile measurement, *Optica Applicata*, **14** (1984), 31–37.

[5.54] Bell S., Nondestructive refractive index characterization of optical fibres using a phase contrast scanning optical microscope, *Opt. Eng.*, **24** (1985), 518–521.

[5.55] Knight P. and Parsons N., Measurement of the width of thin, cylindrical, transparent objects by phase contrast microscopy, *J. Microscopy*, **137** (1985).

[5.56] Niklowitz W., Über geeignete Präparationsverfahren zur Auswendung der Phasenkontrastmikroskopie in der Histologie; in: *Beiträge zur Licht- und Elektronenmikroskopie*, Verlag Leben im Bild, Dr. Konrad Theiss and Co., Aalen/Württ. 1966, pp. 66–68.

[5.57] Hansen H. G., Rominger A., and Michel K., *Das Phasenkontrastverfahren in der Medizin*, Göttingen 1952.

[5.58] Fröhlich R. O., *Phasenkontrastmikroskopie in der Medizin*, Jena 1955.

[5.59] Franke H., *Phasenkontrasthämatologie*, Stuttgart 1954.

[5.60] Evtodeva M. Ya. *et al.*, *Atlas of Phase Contrast Microscopical Images of Blood Cells of Children*, Izdatel'stvo Meditsina, Moscow 1967 (in Russian).

[5.61] Jenny J., *Die Phasenkontrastmikroskopie in der täglichen Praxis*, Verlag Jenny und Artusi, Schaffhausen 1977.

[5.62] Barer R., Phase contrast and interference microscopy in cytology; in: Pollister A. W. (ed.), *Physical Techniques in Biological Research*, Vol. 3, Pt. A, Academic Press, New York 1966, pp. 1–56.

[5.63] Zollinger H. U., Cytologic studies with the phase microscope, *Am. J. Path.*, **24** (1948), 545–567, 569–589, 797–811, 1039–1053.

[5.64] Silva-Inzunza E., Cytological demonstration of sex in fresh unstained preparations under the phase contrast microscope, *Exp. Cell Res.*, **13** (1947), 405–406.

[5.65] Kruszyński J. and Ostrowski K., Phase contrast microrefractometry of fixed cells and tissues, *Folia Morphologica*, **10** (1959), 391–402 (in Polish).

[5.66] Grzycki S., Refractometry and phase contrast microscopic observations on the transitional epithelium of the rat urinary bladder, *Z. mikr. anat. Forsch.*, **70** (1963), 10–20.

[5.67] Bajer A., Note on the behaviour of spindle fibres at mitosis, *Chromosoma*, **12** (1961), 64–71.

[5.68] Neubert W., Die Kernteilung bei der vegetativen Vermehrung der Bäckerhefe. Lebenbeobachtung im Phasenkontrast, *Mikrokosmos*, **70** (1981), 228–230.

[5.69] Schönbohm E., Untersuchungen zur Mechanik der lichtorientierten Chloroplastenbewegung unter besonderer Berücksichtigung der Plasmastrukturen, *Leitz-Mitt. Wiss. u. Techn.*, **6** (1974), 98–109.

[5.70] Sandström B., Phase microscopy of the endoplasmic reticulum of living cell in tissue cultures, *Mikroskopie*, **26** (1970), 313–316.

[5.71] Singer I. I., Use of peroxidatic-enzyme staining to enhance resolution of cultured mammalian cells under phase microscopy, *Stain Technology*, **50** (1975), 11–17.

[5.72] Barer R., Phase-contrast microscopy of viruses, *Nature*, **162** (1948), 251.

[5.73] Orsi E. V., A phase contrast microscopy chamber for virus-infected tissue cultures, *Exp. Cell Res.*, **20** (1960), 139–149.

[5.74] Spencer E. and Pedersen J., Die Phasenkontrastmikroskopie des Harnsediments, *Leitz-Mitt. Wiss. u. Techn.*, **6** (1974), 110–112.

[5.75] Graf J., Poser B. and Rudolf G., Erste Erfahrungen mit Harnsedimentuntersuchungen nach der Methode von Froreich und im Phasenkontrast, *Leitz-Mitt. Wiss. u. Techn.*, **7** (1977), 18–19.

[5.76] Garnett W. J., *Freshwater Microscopy*, Constable and Co., Ltd., London 1965.

[5.77] Beyer H. and Schöppe G., Die Anwendung der Farbimmersionsmethode in Phasenkontrast und Dunkelfeld bei der Untersuchung von Mineralstauben, *Jenaer Rundschau*, **14** (1969), 228–233.

[5.78] Luster E. A., Phase-microscope technique for refractive index determination of anisotropic particles at high magnification, *The Microscope*, **13** (1963), 363–374.

[5.79] Seyfarth H. H. and Hahne B., Die mikroskopische Phasenanalyse von Festkörpergemengen durch Lichtbrechungsmessungen, *Jenaer Rundschau*, **16** (1971), 237–240.

[5.80] Stoves J. L., *Fibre Microscopy*, National Trade Press, London 1957.

[5.81] Loske T., *Methoden der Textilmikroskopie*, Kosmos, Gesellschaft der Naturfreude Franckh'sche Verlagshandlung, Stuttgart 1963.

[5.82] Beyer H. and Schöppe G., Phasenkontrastmikroskopische Untersuchungen geometrischer und physikalischer Oberflächenstrukturen, *Jenaer Jahrbuch 1965*, VEB Gustav Fischer, Jena 1965, p. 31.

[5.83] Beyer H. and Schöppe G., Theorie und Anwendungen der Farbimmersionsmethode im Dunkelfeld und Phasenkontrast, *Jenaer Jahrbuch 1965*, VEB Gustav Fischer, Jena 1965, pp. 21–31.

[5.84] Dione G., Discrepancies in the number of stacking faults revealed by various methods in epitaxial silicon, *J. Appl. Phys.*, **38** (1967), 3417–3418.

[5.85] Dudley R. H., Non-destructive method for revealing stacking faults in epitaxial silicon, *J. Appl. Phys.*, **35** (1964), 1360–1361.

[5.86] Lenie C. A., Characterization of film defects in silicon epitaxial wafers, *SCP and Solid State Technology*, August 1964.

[5.87] Ogino N., Observation of stacking faults in epitaxially grown Si film with phase contrast microscope, *Japan. J. Appl. Phys.*, **3** (1964), 797–798.

[5.88] Lang de H. and Dekkers N. H., Microscope adapted to the generation of various contrast types: theory and experiment, *Appl. Opt.*, **16** (1977), 2215–2222.

[5.89] Sommargren G. E. and Thompson B. J., Linear phase microscopy, *Appl. Opt.*, **12** (1973), 2130–2138.

[5.90] Thompson B. J., New approaches to phase microscopy, *Proc. of the SPIE*, **104** (1977), 6–15.

[5.91] Sprague R. A. and Thompson B. J., Quantitative visualisation of large variation phase objects, *Appl. Opt.*, **11** (1972), 1469–1479.

[5.92] Khan M. A. W. and Rao V. V., Diffraction images of partially coherent phase discs with variable phase plates, *Atti. Fond. Giorgio Ronchi*, **27** (1972), 521–527.

Chapter 6

[6.1] Pluta M., Properties of the amplitude-contrast microscope with soot amplitude rings, *The Microscope*, **16** (1968), 211–226.

[6.2] Pluta M., Improvement of microscopic observation of insufficiently stained specimens by using an amplitude-contrast device, *Folia Histochemica et Cytochemica* (Cracow), 7 (1969), 27–31.

[6.3] Göke G., Das Amplitudenkontrastverfahren, *Mikrokosmos*, **62** (1973), 344–347.

[6.4] Göke G., Phasenkontrast und Amplitudenkonstrast, *Mikrokosmos*, **65** (1976), 261–262.

[6.5] Popielas M., Phase-contrast and amplitude-contrast microscopy, its applications to the study of opaque specimens, *Optyka*, **6** (1971), 67–75 (in Polish).

[6.6] Nomarski G., Perfectionnements aux instruments optiques pour la projection et la reproduction agrandie ou réduite ainsi que pour l'observation usuelle, Franch Patent No. 1,517,701, Paris 1968.

[6.7] Molesini G., Bertani D. and Cetica M., Dark ground microscopy with detuned interference filters, *Opt. Eng.*, **21** (1982), 1061–1063.

[6.8] Wolter H., Schlieren-, Phasenkontrast- und Lichtschnittverfahren; in: Flügge S. (ed.), *Encyclopedia of Physics*, Vol. 24. Springer-Verlag, Berlin–Gottingen–Heidelberg 1956, pp. 555–645.

[6.9] Garrett H. L. and Traylor P. A., Simple oblique transmitted light techniques for photomacrography and photomicrography, *The Microscope*, **25** (1977), 147–160.

[6.10] Hartley W. G., The usefulness of oblique illumination, *Proc. Roy. Micr. Soc.*, **15** (1980), 422–424.

[6.11] Zselyonka L. and Kiss F., Der dreidimensionale (Super-) Kondensor, *Mikroskopie*, **15** (1960), 3–23.

[6.12] Eahlstron E. E., *Optical Crystalography* (2nd ed.), John Wiley and Sons, New York 1951.

[6.13] Holzgraefe M. and Wolfe J. R., A device for dark-field microphotography at low magnification, *J. Microscopy*, **111** (1977), 225–227.

[6.14] Martin H., Die Anwendung der Dunkelfeldbeleuchtung im Makro- und Mikrobereich bei der Untersuchung von Münzen und Medaillen, *Leitz-Mitt. Wiss. u. Techn.*, **7** (1978), 39–45.

[6.15] Schwalbach G., Die Anwendung von Hellfeld, Phasenkontrast und Dunkelfeld in der histologischen Diagnostik von Leberbiopsien, *Leitz-Mitt. Wiss. u. Techn.*, **7** (1978), 48–52.

[6.16] Birk G., Dunkelfeld-Mikroskopie für schwache Vergrösserungen, *Leitz-Mitt. Wiss. u. Techn.*, **7** (1978), 58–59.

[6.17] Stumm C., de Jong M. H and van der Drift C., Untersuchungen über die freie Beweglichkeit von Bakterien unter Anwendung der Dunkelfeld-Mikroskopie, *Leitz-Mitt. Wiss. u. Techn.*, **7** (1978), 61–63.

[6.18] Hausmann K., Die Dunkelfeld-Langzeitbelichtung: Eine Methode zur Darstellung schneller Bewegungvorgänge, *Mikrokosmos*, **73** (1984), 161–165.

[6.19] Toepler A., Über die Methode der Schlierenbeobachtung als mikroskopisches Hilfsmittel, nebst Bemerkungen zur Theorie der schiefen Beleuchtung, *Poggendorffs Ann.*, **127** (1866), 556 (see also **131** (1867), 180).

[6.20] Skotnikov M. M., *Quantitative Schlieren Methods in Gas Dynamics*, Nauka, Moscow 1976 (in Russian).

[6.21] Appelt H., Farb- und Schwarzweisskontraste mit Hilfe der Schlierenmikroskopie, *Mikrokosmos*, **67** (1978), 76–78.

[6.22] Meyer-Arendt J. R. and Appelt H., Microscopic color schlieren system using a wedge-type interference filter, *Appl. Opt.*, **15** (1976), 2017.

[6.23] Dodd J. G., Observation with a schlieren microscope, *The Microscope*, **17** (1969), 1–14.

[6.24] McCrone W. C., A compact schlieren microscope, *The Microscope*, **20** (1972), 309–318.

[6.25] Dodd J. G. and McCrone W. C., A schlieren eyepiece, *The Microscope*, **23** (1975), 89.

[6.26] Schardin H., Das Schlierenmikroskop und seine Anwendung in der Glastechnik, *Glastech. Berichte*, **27** (1954), 3, 70–76.

[6.27] Abramowitz M., Rheinberg illumination, *International Laboratory*, **13** (1983), 8, 76–80.

[6.28] Taylor R. B., Rheinberg updated, *Proc. Roy. Micr. Soc.*, **19** (1984), 253–256 (see also: MacConail M. A., Double illumination microscopy, *Nature*, **176** (1955), 877).

[6.29] Wilson S. D., A reflection-diffraction microscope for observing diatoms in color, *Appl. Opt.*, **5** (1966), 1683–1684.

[6.30] Crossmon G. C., The dispersion staining method for the selective coloration of tissue, *Stain Technology*, **24** (1949), 61.

[6.31] Schmidt K. G., Phase contrast microscopy and dispersion staining, *Staub*, **41** (1955) 436–467 (see also *Z. Staub*, **18** (1958), 247).

[6.32] Cherkasov Yu. A., *Application of focal screening to the measurement of indices of refraction by the immersion method*, Gos. Nauch.-Tekhn. Izdat., Moscow 1957 (in Russian).

[6.33] Brown K. M. and McCrone W. C., Dispersion staining. Part I: Theory, method and apparatus, *The Microscope*, **13** (1963), 311–322.

[6.34] Brown K. M., McCrone W. C., Kuhn R., and Forlini L., Dispersion staining. Part II: The systematic application to the identification of transparent substances, *The Microscope*, **14** (1963), 39–54.

[6.35] McCrone W. C., Dispersion staining as applied to the study of biological tissue, *J. Roy. Micr. Soc.*, **83** (1964), 217–219.

[6.36] McCrone W. C. and Martin J. S., Identifying colourless transparent particles by microscopy, *Research and Development Magazine*, Nov. 1964, 26–31.

[6.37] Forlini L., Expanded and revised tables for the determination of unknowns by dispersion staining, *The Microscope*, **17** (1969), 29–54.

[6.38] Julian Y. and McCrone W. C., Identification of asbestos fibres by microscopical dispersion staining, *The Microscope*, **18** (1970), 1–10.

[6.39] Forlini L. and McCrone W. C., Dispersion staining of fibres, *The Microscope*, **19** (1971), 243–254.

[6.40] McCrone W. C., Routine detection and identification of asbestos, *The Microscope*, **33** (1985), 273–286.

[6.41] McCrone W. C., Darkfield and dispersion staining, *The Microscope*, **23** (1975), 110–117.

[6.42] McCrone W. C., Determination of n_D, n_F and n_C by dispersion staining, *The Microscope*, **23** (1975), 213–220.

[6.43] McCrone W. C., A new dispersion staining objective, *The Microscope*, **23** (1975), 221–226.

[6.44] *McCrone Dispersion Staining Objective*. A technical publication available from the Mc Crone Research Associates Ltd., Belsize Lane, London NW3 5BG.

[6.45] McCrone W. C., Characterization of human hair by light microscopy, *The Microscope*, **25** (1977), 15–30.

[6.46] McCrone W. C. and Delly J. G., *The Particle Atlas* (2nd ed.), Vols. 1 to 4, Ann Arbor Science Publishers, Ann Arbor 1973.

[6.47] Speight R. G., An alternative dispersion staining technique, *The Microscope*, **25** (1977), 215–225.

[6.48] Wright W. D., *The Measurement of Colour*, Adam Hilger Ltd., Letchworth, Herts. (England) 1969.

[6.49] Dick J. and Franke G., *Haarmikroskopie*, VEB Fachbuchverlag, Leipzig 1974.

[6.50] Crossmon G. C., Dispersion staining microscopy as applied to industrial hygiene, *Am. Industr. Hygiene Assoc. Quarterly*, **18** (1957), 341–344.

[6.51] Crossmon G. C. and Vandemark W. C., Microscopic observations correlating toxicity of beryllium oxide with crystal structure, *A.M.A. Arch. Industr. Hyg. and Occup. Med.*, **9** (1954), 481–487.

[6.52] Crossmon G. C., Some macro and microscopic applications of dispersion staining as applied to industrial hygiene and air pollution, *Microchemical J.*, **10** (1966), 273–285.

[6.53] Hall R. H. *et al.*, Acute toxicity of inhaled beryllium, *A.M.A. Arch. Industr. Hyg. and Occup. Med.*, **2** (1950), 25–48.

[6.54] Goodman R. A., Expanded uses and applications of dispersion staining, *The Microscope*, **18** (1970), 41–50.

[6.55] Skirius S. A., Polymer identification by microscopical dispersion staining, *The Microscope*, **34** (1986), 29–43.

[6.56] Douglass P. C. and Crossmon G. C., Examination of textile fibres by the dispersion staining method, *Textile Research Journal*, **19** (1949), 644–646.

[6.57] Hamer D. H., Identification of crystals in human tissue, *The Microscope*, **15** (1966), 230–238.

[6.58] Crossmon G. C., The dispersion staining method for the selective coloration of tissue, *Stain Technology*, **24** (1949), 61–65.

[6.59] Wiemann M. R. (Jr.) and Besic F. C., Dissolution and recrystallisation in dental enamel, *The Microscope*, **21** (1973), 81–99.

[6.60] Grabar D. G. and Principe A. H., Identification of glass fragments by measurement of refractive index and dispersion, *J. Forensic Sci.*, **8** (1963), 54–67.

[6.61] McCrone W. C., *Fusion Methods in Chemical Microscopy*, Interscience Publishers Ltd., London 1957.

[6.62] Palenik S., *Particle Atlas of Illicit Drugs*, A report from McCrone Associates, issued July, 1974.

[6.63] Kirchgessner W. G. and Gaisser A. R., Application of dispersion staining to fiber and plastics examination and identification, *Textile Research Journal*, **35** (1965), 78–80.

[6.64] Kirchgessner W. G., Comparative dispersion staining using a modified comparison microscope, *The Microscope*, **15** (1967), 511–515.

[6.65] Jones F. T., Dispersion and refractive index determinations by means of an annular wedge interference filter, *The Microscope*, **14** (1965), 440.

[6.66] Jones F. T., Some observations made with a spectrum projected into the plane of the object, *The Microscope*, **20** (1972), 319–326.

[6.67] Pepperhoff W., Gefügeentwicklung durch Interferenz-Aufdampfschichten, *Arch. Eisenhüttenwes*, **32** (1961), 269 and 651.

[6.68] Pepperhoff W. and Ettwig H.-H., *Interferenzschichten-Mikroskopie*, Dr. Dietrich Steinkopff Verlag, Darmstadt 1970.

[6.69] Bühler and Hougardy, *Atlas der Interferenzschichten-Metallographie*, 1979.

[6.70] Hasson R., Metallography of molybdenum in color, *The Microscope*, **16** (1968), 329–334.

[6.71] Hasson R. *et al.*, Microscopic study of the nucleation phenomenon in the fluorine-copper reaction, *The Microscope*, **19** (1971), 307–319.

[6.72] Hasson R., *Metallographie en couleurs*, Commissariat à l'Energie Atomique, Centre d'Etudes Nucléaires de Saclay, Rapport CEA-R-4736 (1976).

[6.73] Kubota H. and Ose T., Further study of polarization and interference colours, *J. Opt. Soc. Am.*, **45** (1955), 89–97.

[6.74] Bescós J. and Straud T. C., Optical pseudocolor encoding of spatial frequency information, *Appl. Opt.*, **17** (1978), 2524–2531.

[6.75] Bartelt H., Spatial frequency coloring and feature coloring, *J. Opt.* (Paris), **12** (1981), 169–172.

[6.76] Nowak R. and Zając M., Pseudocolouring of directional structures by spatial frequency filtering, *Optica Applicata*, **13** (1983), 1, 39–46.

[6.77] Levkovich Yu. I. and Lozovskii V. B., New method of producing colour images in microcinematography of colourless transparent biological objects, *Zhurnal Nauchnoi i Prikladnoi Fotografii i Kinematografii*, **20** (1975), 442–443 (in Russian).

[6.78] Nomarski G., Optical differentiation using oblique coherent illumination and pure amplitude spatial filter, *Einladung und Programm zur 76. Tagung der GDaO*, 20. bis 24. Mai 1975 in Bad Ischl (Austria), p. 80.

[6.79] Hoffman R. and Gross L., Modulation contrast microscope, *Appl. Opt.*, **14** (1975), 1169–1176.

[6.80] Hoffman R. and Gross L., The modulation contrast microscope, *Nature*, **254** (1975), 586–588.

[6.81] Hoffman R., The modulation contrast microscope: principles and performance, *J. Microscopy*, **110** (1977), 205–222.

[6.82] De Santis P. *et al.*, Modulation contrast and coherence theory, *Appl. Opt.*, **15** (1976), 2385–2390.

[6.83] Kingslake R. (ed.), *Applied Optics and Optical Engineering*, Vol. 4, pt. 1: *Optical Instruments*, Academic Press, New York–London 1967.

[6.84] Sojecki A., Stereoscopic eyepiece cap for microscopes, *Optica Applicata*, **3** (1973), 1, 33–34.

[6.85] Göke G., Die stereoskopische Beobachtung mit dem monoobjektiven Mikroskop. Stereo-Mikroskopie bei starker Vergrösserung, *Mikrokosmos*, **64** (1975), 112–115.

[6.86] Lawrence M. J., A fibre optic ring illumination system for use in low-powered dark field (including Rheinberg method) video, cine and photomicrography, *J. Microscopy*, **120** (1980), 65–72.

[6.87] Lawrence M. J., The use of fibre optics in video micrography in BBC Open University Productions, *Proc. Roy. Micr. Soc.*, **15** (1980), 425–429.

[6.88] Resua R. and Petraco N., Fiber optics illumination for use in dispersion staining, *The Microscope*, **28** (1980), 51–55.

[6.89] Goldberg O., An obvious illuminator for dispersion staining, *The Microscope*, **24** (1976), 291–293.

[6.90] Freyer F. E. and Goldberg O., Köhler illumination using fiber optics, *The Microscope*, **26** (1978), 69–71.

Chapter 7

[7.1] Nomarski G., Dispositif interférentiel à polarisation pour l'étude des objects transparents ou opaques appartenant à la classe des objects de phase, French patent, No. 1.059.124 (1953).

[7.2] Nomarski G. and Weill A. R., Application à la métallographie des méthodes interférentielles à deux ondes polarisées, *Rev. Métallurgie*, **50** (1955), 121–134.

[7.3] Allen R. D., David G. B. and Nomarski G., The Zeiss–Nomarski differential interference equipment for transmitted-light microscopy, *Z. wiss. Mikroskopie*, **69** (1969), 193–221.

[7.4] Lessor D. L., Hartman J. S., and Gordon R. L., Quantitative surface topography determination by Nomarski reflection microscopy. I. Theory, *J. Opt. Soc. Am.*, **69** (1979), 357–366.

[7.5] Hartman J. S., Gordon R. L., and Lessor D. L., Quantitative surface topography deter-
       mination by Nomarski reflection microscopy. 2. Microscope modification, calibration,
       and planar sample experiments, *Appl. Opt.*, **19** (1980), 2998–3009.

[7.6] Hartman J. S., Gordon R. L., and Lessor D. L., Development of Nomarski microscopy
       for quantitative determination of surface topography, *Proc. SPIE*, **192** (1979), 223–230.

[7.7] Pluta M., *Phase contrast and Interference Microscopy*, PWN, Warsaw 1965 (in Polish).

[7.8] Wilk J., Localization of the apparent point of beam shearing in birefringent prisms,
       Optyka, **11** (1976), 4, 187–192 (in Polish).

[7.9] Lang W., Differential-Interferenzkontrast-Mikroskopie nach Nomarski. I. Grundlagen
       und experimentelle Ausführung. II. Entstehung des Interferenzbildes. III. Vergleich
       mit dem Phasenkontrastverfahren. IV. Anwendungen, *Zeiss-Informationen*, **16** (1968),
       114–120; **17** (1969), 12–16; **18** (1970), 88–93.

[7.10] Padawer J., The Nomarski interference-contrast microscope. An experimental basis
       for image interpretation, *J. Roy. Micr. Soc.*, **88** (1968), 305–349.

[7.11] Rienitz J., Der Bildcharakter beim differentiallen Interferenzkontrast, *Mikroskopie*, **24**
       (1969), 206–228.

[7.12] Wolf R., Differentieller Durchlich-Interferenzkontrast nach Nomarski mit Planapochro-
       maten, Kondensoren langer Schnittweite und anderen Modifikationen, *Mikroskopie*,
       30 (1974), 17–30.

[7.13] Galbraith W. and David G. B., An aid to understanding differential interference contrast
       microscopy: computer simulation, *J. Microscopy*, **108** (1976), 147–176.

[7.14] Galbraith W., The image of a point of light in differential interference contrast microscopy:
       Computer simulation, *Microscopica Acta*, **85** (1982), 233–254.

[7.15] Bessis M. and Thiéry J. P., Les cellules du sang vues au microscope à interférences (System
       Nomarski), *Rev. Hématology*, **12** (1957), 518–528.

[7.16] Duitschaever C. L., The use of the Nomarski interference system in microscopic studies
       of somatic cells in bovine milk and other body fluids, *Mikroskopie*, **23** (1969), 345–347.

[7.17] Geissinger H. D. and Bond E. F., Nomarski differential interference contrast and scanning
       electron microscopy of tissue section and fibroblast cell culture monolayers, *Mikroskopie*
       27 (1971), 32–39.

[7.18] Geissinger H. D. *et al.*, Qualitative and quantitative Nomarski differential interference
       contrast-fluorescence microscopy on cell culture monolayers, *Mikroskopie*, **27** (1971),
       129–135.

[7.19] Geissinger H. D. and Duitschaever C. L., Nomarski differential-interference contrast
       microscopy in transillumination, its use on unstained or stained sections, smears, and wet
       mounts, or on fluorochromed sections and cell culture monolayers, *J. Microscopy*, **94**
       (1971), 107–124.

[7.20] Geissinger D. *et al.*, Nomarski interference contrast—fluorescence microscopy: a new
       technique designed to improve resolution in fluorescent specimen, *Mikroskopie*, **24** (1970),
       321–326.

[7.21] Derry D. M. and Wolfe L. S., Gangliosides in isolated neurons and glial cells, *Science*,
       158 (1967), 1451–1452.

[7.22] Stockem W., Untersuchungen mit dem Differential-Interferenzkontrast über Morpho-
       logie und Cytosen vom Amoeba proteus, *Mikroskopie*, **24** (1970), 332–344.

[7.23] Netzel H., Differential-Interferenzkontrast-Untersuchungen an Protozoen, *Zeiss-Informationen*, **20** (1973), 57–58.

[7.24] Gundlach H. and Heunert H.-H., Zur Anwendung der Interferenzkontrast-Mikroskopie in der Biologie, *Microscopica Acta*, **76** (1975), 305–315.

[7.25] Bajer A. and Molè-Bajer J., Formation of spindle fibres, kinotechore orientation, and behavior of the nuclear envelope during mitosis endosperm. Fine structural and in vitro studies, *Chromosoma* (Berlin), **27** (1969), 448–484.

[7.26] Bajer A., Fine structure studies on phragmoplast and cell plate formation, *Chromosoma* (Berlin), **24** (1968), 383–417.

[7.27] Bajer A. and Allen R. D., Structure and organization of the living mitotic spindle of Haemanthus endosperm, *Science*, **151** (1966), 572–574.

[7.28] Url W. and Gabler F., Beobachtungen an Pflanzenzellen im differentiellen Interferezkontrast, *Mikroskopie*, **22** (1967), 121–132.

[7.29] Gruber F. *et al.*, Hinweise auf eine Kern-Mantelstruktur in Korn der Kartoffelnknollen-Stärke, *Microscopica Acta*, **72** (1972), 189–200.

[7.30] McQuire A. J., Interference microscopy as a technique for studying the cross field pits of softwoods, *J. Inst. Wood Sciences*, **22** (1969), 48–55.

[7.31] Lee D. R., Arnold D. C. and Fensom D. S., Untersuchung lebender Siebröhren im Nomarski-Interferenzkontrast, *J. Exp. Bot.*, **22** (1971), 25.

[7.32] Serentz M. and Schmidt H., Interferenzmikroskopische Bestimmung des Brechungsindex des Inhalts von Nessel-Kapseln aus den Aktionen von Aiptosia mutabilis, *Microscopica Acta*, **73** (1973), 67–81.

[7.33] Graeves H., Neuartige Anwendungsmöglichkeiten des Photomikroskop II in der Holzforschung, *Zeiss-Informationen*, **19** (1971), 4–7.

[7.34] Howells K. and Evans J., The use of Nomarski differential interference contrast optics for visualization of myofibrils in muscle fibres, *Mikroskopie*, **33** (1977), 64–67.

[7.35] Meyer J., Koch H.-J., and Danz R., Interferenzkontrast nach Nomarski am Vertival—ein neues Verfahrens zur Typisierung von Neisseria gonorrhoeae, *Jenaer Rundschau*, **24** (1979), 165–166.

[7.36] Smith M. W. and Jarvis L. G., Use of differential interference contrast microscopy to determine cell renewal times in mouse intestine, *J. Microscopy*, **118** (1980), 153–159.

[7.37] Horster M. and Gundlach H., Application of differential interference contrast with inverted microscopes to the in vitro perfused nephron, *J. Microscopy*, **117** (1979), 375–379.

[7.38] Mitchell J. R., Hanson R. D., and Fleming W. N., Utilizing differential interference contrast microscopy for evaluating abnormal Spermatozoa, *Proc. of the Seventh Technical Conference on Artificial Insemination and Production*, April 14–15, 1978, Madison, Wisconsin. Ed. by National Association of Animal Breeders, Inc., Columbia, Missouri, pp. 64–68.

[7.39] Zalewski W., Differential interference contrast microscopy of fox spermatoza for artificial insemination, Personal communication, Centre of the Medicine Technology, Warsaw, 1984.

[7.40] Allen R. D., Strömgren Allen N., and Travis J. L., Video-enhanced contrast, differential interference contrast (AVEC-DIC) microscopy: a new method capable of analyzing microtubule-related motility in the reticulpodial network of Allogromia laticollaris, *Cell Motility*, **1** (1981), 291–302.

[7.41] Małecki M., The role of axoplasmatic fibrous structures in changes of exoplasma consistency, Doctoral thesis, available from the Institute of Experimental Biology, Polish Academy of Sciences, Warsaw 1982 (in Polish).

[7.42] Kowaliński S. and Weber J., The application of the polarization-interference microscope in soil micromorphological studies, *Soil Mineralogy*, **12** (1979), 65–69.

[7.43] Piesch E., Anwendung der Interferenzkontrast-Mikroskopie zur Kernspurregistrierung in Festkörpern, *Zeiss-Informationen*, **18** (1970), 58–60.

[7.44] Gribnau T. C. J. and Stumm C., Differentielle Interferenzkontrast-Mikroskopie an polymeren Trägermaterialien für die Affinitätschromatographie, *Leitz-Mitt. Wiss. u. Techn.*, **7** (1978), 80–83.

[7.45] Pampuch R., Librant Z., and Piekarczyk J., Texture and sinterability of MgO powders, *Ceramurgia International*, **1** (1975), 14–18.

[7.46] Haberko M. and Haberko K., Effect of glaze on strength of high-tension porcelain, *Ceramurgia International*, **1** (1975), 28–32.

[7.47] David G. B. and Williamson B. S., Amplitude-contrast microscopy in histochemistry, *Histochemie*, **27** (1971), 1–20.

[7.48] Jaquet P. A., Optische Untersuchung metallischer Oberflächen mit dem Nomarski-Gerät, *Jb. Oberflächentechnik*, **13** (1957), 155–169.

[7.49] Bertocci U. and Noggle T. S., Interference contrast employed to measure slopes on metallographic specimens, *Rev. Sci. Instr.*, **37** (1966), 1750–1751.

[7.50] Gham J., Ein neuer Mikrohärtprüfer. *Beiträge zur Licht- und Elektronenmikroskopie*, Verlag Leben im Bild, Dr. Konrad Theiss und Co., Aalen/Württ. 1966.

[7.51] Gham J., A new microhardness tester, *Zeiss-Informationen*, **37** (1966), 120.

[7.52] Demarq J. and Rosch J, in: Van Heel A. C. S. (ed.), *Advanced Optical Techniques*, North Holland, Amsterdam 1967, p. 393.

[7.53] Budevski E., The use of interference contrast in metallographic cinemicrography, *Res. Film*, **8** (1974), 203–214.

[7.54] Ballingall R. A. and Shersby-Harovic R. B., The measurement of small wedge angles using Nomarski interferometry and its application to small semiconductor samples, *J. Phys. E (Sci. Instr.)* **7** (1974), 345–347.

[7.55] Dowell M. B., Hultman C. A., and Rosenblat G. M., Determination of slopes of microscopic surface features by Nomarski polarization interferometry, *Rev. Sci. Instr.*, **48** (1977), 1491–1497.

[7.56] Hard S. and Nilsson O., Laser heterdyne apparatus for roughness measurements of polished surfaces, *Appl. Opt.*, **17** (1978), 3827–3836.

[7.57] De Korte P. A. J. and Lainé R., Assessment of surface roughness by X-ray scattering and differential interference contrast microscopy, *Appl. Opt.*, **18** (1979), 236–242.

[7.58] Bennett H. E., Proposed use of visible optical techniques to evaluate the surface finish of X-ray optics, *Opt. Eng.*, **19** (1980), 610–615.

[7.59] Church E. L. and Takacs P. Z., Survey of the finish characteristics of machined optical surfaces, *Opt. Eng.*, **24** (1985), 396–403 (see also pp. 388–395).

[7.60] Allen R. D. and Stromgren Allen N., Video-enhanced microscopy with a computer frame memory, *J. Microscopy*, **129** (1983), 3–17.

[7.61] Parpart A. K., Televised microscopy in biological research, *Science*, **113** (1951), 483–484.

[7.62] Flory L. E., The television microscope, *Cold Spring Harbor Symp.*, **16** (1951), 505–509.

[7.63] Inoué S., Video image processing greatly enhances contrast quality, and speed in polarization-based microscopy, *J. Cell Biol.*, **89** (1981), 346–356.

[7.64] Pluta M., Differential interference contrast microscope with continuously variable wavefront shear and pupilar compensation, *Optica Acta*, **19** (1972), 1015–1026.

[7.65] Pluta M., Incident-light double refracting interference microscope with variable wavefront shear, *Optica Acta*, **20** (1973), 625–639.

[7.66] Müller J. and Kozłowski T., Microscope for optical fibre end face control; in: *Optics in Science and Practice, Abstracts of the 1983 European Optical Conference*, Rydzyna, Poland, p. 156.

[7.67] Françon M., Interférences par double refraction en lumière blanche. Nouvelle methode d'observation des object transparents, *Rev. d'Opt.*, **31** (1952), 65–86.

[7.68] Françon M., Interférométrie par polarisation en lumière réfléchie, *Rev. d'Opt.*, **32** (1953), 349.

[7.69] Françon M. and Yamamoto T., Un nouveau et très simple dispositif interférentiel applicable au microscope, *Optica Acta*, **9** (1962), 395–408.

[7.70] Hoffman R. and Gross L., Reflected-light differential-interference microscopy: principles, use and image interpretation, *J. Microscopy*, **91** (1970), 149–172.

Chapter 8

[8.1] Curtis A. S. G., The mechanism of adhesion of cells to glass, a study by interference reflection microscopy, *J. Cell Biol.*, **20** (1964), 199–215.

[8.2] Abercrombie M. and Dunn G. A., Adhesions of fibroblasts to substratum during contact inhibition observed by interference reflection microscopy, *Exp. Cell Res.*, **92** (1975), 57–62.

[8.3] Izzard C. S. and Lochner L. R., Cell-to-substrate contacts in living fibroblasts: an interference reflexion study with an evaluation of the technique, *J. Cell Sci.*, **21** (1976), 129–159.

[8.4] Izzard C. S. and Lochner L. R., Formation of cell-to-substrate contacts during fibroblast motility: an interference-reflexion study, *J. Cell Sci.* **42** (1980), 81–116.

[8.5] Bereiter-Hahn J., Strohmeier R., and Beck K., Bestimmung des Dickenprofils von Zellen mit dem Reflexionskontrastmikroskop, *Leitz-Mitt. Wiss. u. Techn.*, **8** (1983), 147–150.

[8.6] Opas M. and Kalnins V. I., Reflection interference contrast microscopy of microfilaments in cultured cells, *Cell Biol. Int. Rep.*, **6** (1982), 1041–1046.

[8.7] Ambrose E. J., Surface contact microscope for the study of cell movements, *Nature*, **178** (1956), 1194.

[8.8] Ambrose E. J. and Jones P. C. T., Surface-contact microscopy studies in cell movements, *Med. Biol. Illust.*, **11** (1961), 104–110.

[8.9] Opas M. and Kalnins V. I., Surface reflection interference microscopy: a new method for visualizing cytoskeletal components by light microscopy, *J. Microscopy*, **133** (1983), 291–306.

[8.10] Ploem J. S., Reflection-contrast microscopy as a tool for investigation of the attachment of living cells to a glass surface; in: van Furth R. (ed.), *Mononuclear Phagocytes in Immunity, Infection, and Pathology*, pp. 405–421, Blackwell Scientific Publications, Oxford 1975.

[8.11] Patzelt W. J., Reflexionskontrast—Eine neue lichtmikroskopische Technik, *Mikrokosmos*, **66** (1977), 78–81.

[8.12] Patzelt W. J., Reflexionskontrast, ein neues lichtmikroskopisches Verfahren, *Leitz-Mitt. Wiss. u. Techn.*, **7** (1979), 141–143.

[8.13] Pera F., Anwendungsmöglichkeiten der Leitz-Reflexionskontrast-Einrichtung in Histologie und Cytologie, *Leitz-Mitt. Wiss. u. Techn.*, **7** (1979), 147–150.

[8.14] Pera F., Effekte der Reflexionskontrastmikroskopie in gefärbten histologischen, hämatologischen und Chromosomenpräparaten, *Mikroskopie*, **35** (1979), 93–100.

[8.15] Piper J. and Pera F., Rekonstruction des Oberflächenreliefs von Erythrocyten mit Hilfe der Leiz-Reflexionskontrast-Einrichtung, *Leitz-Mitt. Wiss. u. Techn.*, **7** (1980), 230–234.

[8.16] Bereiter-Hahn J., Fox C. H., and Thorell B., Quantitative reflection contrast microscopy of living cells, *J. Cell Biol.* **82** (1979), 767–779.

[8.17] Ploem J. S., Enkele methoden voor toxiciteitsonderozoek bij weefselkweekcellen, Thesis, Amsterdam 1967.

[8.18] Opas M., Interference reflection microscopy of adhesion of Amoeba proteus, *J. Microscopy*, **112** (1978), 215–221.

[8.19] Heavens O. S. and Yuan Y. F., Interference effects in observation of cells, *Phys. Med. Biol.*, **24** (1979), 810–814.

[8.20] Gingell D. and Todd J., Interference reflection microscopy: A quantitative theory for image interpretation and its application to cell-substratum separation measurement, *Biophys. J.*, **26** (1979), 507–526.

[8.21] Gingell D., The interpretation of interference reflection images of spread cells: significant contribution from thin peripheral cytoplasm, *J. Cell Sci.*, **49** (1981), 237–247.

[8.22] Beck K. and Bereiter-Hahn J., Evaluation of reflection interference contrast microscope images of living cells, *Microscopica Acta*, **84** (1981), 153–178.

[8.23] Heath J. P. and Dunn G. A., Cell to substratum contacts of thick fibroblasts and their relation to the microfilament system. A correlated interference-reflection and high-voltage electron-microscope study, *J. Cell Sci.* **21** (1976), 129–159.

[8.24] Cottler-Fox M. *et al.*, The process of epithelial cell attachment to glass surfaces studies by reflection contrast microscopy, *Exp. Cell Res.*, **118** (1979), 414–418.

[8.25] Haemmerli G. and Ploem J. S., Adhesion patterns of cell interactions revealed by reflection contrast microscopy, *Exp. Cell Res.*, **118** (1979), 438–443.

[8.26] Haemmerli G., Sträuli P., and Ploem J. S., Cell-to-substrate adhesions during spreading and locomotion of carcinoma cells. A study by microcinematography and reflection contrast microscopy, *Exp. Cell Res.*, **128** (1980), 249–256.

[8.27] Haemmerli G. and Sträuli P., In vitro motility of cells from human epidermoid carcinomas. A study by phase contrast and reflection contrast cinematography, *Int. J. Cancer*, **27** (1981), 603–610.

[8.28] Haemmerli G., Die Anwendung kombinierter Phasenkontrast- und Reflexionskontrastmikroskopie zur Analyse der zellulären Kontaktbildung und Motilität, *Leit-Mitt. Wiss. u. Techn.*, **8** (1982), 39–42.

Chapter 9

[9.1] Haitinger M., *Fluoreszenz-Mikroskopie*, Akademische Verlagsgesellschaft, Geest und Portig K.-G., Leipzig 1959.

[9.2] Coons A. H., Creech H. J., and Jons R. N., Immunological properies of an antibody containing a fluorescence group, *Proc. Soc. Exp. Biol. Med.*, **47** (1941), 200–202.

[9.3] Coons A. H. and Kaplan N. H., Localization of antigen in tissue cells, *J. Exp. Med.*, **91** (1950), 1–13.

[9.4] Coons A. H., The diagnostic application of fluorescent antibodies, *Schweiz. Z. Allg. Pathol. Bakt.*, **22** (1959), 700–723.

[9.5] Brumberg E. M. and Krylova T. N., Application of dielectric interference mirrors in fluorescence microscopy, *Zhurnal Ob. Biologii*, **14** (1953), 461–464 (in Russian).

[9.6] Brumberg E. M., On fluorescence microscopes, *Zhurnal Ob. Biologii*, **16** (1953), 461–464 (in Russian).

[9.7] Ploem J. S., The use of a vertical illuminator with interchangeable dielectric mirrors for fluorescence microscopy with incident light, *Z. wiss. Mikrosk.*, **68** (1967), 129–142.

[9.8] Ploem J. S., Die Möglichkeit der Auflichtfluoreszenzmethoden bei Untersuchungen von Zellen in Durchströmmungskammern und Leighton-Röhren, *Acta Histochemica*, **7** (1967), 339.

[9.9] Ploem J. S., A study of filters and light sources in immunofluorescence microscopy, *Ann. N.Y. Acad. Sci.*, **177** (1971), 414–429.

[9.10] Ploem J. S., Ein neuer Illuminator-Typ für die Auflicht-Fluorezenzmikroskopie, *Leitz-Mitt. Wiss. u. Techn.*, **4** (1969), 225–238.

[9.11] Coons A. H., Fluorescent antibodies as histochemical tools, *Fed. Proc.*, **10** (1951), 558–559.

[9.12] Coons A. H., Labelled antigens and antibodies, *Ann. Rev. Microbiol.*, **8** (1954), 333–352.

[9.13] Coons A. H., Histochemistry with labelled antibody, *Int. Rev. of Cytology*, **5** (1956), 1–23.

[9.14] Coons A. H., The application of fluorescent antibodies to the study of naturally occurring antibodies, *Ann. N.Y. Acad. Sci.*, **69** (1957), 658–662.

[9.15] Coons A. H., Fluorescent antibody methods, *Gen. Cytochem. Meth.*, **1** (1958), 339–422.

[9.16] Coons A. H., Antibodies and antigens labelled with fluorescein, *Schweiz. Z. Path. Bakt.*, **22** (1959), 693–699.

[9.17] Coons A. H., Immunofluorescence, *Pub. Hlth. Rep.* (*Wash.*), **75** (1960), 937–943.

[9.18] Coons A. H., The beginnings of immunofluorescence, *J. Immunol.*, **87** (1961), 499.

[9.19] Coons A. H., Labelling techniques in the diagnoses of viral diseases, *Bact. Rev.*, **28** (1964), 397–399.

[9.20] Coons A. H. *et al.*, The demonstration of pneumococcal antigen in tissues by the use of fluorescent antibody, *J. Immunol.*, **45** (1942), 159–170.

[9.21] Alexander W. R. M. and Potter J. L., Rhodamine-conjugated papain as counterstain in fluorescent microscopy, *Immunology*, **6** (1963), 450–452.

[9.22] Bereznitskaya E. H., Method of differential abolition of non-specific fluorescence of cocci in smears stained with antipertussis and antiparapertussis fluorescent sera, *Z.N. Microb. Epid. Immunobiol.*, **8** (1964), 71–74.

[9.23] Beutner E. H. (ed.), Defined immunofluorescent staining, *Ann. N. Y. Acad. Sci.*, **177** (1971).

[9.24] Chadwick C. S., McEntegart M. B., and Nairn R. C., Fluorescent protein tracers, a simple alternative to fluorescein, *The Lancet*, **1** (1958), 412–414.

[9.25] Chadwick C. S., McEntegart M. B., and Nairn R. C., Fluorescent protein tracers, a trial of new fluorochromes and the development of an alternative to fluorescein, *Immunology*, **1** (1958), 315–327.

[9.26] Chadwick C. S. and Nairn R. C., Fluorescent protein tracers: the unreacted fluorescent material in fluorescein conjugates and studies of conjugates with other green fluorochromes, *Immunology*, **3** (1960), 363–370.

[9.27] Cherry W. B., Goldmann M., and Carski T. R., *Fluorescent antibody techniques in the diagnosis of communicable diseases*, U.S. Dept. of Health, Atlanta, Georgia 1960.

[9.28] Dandliker W. B. *et al.*, Application of the fluorescence polarization on the antigen-antibody reaction, *Theory and Exp. Method Immunochem.*, **1** (1964), 165–195.

[9.29] Dowdle W. R. and Hansen R. A., A phage-fluorescent antiphage staining system for Bacillus anthracis, *J. Infect. Dis.*, **108** (1961), 125–135.

[9.30] Hachmeister U. and Kracht J., Grundzüge der immunhistologischen Technik, *Microscopica Acta*, **77** (1975), 213–220.

[9.31] Goldmann M., *Fluorescent Antibody Methods*, Academic Press, New York and London 1968.

[9.32] Goldwasser R. A. and Shepard C. C., Staining of complement and modifications of fluorescent antibody procedures, *J. Immunol.*, **80** (1958), 122–131.

[9.33] Holborow E. J. *et al.*, Cytoplasmic localization of "complement-fixing" auto-antigen in human thyroid epithelium, *Brit. J. Exp. Path.*, **40** (1959), 583–588.

[9.34] Huang S.-H. and Neurath A. R., Immunohistologic demonstration of hepatitis B viral antigens in liver with reference to its significance in liver injury, *Lab. Invest.*, **40** (1979), 1–17.

[9.35] Jentsch K. D., *Immunofluoreszenz in der Medizinischen Mikrobiologie, Technik and Anwendung*, Barth. Leipzig 1967.

[9.36] Jaksch G. and Ryvarden G., Sicherung des Immunofluoreszenz-Nachweises von Neiseria gonorrhoeae durch Mischlichtbestrahlung, *Mikroskopie*, **25** (1969), 182–185.

[9.37] Jurd R. D. and Maclean N., Detection of haemoglobin in red cells of Xenopus lavis by immunofluorescent double labelling, *J. Microscopy*, **100** (1974), 213–217.

[9.38] Kaplan M. H., Immunologic studies of heart tissues. I. Production in rabbits of antibodies reactive with an autologous myocardial antigen following immunization with heterologous heart tissues, *J. Immunol.*, **80** (1958), 254–267.

[9.39] Kawanishi H., Hepatitis B antigen in hepatocytes of chronic active liver disease. Studies in organ culture, *Archs. Pathol. Lab. Med.*, **103** (1979), 157–164.

[9.40] Klein P. and Burkholder P., Die Darstellung von fixierten Komplement mit markierten Antikomplement, *Schweiz. Z. Path. Bakt.*, **22** (1959), 729–731.

[9.41] Klotz I. M., Protein interactions; in: Neurath H. and Baily K. (eds.), *The Proteins*, Academic Press, New York 1953, pp. 727–806.

[9.42] Kubica J. F. (ed.), *Immunofluorescence*, PZWL, Warsaw 1967 (in Polish).

[9.43] Lachmann P. J., The reaction of sodium acid with fluorochromes, *Immunology*, **7** (1964), 507–510.

[9.44] Linda H., Application of immunofluorescence in clinical bacteriology, *Proc. of the Conference "Microscopy in Science and Practice"—Polmic' 76*, SIMP-ZORPOT, Warsaw 1976, pp. 189–190 (in Polish).

[9.45] Liu Chien, The use of fluorescent antibody in the diagnosis and study of viral and ricketsial infections, *Ergebnisse der Mikrobiologie, Immunitätsforschung und experimentelle Therapie*, **33** (1966), 242–258.

[9.46] Mayersbach H., Immunhistologische Methoden, *Acta Histochem.*, **4** (1957), 260–275; **10** (1960), 44–60.

[9.47] Mayersbach H., Unspecific interactions between serum and tissue sections in the fluorescent-antibody techniques for tracing antigens in tissues, *J. Histochem. Cytochem.*, **7** (1959), 427.

[9.48] Michalak T. *et al.*, Hepatitis B surface antigen and albumin in human hepatocytes. An immunofluorescent and immunoelectron microscopic study, *Gastroenterology*, **79** (1980). 1151–1158.

[9.49] Michalak T. and Nowosławski A., Crystalline aggregates of hepatitis B core particles in cytoplasm of hepatocytes, *Intervirology*, **17** (1982), 247–252.

[9.50] Nairn R. C. (ed.), *Fluorescent Protein Tracing* (4th ed.), Churchill Livingstone, Edinburgh 1976.

[9.51] Nairn R. C. and Ploem J. S., Moderne Immunofluoreszenzmikroskopie und ihre Anwendung in der klinischen Immunologie, *Leitz-Mitt. Wiss. u. Techn.*, **6** (1974), 91–95.

[9.52] Nairn R. C., Richmond H. G., and Forthergill J. E., Differences in staining of normal and malignant cells by non-immune fluorescent protein conjugates, *Brit. Med. J.*, **2** (1960), 1342–1343.

[9.53] Nakane P. K., Localization of hormones with the peroxidase labelled antibody method, *Meth. Enzymol.*, **37** (1975), 133.

[9.54] Nowosłowski A., Immunofluorescence in studies on the etiopathogenesis of viral diseases, *Proc. of the Conference "Microscopy in Science and Practice"—Polmic' 76*, SIMP-ZORPOT, Warsaw 1976, pp. 165–188 (in Polish).

[9.55] Pearse A. G., *Histochemistry, Theoretical and Applied* (2nd ed.), Churchill, London 1959.

[9.56] Petuely F. and Lindner G., Eine einfache Schnellmethode zur Erkennung von pathgenen Colikeimen in Stuhlausstrichen mit Hilfe von fluoreszierenden Antikörpern (Markierung mit 1-dimethylaminonaphthalinsulfosäure-5 als Fluoreszenzfarbstoff), *Arch. Kinderheilk.*, **158** (1958), 248–252.

[9.57] Poetschke G., Der Nachweis von Viren und Virusantigen mit Hilfe fluoreszenzmarkierter Antikörper, *Progr. Med. Virol.*, **3** (1961), 79–157.

[9.58] Poetschke G., Fluoreszenzmarkierte Antikörper in Forsching und Diagnostik. I. Allgemeines und Methodisches, *Das ärztl. Lab.*, **10** (1964), 353–360.

[9.59] Poetschke G., Grundlagen und Möglichkeiten der Anwendung fluoreszenzmarkierter Antikörper, *Ergebn. Lab.-Med.*, **3** (1967), 194–211.

[9.60] Poetschke G., Fluoreszenzmarkierte Antikörper, *Zeitschr. der Karl-Marx-Universität Leipzig*, **15** (1966), 611–624.

[9.61] Rappaport B. Z., Walker J. M., and Booker B. F., Purification of highly conjugated precipitins and globulins containing reagins with silk hydrolysate, *J. Immunol.*, **93** (1964) 782–791.

[9.62] Riggs J. L. *et al.*, Isothiocyanate compounds as fluorescent labelling agent for immune serum, *Am. J. Clin. Path.*, **34** (1958), 1081–1097.

[9.63] Schiller A. A., Schayer R. W., and Hess E. L., Fluorescein-conjugated bovine albumine— physical and biological properties, *J. Gen. Physiol.*, **36** (1953), 489–506.

[9.64] Shimizu F. *et al.*, Application of double staining technique and incident light fluorescence microscopy to immunofluorescence studies on renal diseases, *Japan. J. Exp. Med.*, **45** (1975), 25–32.

[9.65] Silverstein A. M., Contrasting fluorescent labels for two antibodies, *J. Histochem. Cytochem.*, **5** (1957), 94–95.

[9.66] Sternberger L. A., *Immunocytochemistry*, Englewood Cliffs, Prentice Hall, N.J., 1974.

[9.67] Storch W., Immunfluoreszenzmikroskopie—Anwendungsbeispiele aus der klinischen Immunologie zum Nechweis humoraler Antikörper, *Jenaer Rundschau*, **25** (1980), 107–110.

[9.68] Taylor C. R., Immunoperoxidase techniques, *Arch. Pathol. Lab. Med.*, **102** (1978), 113.

[9.69] Suter-Kopp V. and Geigy R., Bedeutung der Immunfluoreszenzmethode für den Nechweis von parasitären, insbensondere Tropenerkrankungen, *50 Jahre Wild Heerbrugg 1921–1971. Festschrift Mikroskopie*, Herausgegeben von Wild Heerbrugg AG, Heerbrugg/Schweiz 1971, pp. 63–66.

[9.70] Voss H. *et al.*, Die Kopplung von Immunseren mit Fluoresceinisothiocyanat, *Zentralbl. Bakteriol.* **203** (1967), 13–40.

[9.71] Wachsmuth E. D., The localization of enzymes in tissue sections by immuno-histochemistry. Conventional antibody and mixed aggregation techniques, *Histochem. J.*, **8** (1976), 253.

[9.72] Uehleke H., Untersuchungen mit fluoreszenz-markierten Antikörpern, *Schweiz Z. Allg. Path.*, **22** (1959), 724–729.

[9.73] Warweg U., Doppelmarkierung von Antikörpern mit Fluoreszeinisothiocyanat (FITC) und Meerrettichperoxidase (HPOD), *Microscopica Acta*, **87** (1983), 15–18.

[9.74] Weller T. H. and Coons A. H., Fluorescentti antibody studies with agents of varicella and herpens zoster propagated in vitro, *Proc. Soc. Biol. Med.*, **86** (1954), 789–794.

[9.75] Yamada G. and Nakane P. K., Hepatitis B core and surface antigens in liver tissue. Light and electron microscopic localization by the peroxidase-labelled antibody method, *Lab. Invest.*, **36** (1977), 649–659.

[9.76] Holz H. M., *Worthwhile facts about fluorescence microscopy*, Brochure No. JG XII/75 Poo, published by C. Zeiss Oberkochen, 1975.

[9.77] Wachsmuth E. D., Principles of immunocytochemical assays, *Proc. Roy. Micr. Soc.*, **14** (1979), 252–255.

[9.78] Ploem J. S., *Standards for fluorescence microscopy. Conference on Standardization in Immunofluorescence*, London, Oct. 1968.

[9.79] Halbrow H. J. (ed.), *Standardization in Immunofluorescence*, Blackwell Scientific Publications, Oxford–Edinburgh 1970.

[9.80] Jongsma A. P. M., Hijmans W., and Ploem J. S., Quantitative immunofluorescence— standardization and calibration in microfluorometry, *Histochemie*, **25** (1971), 329–343.

[9.81] Ploem J. S., General introduction, *Ann. N.Y. Acad. Sci.*, **254** (1975), 4–20.

[9.82] Koch K.-F., Lichtquellen für die Fluoreszenzmikroskopie. 1. FITC-Immunfluoreszenz, *Leitz-Mitt. Wiss. u. Techn.*, **5** (1971), 146–148.

[9.83] Koch K.-F., Lichtquellen für die Fluoreszenzmikroskopie. 2. Allgemeine Fluoreszenz, *Leitz-Mitt. Wiss. u. Techn.*, **5** (1972), 206.

[9.84] Young M. R. and Armstrong A., Fluorescence microscopy with the quartz-iodine-lamp, *Nature*, **213** (1967), 649.

[9.85] Tomlinson A. H., Tungsten halogen lamps and interference filters for immunofluorescence microscopy, *Proc. Roy. Micr. Soc.*, **7** (1972), 27–37.

[9.86] Fey H. and Braun K., Erfahrungen mit der Quartzjodlampe für Fluoreszenzmikroskopie, *50 Jahre Wild Heerbrugg 1921–1971*, Herausgegeben von Wild Heerbrugg AG, Heerbrugg/Schweitz 1971.

[9.87] Trapp L., Über Lichtquellen und Filter für die Fluoreszenzmikroskopie und über die Auflichtfluoreszenzmethode bei Durchlichtpräparaten, *Acta Histochemica, Suppl. VII* (1967), 327–338.

[9.88] Young M. R., Flash-fluorescence microscopy, *The Microscope*, **17** (1969), 15–18.

[9.89] Richards O. W. and Waters P., A new interference exciter filter for fluorescence microscopy of fluorescein-tagged substances, *Stain Technology*, **42** (1967), 320.

[9.90] Pluta M., Fluorescence equipment for PZO microscopes, *Optyka*, **4** (1969), 17–25 (in Polish).

[9.91] Rygaard J. and Olsen W., Interference filters for improved immunofluorescence microscopy, *Acta Path. Microbiol. Scand.*, **76** (1969), 146–148.

[9.92] Rygaard J. and Olsen W., Toward quantitation of excitation, *Ann. N.Y. Acad. Sci.*, **177** (1971), 430–433.

[9.93] Rygaard J. and Olsen W., Determination of characteristics of interference filters, *Ann. N.Y. Acad. Sci.*, **177** (1971), 430–433.

[9.94] Kraft W., Ein neues FITC-Erregerfilter für die Routinefluoreszenz, *Leitz-Mitt. Wiss. u. Techn.*, **5** (1970), 41–44.

[9.95] Popielas M., Fluorescence microscopy with dichroic mirrors, *Optyka*, **4** (1969), 168–177 (in Polish).

[9.96] Popielas M., Equipment for fluorescence epi-microscopy, *Optyka*, **6** (1971), 173–183 (in Polish).

[9.97] Shiffers L. A., The use of a interference beam-splitter for the study of polarizing fluorescence of microobjects, *Opt.-Mekh. Prom.*, **39** (1972), 11, 3–6 (in Russian).

[9.98] Kraft W., Die Fluoreszenzmikroskopie und ihre gerätetechnischen Anforderungen, *Leitz-Mitt. Wiss. u. Techn.*, **5** (1972), 193–206 (see also **4** (1969), 239–242).

[9.99] Walter F., Eine Auflich-Fluoreszenz für die Routine Diagnose, *Leitz-Mitt. Wiss. u. Techn.*, **4** (1969), 186–187.

[9.100] Kraft W. and Koch K.-F., Ein neuer Mehrwellenlängen-Fluoreszenzilluminator für Forschung und Routine, *Microscopica Acta*, **75** (1974), 249–257.

[9.101] Kraft W., The technology of new fluorescence illumination system, *Mikroskopie*, **31** (1975), 129–146.

[9.102] Otto L., *et al.*, Das neue Fluoreszenzmikroskop Fluoval 2, *Jenaer Rundschau*, **23** (1978), 170–176 (see also **17** (1972), 72–74).

[9.103] Otto L., Einige gerätetechnische Erfahrungen aus der Fluoreszenzmikroskopie, *Acta Histochemica, Suppl. VII* (1967), 345–349.

[9.104] Borkowski W., Equipment for routine fluorescence microscopy—version "dia", *Optyka*, **9** (1974), 166–174 (in Polish).

[9.105] Ploem J. S. and Tanke H. J., *Fluorescence Microscopy*. Oxford University Press and Royal Microscopical Society, Oxford 1985.

[9.106] Storz H., Anmerkungen zu gerätetechnischen Fragen der Fluoreszenzmikroskopie unter besonderer Berücksichtigung der Immunfluoreszenz, *Jenaer Rundschau*, **25** (1980), 106–107.

[9.107] Göke G., Methoden der Durchlicht-Fluoreszenzmikroskopie. 1. Einführung in die Technik und Problematik, *Mikrokosmos*, **73** (1984), 167–174.

[9.108] Bruch H. and Röhler B., Das Laboval 2a-fl; ein neues Auflicht-Fluoreszenzmikroskop für Routine und Forschung, *Jenaer Rundschau*, **24** (1979), 160–164.

[9.109] Reynolds G. T., Image intensification applied to microscope systems; in: Barer R. and Cosslett V. E. (eds.), *Advances in Optical and Electron Microscopy*, Vol. 2. Academic Press, London and New York 1968, pp. 1–40.

[9.110] Storz H., Anmerkungen zur Fluoreszenzmikrofotografie, *Jenaer Rundschau*, **28** (1983) 79–81.

[9.111] Haselmann H. and Wittekind D., Phasenkontrast-Fluoreszenzmikroskopie, *Z. Wiss. Mikroskopie*, **63** (1957), 216–226.

[9.112] Gabler F. and Herzog F., Über die Kombination des Phasenkontrastverfahrens mit der Fluoreszenzmikroskopie, *Appl. Opt.*, **4** (1965), 469–472.

[9.113] Moncel C., Nouveau dispositif pour la microcinématographie vue par vue an fluorescence secondaire et contraste de phase, *Z. wiss. Mikrosk.*, **70** (1970), 23–28.

[9.114] Pluta M., Kombination des Phasenkontrastverfahrens mit der Fluoreszenzmikroskopie mittels aus dünnen Interferenzschichten ausgeführter Kondensor-Ringblenden, *Mikroskopie*, **27** (1971), 121–128.

[9.115] Eger W., Kämmerer H. and Trapp L., Simultane Fluoreszenz- und polarisationsmikroskopische Untersuchungen an unentkalkten Dünnschliffen von Knochengewebe; in: *Beiträge zur Licht- und Elektronenmikroskopie*, Verlag Leben im Bild, Dr. Konrad Theiss and Co., Aalen/Württ. 1966, pp. 24–28.

[9.116] Kaufman G. I., Mester J. F., and Wasserman D. E., An experimental study of lasers as excitation light sources for automated fluorescent antibody instrumentation, *J. Histochem. Cytochem.*, **19** (1971), 469–476.

[9.117] Wayland H., Intravital microscopy; in: Barer R. and Cosselett V. E. (eds.), *Advances in Optical and Electron Microscopy*, Vol. 6, Academic Press, London and New York 1975, pp. 1–47.

[9.118] Harris R. B., The laser in biomedical applications, *Electro-Optical Systems Design*, **5** (1973), 5, 20–24.

[9.119] Steinkamp J. A. *et al.*, A new multiparameter separator for microscopic particles and biological cells, *Rev. Sci. Instr.*, **44** (1973), 1301–1310.

[9.120] Bergquist R., The laser in fluorescent microscopy or watch that lymphocyte! *Electro-Optical Systems Design*, **6** (1974), 7, 24–27.

[9.121] Pluta M., Laser fluorescence microscopy, *Optyka*, **10** (1975), 6–11 (in Polish).

[9.122] Marowsky G., Cornelius G., and Rensing L., Laser-induced fluorescence studies of intact biological membranes, *Opt. Comm.*, **22** (1977), 361–364.

[9.123] Geel van F. *et al.*, Fluorescence microscopy with pulsed nitrogen tunable laser source, *J. Microscopy*, **133** (1983), 141–148.

[9.124] Ware B. R., Fluorescence photobleaching recovery. *International Laboratory*, **14** (1984), 6, 12–21.

[9.125] McGregor G. N. *et al.*, Laser-based fluorescence microscopy of living cells, *Laser Focus/Electro-Optics*, Nov. 1984, 85–93.

[9.126] Axelrod D. *et al.*, Total internal reflection fluorescent microscopy, *J. Microscopy*, **129** (1983), 19–28.

[9.127] Stolyarov V. J. *et al.*, Contact fluorescence rectomicroscope and its application in clinical diagnostics, *Opt.-Mekh. Prom.*, **39** (1972), 4, 24–26 (in Russian).

[9.128] Yakubenas A. V., *et al.*, Contact microscope for medical investigation. *Opt.-Mekh. Prom.*, **38** (1971), 1, 29–34 (in Russian; see also: **38** (1971), 12, 67–68 and **43** (1976), 3, 29–33).

[9.129] Brumberg E. M. and Khomenkova S. A., A simple contact microscope for anatomo-pathological studies. *Opt.-Mekh. Prom.*, **42** (1975), 4, 71–72 (in Russian; see also **38** (1971), 12, 26–30).

[9.130] Kozłowski T., Surgical microscope for fluorescence investigation, *Przegląd Techniki Medycznej*, No. 1/1981, 33–43 (in Polish).

[9.131] Kazanowska W., *et al.*, Usefulness of the Klp2 colposcope for fluorocolposcopic studies, *Gin. Pol.*, 4 (1970), 297–300 (in Polish).

[9.132] Braun K., Fluorescence microscopy in medical research and diagnostics, *Microskopion* (Wild Heerbrugg), No. 8/9, 21–27.

[9.133] Walter F., Fluoreszenzmikroskopie in Biologie und Medizin, *Leitz-Mitt. wiss. u. Techn.*, **5** (1970), 33–40.

[9.134] Thaer A. A. and Sernetz M. (eds.), *Fluorescence Techniques in Cell Biology*, Springer-Verlag, Berlin–Heidelberg–New York 1973.

[9.135] Kohen E. *et al.*, 32-ms scan of the NAD/P/H fluorescence spectrum in single living cells, *Rev. Sci. Instr.*, **44** (1973), 1784–1785.

[9.136] Stockinger L., Vitalfärbung und Vitalfluorochromierung tierischer Zellen, *Protoplasmologia. Handbuch der Protoplasmaforschung*, Vol. 2: *Cytoplasma D1*, Springer-Verlag, Wien 1964.

[9.137] Gruhn I. and Brunnemann H., Die Anwendung des Fluorochroms Akridinorange in der Medizinischen Mikrobiologie, *Jenaer Rundschau*, **18** (1973), 185–187.

[9.138] Grossgebauer K. and Rolly H., Fluorescent, DNA-binding dyes for rapid detection of Chlamydia trochomatis, *Microscopica Acta*, **86** (1982), 1–11.

[9.139] Janowiec M., Pichulowa K., and Pelczarska B., The value of fluorescence method in microbiological diagnostics of tuberculosis, *Proc. of the Conference "Microscopy in Science and Practice"—Polmic' 76*. SIMP-ZORPOT, Warsaw 1976, pp. 182–188 (see also: Selibórska Z. and Zalewska L., The application of the fluorescence microscopy in the laboratory diagnosis of veneral diseases, *ibid.*, p. 191).

[9.140] Faissner W., Losert A., and Steiner E., Beiträge zur Vitalfluorochromierbarkeit von Ciliaten, *Mikroskopie*, **31** (1975), 233–240.

[9.141] Mdzewski B., Fluorescence studies of lysosomes in stimulated and non-stimulated lymphocytic cultures, *Proc. of the Conference Microscopy in Science and Practice—Polmic' 76*, SIMP-ZORPOT, Warsaw 1976, pp. 188–189.

[9.142] Eder H. *et al.*, Fluoreszenzmikroskopische Darstellung eosinophiler Leukocyten mit einer Reihe von Fluorochromen, *Acta Histochem.*, **59** (1977), 308–313.

[9.143] Eder H. and Kayser G., Zur fluoreszenzmikroskopischen Darstellung von Erythrozyten, *Leitz-Mitt. Wiss. u. Techn.*, **6** (1973), 14–17.

[9.144] Hamperl H., Über fluoreszierende Messenchymzellen (Fluorozyten), *Leitz-Mitt. Wiss. u. Techn.*, **4** (1969), 243–246.

[9.145] Lin C. C. and van de Sande J. H., Differential fluorescent staining of human chromosomes with daunomycin and adriamycin, *Science*, **190** (1975), 61–63.

[9.146] Vagner-Capadano A. M. *et al.*, Study of human chromosomes by fluorescent techniques, *Pathol. Biol.*, **23** (1975), 119–132.

[9.147] Hauge M. *et al.*, The value of fluorescence markers in the distinction between maternal and fetal chromosomes, *Humangenetik*, **26** (1975), 187–192.

[9.148] Fabricius H. A., Eine einfache Methode zur Auszählung von Mitosen in Gewebekulturen, *Microscopica Acta*, **71** (1972), 199–201.

[9.149] Kałużewski B. *et al.*, Report of a case of true hermaphroditism with karyotype 45, X/46, XX/46, X, mar/47, XX, mars, *J. Génét. hum.*, **25** (1977), 195–203.

[9.150] Eichler J. and Walter F., Ein Beitrag zur Fluoreszenzmikroskopie des Knochengewebes, *Leitz-Mitt. Wiss. u. Techn.*, **4** (1967), 110–114.

[9.151] Rahn B. A., Die polychrome Fluoreszenzmarkierung des Knochenanbaus, *Zeiss Information*, **22** (1976), 36–39.

[9.152] Schmidt K. H. *et al.*, Fluoreszenzmikroskopische Darstellung unentkalkter Knochenschnitte mit unterschiedlichen Filterkombination, *Leitz-Mitt. Wiss. u. Techn.*, **7** (1979), 235–238.

[9.153] Spiekermann H. and Gehlhar P., Die Osteonenausmessung mit Hilfe des Leitz A.S.M.— ein neues Verfahren bei der Auswertung von Knochenpräparaten in der Fluoreszenzmikroskopie, *Leitz-Mitt. Wiss. u. Techn.*, **7** (1980), 258–261.

[9.154] Pilz W. and Radtke G., Fluoreszenzmikroskopische Untersuchungen zum Dentinwachstrum an Nagezähnen von Ratte und Kaninchen, *Jenaer Rundschau*, **25** (1980), 110–112.

[9.155] Becher H., Fluoreszenzhistochemie beginer Amine, *Mikroskopie*, **26** (1970), 153–168.

[9.156] Mootz W. and Müsebeck K., Fluoreszenzmikroskopischer Katecholaminnachweis, *Leitz-Mitt. Wiss. u. Techn.*, **4** (1969), 247–249.

[9.157] Tchacarof E. *et al.*, Sur les possibilités d'application de réactifs UV-sensibles dans l'histochemie luminescente des lipides. II. Detection de lipides non-saturés, *Mikroskopie*, **27** (1971), 24–31.

[9.158] Polak J. M. and Noorden van S., *An Introduction to Immunocytochemistry*, Oxford University Press and Royal Microscopical Society, Oxford 1984.

[9.159] Moroz P. E. and Kobilinsky L., The detection of quinine in tissue sections by fluorescence microscopy, *The Microscope*, **33** (1985), 37–43.

[9.160] Kho O. and Baer J., Die Fluoreszenzmikroskopie in der botanischen Forschung, *Zeiss-Informationen*, **18** (1970), 54–57.

[9.161] Friedrich W. L. and Schaarschmidt F., Epifluorescence of fossil plants, *Zeiss-Information*, No. 1/1980, 3–4.

[9.162] Nordhorn-Richter G., Primäre Fluoreszenz bei Moosen, *Leitz-Mitt. Wiss. u. Techn.*, **8** (1984), 167–170.

[9.163] Floroskaya V. N. and Ovchinnikova L. I., *Luminescence Microscopy of Bituminous Substances*, Izd. Mosk. Univ., Moscow 1970 (in Russian).

[9.164] Garret H. L., Use of fluorescent dye to reveal spray pattern on plant leaves, *The Microscope*, **33** (1985), 115–120.

[9.165] Blitz F., Lumineszenzuntersuchungen der Energieübertragung im Mikrokristallen, *Leitz-Mitt. Wiss. u. Techn.*, **5** (1970), 85–87.

[9.166] Göke G., Fluoreszenzmikroskopie von Minerallen, *Der Aufschluss*, **21** (1971), 321.

[9.167] Jacob H., Lumineszenz-Mikroskopie der organopetrographischen Bestandteile von Sedimentgesteinen, *Leitz-Mitt. Wiss. u. Techn.*, **4** (1969), 250–254.

[9.168] Mörtel H., Fluoreszenzauflichtmikroskopische Untersuchungen an Dünnschliffen basaltischer Gesteine des Vogelsberger, *Microscopica Acta*, **72** (1972), 55–61.

[9.169] Pauli F., Fluorescence microscopy in organic geochemistry, *Mikroskopie*, **30** (1974), 202–207.

[9.170] Crelling J. C., Current uses of fluorescence microscopy in coal petrology, *J. Microscopy*, **132** (1983), 251–266.

[9.171] Johnston W. D., Jr., Microscopic determination of fluorescence from Al_xGa_{1-x}As-GaAs DH material following microscopic physical damage, *Appl. Phys. Lett.*, **24** (1974), 494–496.

[9.172] Beesley J. E., Colloidal gold: a new revolution in marking cytochemistry. *Proc. Roy. Micr. Soc.*, **20** (1985), 187–196 and 255–256.

[9.173] Lucocq J. M. and Roth J., Colloidal gold and colloidal silver-metallic markers for light microscopical histochemistry; in: Bullock G. R. and Petrusz P. (eds.), *Techniques in Immunocytochemistry*, Vol. 3, Academic Press, London and Orlando (Florida) 1985.

Chapter 10

[10.1] Köhler A., Mikrophotographische Untersuchungen mit ultraviolette Licht, *Z. wiss. Mikr.*, **21** (1904), 129 and 274.

[10.2] Heimann W. Elektonenoptische Bildwandler und Bildverstärker, *Z. Instr.*, **73** (1965), 108.

[10.3] Friel H.-I. and Weise H., Der Bildwandler—ein photoelektrisches Gerät zur bildmässigen Erschliessung des ultravioletten und infraroten Spektralbereichs und zur Bildverstärkung, *Jenaer Rundschau*, **4** (1959), 93.

[10.4] Hurwitz C. E., Detectors for the 1.1 to 1.6 wavelength region, *Opt. Eng.*, **20** (1981), 658–664.

[10.5] Andre Y. V. *et al.*, Infrared video camera at 10 μm, *Appl. Opt.*, **18** (1979), 2607–2608.

[10.6] Mitin V. P. and Muraveyskaya A. A., Correction of the entrance aperture of a photodetector in a scanning IR microscope, *Opt.-Mekh. Prom.*, **46** (1979), 6, 38–39 (in Russian).

[10.7] Arapova E. Ya., Laser projection infrared microscope, *Kvantovaya Elektronika*, **2** (1975), 1568–1570 (in Russian).

[10.8] Brumberg E. M., Colour microscopy in ultraviolet rays, *Nature*, **152** (1943), 357.

[10.9] Swift H., Analytical microscopy of biological materials; in: Wied G. L. (ed.), *Introduction to Quantitative Cytochemistry*, Academic Press, New York and London 1966, pp.1–39.

[10.10] Swift H., The quantitative cytochemistry of RNA; in: Wied G. L. (ed.), *Introduction to Quantitative Cytochemistry*, Academic Press, New York and London 1966, pp. 355–386.

[10.11] Svihla G., Ultraviolettmikroskopische Beobachtungen an Hefen und an Entamoeba invadens; in: *Beiträge zur Licht- und Elektronenmikroskopie*, Verlag Leben im Bild, Dr. Konrad Theiss und Co., Aalen/Württ. 1966, pp. 74–78.

[10.12] Wood J. R. and Goring D. A. I., Ultraviolet microscopy at wavelengths below 240 nm, *J. Microscopy*, **100** (1974), 281–292.

[10.13] Mough T. H., A new microscopic tool for biology, *Science*, **206** (1979), 918–919.

[10.14] Hagins W. A., Near-infrared microscopy, *Science*, **207** (1980).

[10.15] Linskens H. F., Moderne Methoden der mikroskopischen Beobachtung: Das Infrarot-Mikroskop, *Mikrokosmos*, **69** (1980), 219–221.

[10.16] Blank R. and Korn H., Infrared photomicrography of neurosecretory cells, *Zeiss-Information*, No. 1/1980, 5–9.

[10.17] Göke G., Anwendung von Infrarot zur mikroskopischen Untersuchung von Mineralien und Mikrofossilien, *Der Aufschluss*, **22** (1971), 167.

[10.18] Swift E. and Taylor W. R., Bioluminescence and chloroplast movement in the dinoflagellate Pyrocystis Lunula, *J. Phycology*, **3** (1967), 77–81.

[10.19] Lussier F. M., Guide to ir-transmissive materials, *Laser Focus*, Dec. 1976, 47–50.

[10.20] Piotrowski K. and Piotrowski T., IR-microscopes for the study of semiconductor materials, *Optyka* **4**, (1969), 155–160 (in Polish).

[10.21] Bellin P. H. and Zwicker W. K., Observation of surface defects in electrolytically etched silicon by infrared microscopy, *J. Appl. Phys.*, **42** (1971), 1216–1221.

[10.22] Schulz M. and Sarodnik R., Defektstrukturuntersuchungen an GaAs-Einkristallen mit Hilfe der IR-Mikroskopie, *Kristal und Technik*, **8** (1973), 1413–1420.

[10.23] Kawakami T. and Saito H., Defect observations of GaAs-Al$_x$Ga$_{1-x}$ as heterostructures by transmission infrared microscopy, *Japan. J. Appl. Phys.*, **14** (1975), 2073–2074.

[10.24] Lihl F. *et al.*, Polarisationsoptische Untersuchungen as Silizium-Einkristallen zum Nachweis von Versetzungen, *Z. ang. Phys.*, **32** (1971), 287–291.

[10.25] Cohen B. G., Infrared microscopy—some applications to solid state device, *Solid State Techn.*, **17** (1974), 36–40.

[10.26] Poness J., Ein Beitrag zur Infrarotmikroskopie an elektronischen Bausteinen, *Feingerätetechnik*, **22** (1973), 208–210.

[10.27] Enoch R. D. and Lambert R. M., An infrared polarizing microscope for observation of domains in thick samples of magnetic oxides, *J. Phys. E. (Sci. Instr.)*, **3** (1970), 728–730.

[10.28] Sunshine R. A. and Goldsmith N., Infrared transmission microscopy utilizing a high resolution video display, *Solid State Techn.*, **16** (1973), 3, 35–38.

[10.29] Simpson W. A. *et al.*, An infrared microscope system for the detection of internal flows in solids, *Mater. Eval.*, **28** (1970), 205–211.

[10.30] Cohen B. G., Infrared microscopy for evaluation of silicon devices and die-attach bonds, *Proc. of the SPIE*, **104** (1977), 125–131.

[10.31] Leftwich R. F. and Lidback C. A., The infrared microimager and integrated circuits, *Proc. of the SPIE*, **104** (1977), 104–110.

[10.32] Ovtharenko G. M. *et al.*, Scanning IR-microscope, *Opt.-Mekh. Prom.*, **42** (1975), 31–33 (in Russian).

[10.33] Ovtharenko G. M. *et al.*, Improvement of the stability of a IR-microscope, *Opt.-Mekh. Prom.*, **48** (1981), 12, 22–23 (in Russian).

[10.34] Soboleva N. F. and Mitin V. P., Analysis of image distortion in a IR-microscope with a single scanning mirror, *Opt.-Mekh. Prom.*, **42** (1975), 4, 6–7 (in Russian).

[10.35] Nowakowski A. and Szwedowski A., Infrared microscope for non-contact measurement of temperature, *Proc. of the Conference "Microscopy in Science and Practice"—Polmic'76*, SIMP-ZORPOT, Warsaw 1976, pp. 197–199 (in Polish).

[10.36] Shephard G., Todays infrared-reading vidicons map clearer pictures, *Electronics*, **50** (1977), 24, 99–105.

[10.37] Kogan A. V., On the theory of the optical pyrometer, *Optika i Spektroskopiya*, **35** (1973), 555–560 (in Russian).

[10.38] Simon I., *Infrared Radiation*, D. Van Nostrand Company, Inc., Princeton, N.J. 1966.

Chapter 11

[11.1]   Gabor D., A new microscope principle, *Nature*, **161** (1948), 777–778.

[11.2]   Gabor D., Microscopy by reconstructed wave-fronts, *Proc. Roy. Soc.*, **A197** (1949), 454–487.

[11.3]   Gabor D., Microscopy by reconstructed wave-fronts, *Proc. Phys. Soc. (Section B)*, **64** (1951), 449–469.

[11.4]   Bragg W. L., A new type of X-ray microscope, *Nature*, **143** (1939), 678.

[11.5]   Bragg W. L., The X-ray microscope, *Nature*, **149** (1942), 470–472.

[11.6]   Wolfke M., Über die Möglichkeit der optischen Abbildung von Molekulargitter, *Physikalische Zeitschrift*, **21** (1920), 495–497.

[11.7]   Boersch H., Z. techn. Phys., 19 (1938), 337–338.

[11.8]   Shushurin S. F., On the history of holography, *Usp. Fiz. Nauk*, **105** (1971), 145–148.

[11.9]   Stroke G. W., *An Introduction to Coherent Optics and Holography*, 2nd ed., Academic Press, New York and London 1969.

[11.10] Bragg W. L., Microscopy by reconstructed wavefronts, *Nature*, **166** (1950), 399–400.

[11.11] Haine, M. E. and Dyson J., A modification to Gabor's proposed diffraction microscope, *Nature*, **166** (1950), 315–316.

[11.12] Rogers G. L., Gabor diffraction microscopy: the hologram as a generalized zone plate, *Nature*, **166** (1950), 237.

[11.13] Rogers G. L., Experiments in diffraction microscopy, *Proc. Roy. Soc. Edinburgh, Section A*, **63** (1950), 193–221.

[11.14] Bragg W. L. and Rogers G. L., Elimination of unwanted image in diffraction microscopy, *Nature*, **167** (1951), 190–191.

[11.15] Baez A. V., A study in diffraction microscopy with special reference to X-rays, *J. Opt. Soc. Am.*, **42** (1952), 756–762.

[11.16] El-Sum H. M. A., Reconstructed wavefront microscopy, Ph. D. thesis, Stanford Univ., Stanford, California 1952.

[11.17] Kirkpatrick P. and El-Sum H. M. A., Image formation by reconstructed wave-fronts, *J. Opt. Soc. Am.*, **46** (1956), 825.

[11.18] Rogers G. L., Two holograms method in diffraction microscopy, *Proc. Roy. Soc.*, **A64** (1956), 209–221.

[11.19] Baez A. V. and El-Sum H. M. A., in: *Microradiogr. Proc. Symp.*, pp. 347–366, Academic Press, New York 1957.

[11.20] Leith E. N. and Upatnieks J., Microscopy by wavefront reconstruction, *J. Opt. Soc. Am.*, **55** (1965), 981–986.

[11.21] Stroke G. W. and Falconer D. G., Attainment of high resolutions in holography by multi-directional illumination and moving scatterers, *Phys. Lett.*, **15** (1965), 238–240.

[11.22] Van Ligten R. F. and Osterberg H., Holographic microscopy, *Nature*, **211** (1966), 282–283.

[11.23] Snow K. and Vandewarker R., An application of holography to interference microscopy, *Appl. Opt.*, **7** (1968), 549–554.

[11.24] Daszkiewicz M., Pawluczyk R., and Pluta M., A microholographic system enabling to obtain holograms with no pseudoscopic real image, *Acta Phys. Polon.*, **36** (1969), 37–41.

[11.25] Toth L. and Collins S. A., Jr., Reconstruction of a three-dimensional microscopic sample using holographic techniques, *Appl. Phys. Lett.*, **13** (1968), 7–10.

[11.26] Thompson B. J., Ward J. H., and Zinky W. R., Application of hologram techniques for particle size analysis, *Appl. Opt.*, **6** (1967), 519–526.

[11.27] Kopylov G. I. Contribution to the theory of holographic image magnification, *JETP Letters*, **5** (1967), 314–315.

[11.28] Stroke G. W. and Halioua M., Attainment of diffraction-limited imaging in high-resolution electron microscopy by "a posteriori" holographic image sharpening, *Optik*, **35** (1972), 50–65.

[11.29] Stroke G. W. and Halioua M., Image deblurring by holographic deconvolution with partially coherent low contrast objects and application to electron microscopy, *Optik*, **35** (1972), 489–505.

[11.30] Stroke G. W. and Halioua M., Image improvement in high-resolution electron microscopy with coherent illumination low-contrast objects using holographic image-deblurring deconvolution, *Optik*, **37** (1973), 192–203; 250–264.

[11.31] Stroke G. W., Halioua M., Thon F., and Willasch D., Image improvement in high-resolution electron microscopy using holographic image deconvolution, *Optik*, **41** (1974), 319–343.

[11.32] Mróz E., Pawluczyk R., and Pluta M., A method for coherent optical noise elimination in optical systems with laser illumination, *Optica Applicata*, **1** (1971), 2, 9–15.

[11.33] Close D. H., High resolution portable holocamera, *Appl. Opt.*, **11** (1972), 376–383.

[11.34] George N. and Jain A., Speckle in microscopy, *Opt. Comm.*, **6** (1972), 253–257.

[11.35] Pawluczyk R., Coherent noise elimination in holographic microscope, *Opt. Comm.*, **7** (1973), 366–370.

[11.36] Van Ligten R. F., Speckle reduction by simulation of partially coherent object illumination in holography, *Appl. Opt.*, **12** (1973), 255–265.

[11.37] Golbach H., Granulationsverminderung in der holographischen Auflichtmikroskopie, *Optik*, **37** (1973), 45–49.

[11.38] Händler E., Haina D., and Waidelich W., Bildverbesserung in der holographischen Mikroskopie, *Laser + Elektrooptik*, **6** (1974), 4, 46–48.

[11.39] Pawluczyk R., Suppression of coherent noise in holographic microinterferometry, Ph. D. thesis, Central Optical Laboratory, Warsaw 1984 (in Polish).

[11.40] Collier R. J., Burckhardt Ch. B., and Lin L. H., *Optical Holography*, Academic Press, New York and London 1971.

[11.41] Butters J. N., *Holography and Its Technology*, Peter Peregrinus Ltd., London 1971.

[11.42] Abramson N., *The Making and Evaluation of Holograms*, Academic Press, London and New York 1981.

[11.43] Soroko L. M., *Holography and Coherent Optics*, Plenum Press, New York and London 1980.

[11.44] Ginzburg V. M. and Stepanov B. M. (eds.), *Optical Holography—Practical Applications*, Sovetskoe Radio, Moscow 1978 (in Russian).

[11.45] Ostrovskii Yu. I., Butusov M. M., and Ostrovskaya G. W., *Holographic Interferometry*. Izd. Nauka, Moscow 1977 (in Russian).

[11.46] Pluta M. (ed. and co-author), *Optical Holography*, PWN, Warsaw 1980 (in Polish).

[11.47] Leith E. N. and Upatnieks J., New techniques in wavefront reconstruction, *J. Opt. Soc. Am.*, **51** (1961), 1469–1475.

[11.48] Denisyuk Yu. N., On imaging of optical properties of an object in its wave field of scattered radiation, *Dokl. Akad. Nauk SSSR*, **144** (1962), 1275 (in Russian).

[11.49] Smith H. M., Holography, in: Thomas W. (ed.), *SPSE Handbook of Photographic Science and Engineering*, John Wiley and Sons, New York 1973.

[11.50] Leith E. N., Upatnieks J., and Haines K. A., Microscopy by wavefront reconstruction, *J. Opt. Soc. Am.*, **55** (1965), 981–986.

[11.51] Meier R. W., Magnification and third-order aberrations in holography, *J. Opt. Soc. Am.*, **55** (1965), 987–992.

[11.52] Smith H. M., *Principles of Holography*, John Wiley and Sons, New York 1969.

[11.53] Pawluczyk R., Holographic interference microscope, *Optica Applicata*, **2** (1972), 1, 27–34.

[11.54] Ginzburg V. M. *et al.*, Application of holography for the examination of crystals, *Metrologiya*, **9** (1971), 11–14 (in Russian).

[11.55] Ginzburg V. M. *et al.*, MGI-1 holographic microscope, *Prib. i Tech. Eksp.* **3** (1975), 212–214 (in Russian).

[11.56] Dodd J. G., A holographic attachment for the light microscope, *The Microscope*, **25** (1977), 55–63.

[11.57] McFee R. H., Holographic interferometry of birefringent crystals growth from the melt, *Appl. Opt.*, **9** (1966), 1834–1837.

[11.58] Smith R. W. and Williams T. H., A depth encoding high resolution holographic microscope, *Optik*, **39** (1973), 150–155.

[11.59] Cox M. E. and Vahala K. J., Image plane holograms for holographic microscopy, *Appl. Opt.*, **17** (1978), 1455–1457.

[11.60] Smith R. W., Holographic methods in microscopy, *J. Microscopy*, **129** (1983), 29–47.

[11.61] Briones R. A., Heflinger L. O., and Wuerker R. F., Holographic microscopy, *Appl. Opt.*, **17** (1978), 944–950.

[11.62] Jeong T. H. and Snyder H., Holographic microscope system using a triangular interferometer, *Appl. Opt.*, **12** (1973), 146–147.

[11.63] Sato T., Ueda M., and Yamagishi G., Superresolution microscope using electrical superresolution of holograms, *Appl. Opt.*, **13** (1974), 406–408.

[11.64] Courjon D. and Bulabois J., Real time holographic microscopy using a peculiar holographic illuminating system and rotary shearing interferometer, *J. Optics* (Paris), **10** (1979), 125–128.

[11.65] Courjon D. and Bulabois J., Noncoherent microholography using a holographic optical element as a beamsplitter, *Opt. Eng.*, **20** (1981), 233–235.

[11.66] Van Renesse R. L. and Bouts F. A. J., Efficiency of bleaching agents for holography, *Optik*, **38** (1973), 156–168.

[11.67] Pawluczyk R. and Pluta M., Application of laser light for testing the quality of microscope objectives, *Optyka*, **3** (1968), 1, 5–8 (in Polish).

[11.68] Dainty J. C. (ed.), *Laser Speckle and Related Phenomena*, Springer-Verlag, Berlin–Heidelberg–New York 1975.

[11.69] Gerritsen H., Hannan W., and Ramberg E., Elimination of speckle noise in holograms with redundancy, *Appl. Opt.*, **7** (1968), 2301–2311.

[11.70] Gabor D., Information processing with coherent light, *Optica Acta*, **16** (1969), 519–533.

[11.71] Lewis R. W., Redundancy modulation for coherent imaging systems, *Opt. Comm.*, **7** (1973), 22–25.

[11.72] Upatnieks J. and Lewis R. W., Noise suppresion in coherent imaging, *Appl. Opt.*, **12** (1973), 2161–2266.

[11.73] Katti P. K. and Mehta P. C., Noise elimination technique in holography, *Appl. Opt.*, **15** (1976), 530–533.

[11.74] Som S. C. and Budhiraja C. J., Noise reduction by continuous addition of subchannel holograms, *Appl. Opt.*, **14** (1975), 1702–1705.

[11.75] Caulfield H. J., Speckle averaging by spatially multiplexed holograms, *Opt. Comm.*, **3** (1971), 322–323.

[11.76] Martienssen W. and Spiller S., Holographic reconstruction without granulation, *Phys. Lett.*, **A24** (1967), 126–127.

[11.77] Mróz E. and Pawluczyk R., An improvement of the holographic imaging quality by the method of noncoherent superposition of images, *Optica Applicata*, **10** (1980), 205–210.

[11.78] Dainty J. C. and Welford W. T., Reduction of speckle in image plane hologram reconstruction by moving pupils, *Opt. Comm.*, **3** (1971), 289–294.

[11.79] Brandt G. B., Image plane holography, *Appl. Opt.*, **8** (1969), 1421–1429.

[11.80] Withrington R. J., in: Robertson, E. R. and Harvey, J. M (eds.), *Engineering Uses of Holography*, Cambridge Univ. Press, London 1970, p. 267.

[11.81] Rosen L., Focused–image holography with extended sources, *Appl. Phys. Lett.*, **9** (1966), 337.

[11.82] Stroke G. W., White light reconstruction of holographic images, *Phys. Lett.*, **23** (1966), 325.

[11.83] Welford W. T., Obtaining increased focal depth in bubble chamber photography by an application of the hologram principle, *Appl. Opt.*, **5** (1966), 872–873.

[11.84] Klimenko J. S. and Matinyan E. G., On certain features of holograms of focalized images, *Optika i Spektroskopiya*, **28** (1970), 556–560.

[11.85] Pawluczyk R. and Mróz E., Unidirectional optical coherent noise elimination in optical systems with laser illumination, *Optica Acta*, **20** (1973), 379–386.

[11.86] Pawluczyk R., Holographic microinterferometry, in: *Proceedings of the Fifth Soviet*

Union School on Holography, Novosibirsk 1973, Izd. LIYaF, 1975, pp. 314–333 (in Russian).

[11.87] Pluta M. and Pawluczyk R., Evaluation of imagery quality of holographic microinterference system, *76. Tagung DGaO*, 20. bis 24. Mai 1975 in Bad Ischl (Austria).

[11.88] Ginzburg V. M. and Stepanov B. M., *Holographic Measurements*, Radio i Svyaz', Moscow 1981 (in Russian).

[11.89] Ellis J. W., Holomicrography. Transformation of image during reconstruction "a posteriori", *Science*, **154** (1966), 1195–1197.

[11.90] Anderson W. L., Carrier suppresion and restoration in hologram microscopy, *J. Opt. Soc. Am.*, **59** (1969), 96–103.

[11.91] Tsuruta T. and Itoh Y., Hologram schlieren and phase-contrast methods, *Japan. J. Appl. Phys.*, **8** (1969), 96–103.

[11.92] Knox C., Holographic microscopy as a technique for recording dynamic microscopic subjects, *Science*, **153** (1966), 989–990.

[11.93] Knox C. and Brooks R. E., Holographic motion-picture microscopy, *J. Soc. Motion Picture Telev. Engrs.*, **79** (1970), 594–598.

[11.94] Felappa E., Jr., Holographic motion-induced-contrast images. *J. Cell Biology*, **40** (1969), 838–842 (see also *Phys. Today*, **22** (1969), 25).

[11.95] Cox M. E., Buckles R. G., and Whitlow D., Cineholomicroscopy of small animal microcirculation, *Appl. Opt.*, **10** (1971), 128–131.

[11.96] Evans A. E., Quantitative reconstruction and superresolution of red-blood-cell image holograms, *J. Opt. Soc. Am.*, **61** (1971), 991–997.

[11.97] Cox M. E., Holographic microscopy—a review, *The Microscope*, **19** (1971), 137–149.

[11.98] Cox M. E., Holographic microscopy—a reassessment, *The Microscope*, **22** (1974), 361–366.

[11.99] Sokolov A. V., *Applications of Holographic Microscopy Methodes to Study Biological Microobjects*, Nauka, Leningrad 1978 (in Russian).

[11.100] Opas M., Holographic microscopy of glycerination of Amoeba proteus, *J. Microscopy*, **112** (1978), 301–305 (see also *Acta Protozoologica*, **15** (1976), 485–499).

[11.101] Baranowski Z., Three-dimensional analysis of movement in Physrum polycephalum plasmodia, *Cytobiology*, 13 (1976), 118–131.

[11.102] Heflinger L. O., Stewart G. L., and Booth C. R., Holographic motion pictures of microscopic plankton, *Appl. Opt.*, **17** (1978), 951–954.

[11.103] Carder K. L., Holographic microvelocimeter for use on studying ocean particle dynamics, *Opt. Eng.*, **18** (1979), 524–525.

[11.104] Rosen A. N., Holographic fundoscopy with fibre optic illumination, *Optics and Laser Technology*, **7** (1975), 127–129.

[11.105] Tokuda A. R., Auth D. C., and Bruckner A. P., Holocamera for 3-D micrography of the alert human eye, *Appl. Opt.*, **19** (1980), 2219–2225.

[11.106] Sharnoff M., Microdifferential holography, *J. Opt. Soc. Am.*, **A2** (1985), 1619–1628.

[11.107] Bedaria F. and Pontiggi C., An application of holography in reflection microscopy, *Acta Crystallogr.*, **24** (1968), 614–615.

[11.108] Rhodes M. B. and Cournoyer R. F., Interferometry with a holographic microscope, *Proc. SPIE*, **104** (1977), 21–28.

[11.109] Pawluczyk R., Kibalczyc W., and Sokołowski T., Application of holographic interference microscope in the investigation of crystal dissolution, *Optica Applicata*, **16** (1986), 25–33.

[11.110] Magill P. and Wilson A. D., Applications of a holographic interference microscope, *J. Appl. Phys.*, **39** (1968), 4717–4725.

[11.111] Van Ligten R. F. and Lawton K. C., Holographic microscopy and integrated circuit inspection, *Ann. N.Y. Acad. Sci.*, **168** (1970), 510–535.

[11.112] Pierattini G., Real-time and double-exposure microholographic interferometry for observing the dynamics of phase variations in transparent specimens, *Opt. Comm.*, **5** (1972), 41–44.

[11.113] Presby H. M., Time-resolved differential holographic microscopy, *Appl. Phys. Lett.*, **21** (1972), 31–32.

[11.114] Rhodes M. B. and Cournoyer R. F., Holographic interferometric microscopy applied to the study of diffusion mechanisms in polymeric systems, *J. Opt. Soc. Am.*, **64** (1974), 1381.

[11.115] Ribbens W. B., Surface roughness meseurement by holographic interferometry, *Appl. Opt.*, **11** (1972), 807–810.

[11.116] Karger A. M. and Holeman J. M., Microscopic holography of small parts, *Appl. Opt.*, **11** (1972), 1641–1647.

[11.117] Attwood D. T., Coleman L. W., and Sweeney D. W., Holographic microinterferometry of laser-produced plasmas with frequency-tripled probe pulses, *Appl. Phys. Lett.*, **26**, (1975), 616–618.

[11.118] Fedosejevs R. and Richardson M. C., Subnanosecond microscopic holographic interferometry of plasmas by 1-nsec CO_2 laser pulses, *Appl. Phys. Lett.*, **27** (1975), 115–117.

[11.119] Pierce E. L., Designing a probe beam and an ultraviolet holographic microinterferometer for plasma probing, *Appl. Opt.*, **19** (1980), 852–861.

[11.120] Webster J. M., The application of holography as a technique for size and velocity analysis of high velocity droplets and particles, *J. Photogr. Sci.*, **19** (1971), 38–44.

[11.121] Stasel'ko D. J. and Kosnikovskii V. A., Holographic recording of three-dimensional media of high velocity particles, *Optika i Spektroskopiya*, **34** (1973), 365–374 (in Russian).

[11.122] Stasel'ko D. J., Properties of holographic recording of high speed processes by using a pulsed ruby laser; in: Denisyuk Yu. N. and Ostrovskii Yu. J. (eds.), *Optical Holography*, Izd. Nauka, Leningrad 1975, pp. 4–70 (in Russian).

[11.123] Thompson B. J., Holographic particle sizing techniques, *J. Phys. E (Sci. Instr.)*, **7** (1974), 781–788.

[11.124] Royer H., Holographic velocimetry of submicron particles, *Opt. Comm.*, **20** (1977), 73–75.

[11.125] Moroz E. V., Presnyakov Yu. P., and Tsarfin V. Ya., Errors in size measurements of droplets using holographic methods, *Izmer. Tekhn.*, 1978, 4, 36–37 (in Russian).

[11.126] Ewan B. C. R., Particle velocity distribution measurement by holography, *Appl. Opt.*, **18** (1979), 3156–3166.

[11.127] Dunn P. and Walls J. M., Improved microimages from in-line absorption holograms, *Appl. Opt.*, **18** (1979), 263–264.

[11.128] Ewan B. C. R., Fraunhofer plane analysis of particle field holograms, *Appl. Opt.*, **19** (1980), 1368–1372.

[11.129] Grabowski W., Measurement of the size and position of aerosol droplets using holography, *Optics and Laser Technology*, **15** (1983), 199–205.

[11.130] Malyak P. H. and Thompson B. J., Particle displacement and velocity measurement using holography, *Opt. Eng.*, **23** (1984), 567–576.

[11.131] Wuerker R. F. and Hill D. A., Holographic microscopy, *Opt. Eng.*, **24** (1985), 480–484.

[11.132] Koechner W., Holography's pulses, *Industrial Research*, April 1973, 44–48.

[11.133] Bexon R. *et al.*, In-line holography and the assessment of aerosols, *Optics and Laser Technology*, **8** (1976), 161–165.

[11.134] Witherow W. K., A high resolution holographic particle sizing system, *Opt. Eng.*, **18** (1979), 249–255.

[11.135] Belz R. A. and Menzel R. W., Particle field holography at Arnold Engineering Development Center, *Opt. Eng.*, **18** (1979), 256–265.

[11.136] Ward J. H. and Thompson B. J., In-line hologram system for bubble chamber recording, *J. Opt. Soc. Am.*, **57** (1967), 275–276.

[11.137] Zachara B., Kraska Z., and Pawluczyk R., Patent Application, OWP-1-P/1389, 1980 (in Polish).

[11.138] Zachara B., Holographic Oil-Mist Droplet Sizing, Ph. D. Thesis, Technical University of Mining and Metallurgy, Cracow 1982 (in Polish).

[11.139] Albe F., Royer H., and Smigielski P., Microholographie ultra-rapide, *Optik*, **39** (1974), 185–194.

[11.140] Royer H., Une application de la microholographie ultra-rapide. La métrologie des brouillards, *Nouv. Rev. Opt.*, **5** (1974), 87–93.

[11.141] Kozikowska A., Investigation of Spatial Distribution of Fog Droplets by a Holographic Method, Ph. D. Thesis, Polish Academy of Sciences, Institute of Geophysics, Warsaw 1981 (in Polish).

[11.142] Belz R. A. and Dougherty N. S., Jr., In-line holography of reacting liquid sprays; in: *Proc. of the SPIE "Engineering Application of Holography"*, Los Angeles, California 1972, pp. 209–218.

[11.143] Jones A. R. *et al.*, Application of in-line holography to drop size measurement in dense fuel sprays, *Appl. Opt.*, **17** (1978), 328–330.

[11.144] Trolinger J. D. and Heap M. P., Coal particle combustion studied by holography, *Appl. Opt.*, **18** (1979), 1757–1762.

[11.145] Thompson B. J. and Zinky W. R., Holographic detection of submicron particles, *Appl. Opt.*, **7** (1968), 2426–2428.

[11.146] Royer H., L'utilisation de la microholographie dans les chambres à bulles, *J. Optics (Paris)*, **12** (1981), 347–350.

[11.147] Budziak A. *et al.*, Laser streamer chamber—high presure targets with small corrective admixtures to filling gas, *European Optical Conference*, Rydzyna, Poland 1983 (in *Abstracts Volume*, p. 151).

[11.148] Tschudi T., Herziger G., and Engel A., Particle size analysis using computer-synthesized holograms, *Appl. Opt.*, **13** (1974), 245–248.

[11.149] Stanton A. C., Caulfield H. J., and Stewart G. W., An approach for automated analysis of particle hologram, *Opt. Eng.*, **23** (1984), 577–582.

[11.150] Caulfield H. J., Automated analysis of particle holograms, *Opt. Eng.*, **24** (1985), 462–463.

[11.151] Vikram C. S. and Billet M. L., Holographic image formation of objects inside a chamber, *Optik*, **61** (1982), 427–432.

[11.152] Vikram C. S., Far-field holography of non-image planes for size analysis of small particles, *Appl. Phys. B* (Springer), **33** (1984), 149–153.

[11.153] Hess C. F. and Trolinger J. D., Particle field holography data reduction by Fourier transform analysis, *Opt. Eng.*, **24** (1985), 470–475.

Chapter 12

[12.1] Rabinowitz P. and Chimenti R., Superradiant-illuminator projector, *J. Opt. Soc. Am.*, **60** (1970), 1577.

[12.2] Zemskov K. I. *et al.*, Investigation of principal characteristics of the laser projection microscope, *Kvantovaya Elektronika*, **3** (1976), 35–43 (in Russian).

[12.3] Zemskov K. I. *et al.*, A transmitted-light laser projection microscope, *Kvantovaya Elektronika*, **6** (1979), 2473–2475 (in Russian).

[12.4] Bunkin F. V. *et al.*, Projection system with brightness amplifier for biology and medicine, *Microscopica Acta*, **82** (1979), 229–233.

[12.5] Savransky V. V. and Sitnikov G. A., Laser projection microscope principle and applications in biology; in: von Bally G. and Greguss P. (eds.), *Optics in Biomedical Sciences*, Springer-Verlag, Berlin–Heidelberg–New York 1982, pp. 21–23.

[12.6] Bunkin F. V. *et al.*, Power self-control and formation of the negative image in an illuminating beam of the laser projection microscope, *Kvantovaya Elektronika*, **8** (1981), 1372–1373 (in Russian).

[12.7] Zemskov K. I. *et al.*, A study of characteristics of negative images in optical systems with copper-vapour intensity amplifiers, *Kvantovaya Elektronika*, **10** (1983), 2278–2283 (in Russian).

[12.8] Roberts F. and Young J. Z., A flying spot microscope, *Nature*, **167** (1951), 231.

[12.9] Montgomery P. O'B., Roberts F., and Bonner W., The flying spot monochromatic ultraviolet television microscope, *Nature*, **177** (1956), 1172 (see also Clark G. L. (ed.), *The Encyclopedia of Microscopy*, Reinhold Publishing Co., New York 1961, pp. 334–338).

[12.10] Box H. C. and Freund H., Flying-spot microscope adapted for quantitative measurements, *Rev. Sci. Instr.*, **30** (1959), 28–30.

[12.11] Cosslett V. E., *Modern Microscopy*, Bell, London 1966.

[12.12] Gregory R. L. and Donaldson P. E. K., A solid image microscope, *Nature*, **182** (1958), 1434.

[12.13] King M. C. and Berry D. H., Vertical mirror technique for video transmission of three-dimensional images, *Appl. Opt.*, **9** (1970), 2035–2039.

[12.14] King M. C. and Berry D. H., A depth scanning microscope, *Appl. Opt.*, **10** (1971), 208–209.

[12.15] Egger M. D. and Petran M., New reflected-light microscope for viewing unstained brain and ganglion cells, *Science*, **157** (1967), 305–307.

[12.16] Petran M. *et al.*, Tandem-scanning reflected-light microscope, *J. Opt. Soc. Am.*, **58** (1968), 661–664.

[12.17] Davidovits P. and Egger M. D., Scanning laser microscope for biological investigations, *Appl. Opt.*, **10** (1971), 1615–1619 (see also *Nature*, **223** (1969), 831).

[12.18] Whitman R. L., Increased depth of focus in scanning microscope, *Proc. of the SPIE*, **104** (1977), 59–69.

[12.19] Black J. F., Summers C. J. and Sherman B., Scanned-laser microscope for photoluminescence studies, *Appl. Opt.*, **11** (1972), 1553–1562.

[12.20] Sawatari T., Optical heterodyne scanning microscope, *Appl. Opt.*, **12** (1973), 2768–2772.

[12.21] Nomarski G., Simple method for reducing the depth of focus, *J. Opt. Soc. Am.*, **65** (1975), 1166.

[12.22] Hadni A. *et al.*, Laser scanning microscope for pyroelectric display in real time, *Appl. Opt.*, **15** (1976), 2150–2158.

[12.23] Sheppard C. J. R. and Choudhury A., Image formation in the scanning microscope, *Optica Acta*, **24** (1977), 1051–1073.

[12.24] Sheppard C. J. R., The use of lenses with annular aperture in scanning optical microscopy, *Optik*, **48** (1977), 329–334.

[12.25] Sheppard C. J. R. and Wilson T., Image formation in scanning microscopes with partially coherent source and detector, *Optica Acta*, **25** (1978), 315–325.

[12.26] Sheppard C. J. R. and Kompfner R., Resonant scanning optical microscope, *Appl. Opt.*, **17** (1978), 2879–2882.

[12.27] Sheppard C. J. R. and Wilson T., The theory of scanning microscopes with Gaussian pupil functions, *J. Microscopy*, **114** (1978), 179–198.

[12.28] Cremer Ch. and Cremer Th., Considerations on a laser-scanning-microscope with high resolution and depth of field, *Microscopica Acta*, **81** (1978), 31–44.

[12.29] Brakenhoff G. J., Blom P., and Barends P., Confocal scanning light microscopy with high aperture immersion lenses, *J. Microscopy*, **117** (1979), 219–232 (see also *Optica hoy y mañana*, *Proc. of the ICO-11*, Madrid 1978, pp. 215–218).

[12.30] Brakenhoff G. J., Imaging modes in confocal scanning light microscopes, *J. Microscopy*, **117** (1979), 232–242 (see also *Optica hoy y mañana*, *Proc. of the ICO-11*, Madrid 1978, pp. 219–222).

[12.31] Sheppard C. J. R. and Wilson T., Effect of spherical aberration on the imaging properties of scanning optical microscopes, *Appl. Opt.*, **18** (1979), 1058–1063.

[12.32] Ash E. A. (ed.), *Scanned Image Microscopy*, Academic Press, London 1980.

[12.33] Wilson T., Imaging properties and applications of scanning optical microscopes, *Appl. Phys.* (Springer), **22** (1980), 119–128.

[12.34] Sheppard C. J. R. and Wilson T., Fourier imaging of phase information in scanning and conventional microscopes, *Phil. Trans. Roy. Soc. London*, **A295** (1980), 513–536.

[12.35] Sheppard C. J. R. and Wilson T., Multiple traversing of the object in the scanning microscope, *Optica Acta*, **27** (1980), 611–624.

[12.36] Sheppard C. J. R. and Wilson T., Image formation in confocal scanning microscopes, *Optik*, **55** (1980), 331–342.

[12.37] Wilson T. and Gannaway J. N., The imaging of periodic structures in partially coherent and confocal scanning microscopes, *Optik*, **54** (1979), 201–210.

[12.38] Sheppard C. J. R. and Wilson T., The theory of the direct-view confocal microscope, *J. Microscopy*, **124** (1981), 107–117.

[12.39] Cox I. J., Sheppard C. J. R., and Wilson T., Improvement in resolution by nearly confocal microscopy, *Appl. Opt.*, **21** (1982), 778–781.

[12.40] Cox I. J., Scanning optical fluorescence microscopy, *J. Microscopy*, **133** (1983), 149–154.

[12.41] Cox I. J. and Sheppard C. J. R., Scanning optical microscope incorporating a digital frame store and microcomputer, *Appl. Opt.*, **22** (1983), 1474–1478.

[12.42] Hamilton D. K. and Wilson T., Three-dimensional surface measurement using the confocal scanning microscope, *Appl. Phys. B*, **27** (1982), 211–213.

[12.43] Deinet W. *et al.*, Laser scanning microscope with automatic focusing, *Microscopica Acta*, **87** (1983), 129–138.

[12.44] Hamed A. M. and Clair J. J., Image and super-resolution in optical coherent microscopes, *Optica Applicata*, **13** (1983), 215–222.

[12.45] Hamed A. M., Compromising resolution and contrast in confocal microscopy using a novel obstruction of circular pupils, *Optica Applicata*, **13** (1983), 265–271.

[12.46] Clair J. J. and Hamed A. M., Theoretical remarks on optical coherent microscopes, *Optica Applicata*, **13** (1983), 141–148 (see also *Optik*, **64** (1983), 133–141).

[12.47] Hamilton D. K. and Sheppard C. J. R., Differential phase contrast in scanning optical microscopy, *J. Microscopy*, **133** (1984), 27–39.

[12.48] Jungerman R. L., Hobbs P. C. D., and Kino G. S., Phase sensitive scanning optical microscope, *Appl. Phys. Lett.*, **45** (1984), 846–848.

[12.49] Magiera A. and Magiera L., Remarks on point spread function in confocal scanning microscope with apodized pupil, *Optica Applicata*, **15** (1985), 107–110.

[12.50] Hamilton D. K. and Wilson T., Two-dimensional phase imaging in the scanning optical microscope, *Appl. Opt.*, **23** (1984), 348–352.

[12.51] Wilson T. and Hamilton D. K., Difference confocal scanning microscopy, *Optica Acta*, **31** (1984), 453–465.

[12.52] Mendez E. R., Speckle statistics and depth discrimination in the confocal scanning optical microscope, *Optica Acta*, **32** (1985), 209–221.

[12.53] Bouwhuis G. and Dekkers N. H., Ultramicroscopy in scanning microscopes, *Optik*, **56** (1980), 233–242.

[12.54] Wilson T. and Sheppard C. J. R., *Theory and Practice of Scanning Optical Microscopy*, Academic Press, London 1983.

[12.55] Petran M. *et al.*, The tandem scanning reflected light microscope. Part 1—The principle, and its design, *Proc. Roy. Micr. Soc.*, **20** (1985), 125–129.

[12.56] Boyde A., The tandem scanning reflected light microscope. Part 2—Pre-Micro'84 applications at UCL, *Proc. Roy. Micr. Soc.*, **20** (1985), 131–139.

[12.57] Boyde A., Petran M., and Hadravsky M., Tandem scanning reflected light microscopy of internal features in whole bone and tooth samples, *J. Microscopy*, **132** (1983), 1–7.

[12.58] Boyde A., Stereoscopic images in confocal (tandem scanning) microscopy, *Science*, **230** (1985), 1270–1272.

[12.59] Howard V. *et al.*, Unbiased estimation of particle density in the tandem scanning reflected light microscope, *J. Microscopy*, **138** (1985).

[12.60] Svishchyov G. M., Microscope for the study of transparent highly scattering objects in incident light, *Optika i Spektroskopiya*, **26** (1969), 313–315 (in Russian).

[12.61] Svishchyov G. M., Image contrast in a microscope with synchronous scanning of sample by slit field diaphragms, *Optika i Spektroskopiya*, **30** (1971), 349–355; Image contrast in a microscope with synchronous scanning of the object by point or raster field diaphragm, *ibid.*, 34 (1973), 393–400 (in Russian).

[12.62] Bogdankevich O. V. *et al.*, Scanning optical microscope, *Izm. Tehk.*, No. 12/1978, 30–32.

[12.63] Fujii Y., Takimoto H., and Igarashi T., Optimum resolution of laser microscope by using optical heterodyne detection, *Opt. Comm.*, 38 (1981), 85–90.

[12.64] McMahon R. E., Laser tests ICs with light touch, *Electronics*, 8 (1971), 92–95.

[12.65] Wilson T., Gannaway J. N., and Sheppard C. J. R., Optical fibre profiling using a scanning optical microscope, *Optical and Quantum Electronics*, 12 (1980), 341–345.

[12.66] Hellwarth R. and Christensen P., Nonlinear optical microscopic examination of structure in polycrystalline ZnSe, *Opt. Comm.*, 12 (1974), 318–322.

[12.67] Hellwarth R. and Christensen P., Nonlinear optical microscope using second harmonic generation, *Appl. Opt.*, 14 (1975), 247–248.

[12.68] Gannaway J. N. and Sheppard C. J. R., Second-harmonic imaging in the scanning optical microscope, *Optical and Quantum Electronics*, 10 (1978), 435–439.

[12.69] Wilson T. and Sheppard C. J. R., Imaging and super resolution in the harmonic microscope, *Optica Acta*, 26 (1979), 761–770.

[12.70] Gannaway J. N. and Wilson T., Imaging properties of the scanning harmonic optical microscope, *Proc. Roy. Micr. Soc.*, 14 (1979), 170–174.

[12.71] Delhaye M. and Dhamelincourt P., Raman microprobe and microscope with laser excitation, *J. Raman Spectr.*, 3 (1975), 33–43.

[12.72] Dhamelincourt P. and Bisson P., Principe et réalisation d'un microscope optique utilisant l'effect Raman, *Microscopica Acta*, 79 (1977), 267–276.

[12.73] Sombert B. *et al.*, Etude par spectroscopie et microscopie Raman de phases oxyde de molybdène déposé par alumines, *J. Raman Spectr.*, 9 (1980), 291–296.

[12.74] Busse G., Imaging and microscopy with optoacoustic and photothermal methods; in: von Bally G. and Greguss P. (eds.), *Optics in Biomedical Sciences*, Springer-Verlag, Berlin–Heidelberg–New York 1982, pp. 34–37.

[12.75] Rosencweig A. and Busse G., High-resolution photoacoustic thermal-wave microscopy, *Appl. Phys. Lett.*, 36 (1980), 725–727.

[12.76] Quimby R. S., Real-time photoacoustic microscopy, *Appl. Phys. Lett.*, 45 (1984), 1037–1039.

[12.77] Busse G., Photothermal transmission imaging and microscopy, *Opt. Comm.*, 36 (1981), 441–443.

[12.78] Mundy W. C. and Hughes R. S., Photothermal deflection microscopy of dielectric thin film, *Appl. Phys. Lett.*, 39 (1983), 985–987.

[12.79] Arendt A. F. and Begley D. L., Scanning photoacoustic microscopy of coal, *Optics and Lasers in Engineering*, 7 (1986/7), 15–24.

[12.80] Zharov V. P. and Letokhov V. S., *Laser Optoacoustic Spectroscopy*, Springer-Verlag, Berlin–Heidelberg–New York 1986.

[12.81] Pluta M., *Laser Microscopy* (in preparation).

[12.82] Duncan M. D., Reintjes J., and Manuccia T. J., Imaging biological compounds using the coherent anti-Stokes Raman scattering microscope, *Opt. Eng.*, 24 (1985), 353–355.

[12.83] Patzelt W. J., Leitz-Akustomikroskop ELSAM, technische Realisierung des Serienmodells, *Leitz-Mitt. Wiss. u. Techn. Sonderheft Achema, Juni* 1985, 3–8.

Epilogue to Volume 2

[E2.1] De Brabander M., Nuydens R., Geuens S., Geerts H., Moeremans M., and De May J., Visualization of submicroscopic colloidal gold particles with video contrast enhancement (Nanovid microscopy), *Proc. Roy. Micr. Soc.*, **22** (1987), 260.

[E2.2] Leuvering J. H. W., SPIA: a new immunological technique, *International Clinical Products Review*, March/April 1986, 22–25.

[E2.3] Janssens Nanovid Microscopy, *Proc. Roy. Micr. Soc.*, **22** (1987), 378.

[E2.4] Glauert A., The use of light microscopy for the study of cultured cells, *Proc. Roy. Micr. Soc.*, **22** (1987), 337–338.

[E2.5] Hamamatsu Ultra-Low-Light-Level Camera, *Proc. Roy. Micr. Soc.*, **22** (1987), 309.

[E2.6] Hamamatsu Video Intensified Microscope, *Proc. Roy. Micr. Soc.*, **22** (1987), 225.

[E2.7] Allen T. D., Time lapse video microscopy using an animation unit, *J. Microscopy*, **147** (1987), 129–135 (see also *Proc. Roy. Micr. Soc.*, **22** (1987), 235).

[E2.8] Brakenhoff G. J., Voort van der, H. T. M., Spronsen van, E. A., Valkenburg J., and Nanninga N., 3-D image formation in confocal scanning microscopy and applications in biology, *Proc. Roy. Micr. Soc.*, **22** (1987), 91.

[E2.9] Carlsson K. and Liljeborg, PHOIBOS, a confocal laser microscope scanner for digital recording of optical serial sections, *Proc. Roy. Micr. Soc.*, **22** (1987), 91.

[E2.10] Dixon A. J. and Pang T. M., 3-D imaging applications of the Lasersharp CLSM, *Proc. Roy. Micr. Soc.*, **22** (1987), 99.

[E2.11] White N. S., Dixon A., Doe N., and Shotton D. M., Scanning optical microscopy (SOM) of biological specimens, *Proc. Roy. Micr. Soc.*, **22** (1987), 99.

[E2.12] Sheppard C. J. R., Scanning optical microscopy, in: Barer R. and Cosslett V. E. (eds.), *Advances in Optical and Electron Microscopy*, Vol. 10, Academic Press, London 1987, pp. 1–98.

[E2.13] Garza F. M., Diller K. R., Bovik A. C., Aggarwal S. J., and Aggarwal J. K., Improvement in the resolution of three-dimensional data sets collected using optical serial sectioning, *J. Microscopy*, **150** (1988), in press.

[E2.14] Pang T. M., SOM-the scanning laser optical microscope, *Laboratory Practice*, December 1986.

[E2.15] Pang T. M., Laboratory applications of the scanning optical microscope, *International Laboratory*, July/August 1987, 64–67.

[E2.16] Boyde A., Jones S. J., Watson I. F., Radcliffe R., Prescott C., and Maroudas N., Stereoscopic tandem scanning reflected light microscopy: application in bone, dental materials and nervous tissue research, *Proc. Roy. Micr. Soc.*, **22** (1987), 91.

[E2.17] Blackie R. A. S., Bagby R., Wright L., Drinkwater J., and Hart S., Reconstruction of three-dimensional images of microscopic objects using holography, *Proc. Roy. Micr. Soc.*, **22** (1987), 98.

[E2.18] Horikawa Y., Yamamoto M., and Dosaka S., Laser scanning microscope: differential phase images, *J. Microscopy*, **148** (1987), 1–10.

[E2.10] Suzuki T. and Horikawa Y., Development of real-time scanning laser microscope for biological use, *Appl. Opt.*, **25** (1986), 4115–4121.

Index of Names

Subject Index

Plates

a)

b)

c)

d)

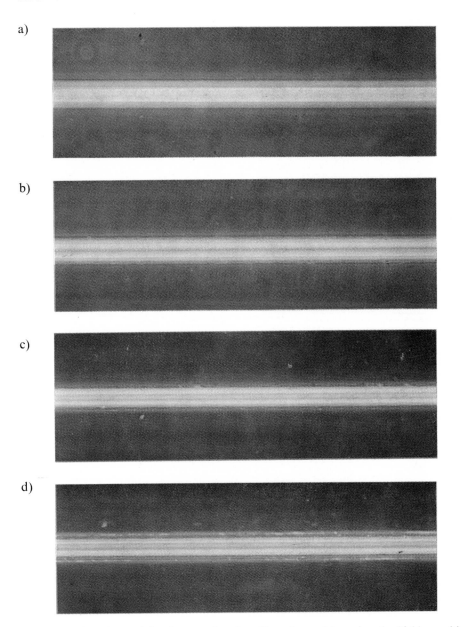

Fig. VIII. Dispersion staining images of a glass fibre observed by using the highly sensitive phase contrast device (KFA): a) refractive index n_D of the HD Cargille liquid is equal to 1.500, b) $n_D = 1.505$, c) $n_D = 1.510$, and d) $n_D = 1.515$. Cross ref.: Subsections 6.3.3 and 5.5.2.

a)

b)

c)

d)

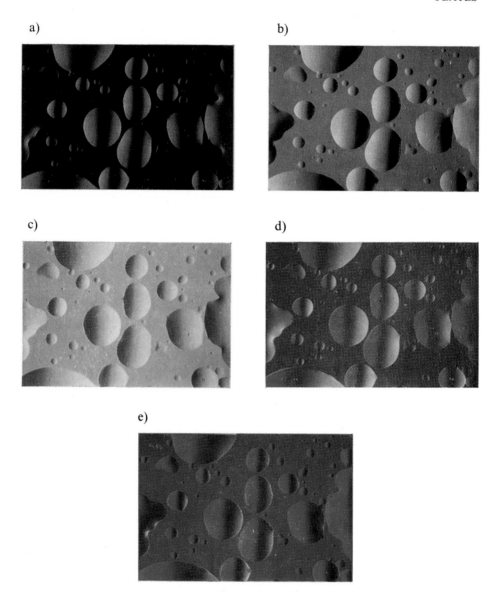

e)

Fig. IX. DIC images of Canada balsam droplets deposited on a microscope slide, then covered by a layer of immersion liquid (water) and cover slip: a) bias retardation $\Delta_b = 0$, b) $\Delta_b = 0.05\,\mu\text{m}$, c) $\Delta_b = 0.3\,\mu\text{m}$, d) $\Delta_b = 0.55\,\mu\text{m}$, e) $\Delta_b = 0.60\,\mu\text{m}$. Objective $10\times/0.24$; print magnification $120\times$. Photomicrographs were taken with the Biolar PI double-refracting interference microscope manufactured by the Polish Optical Works (PZO), Warsaw. Cross ref.: Subsection 7.1.4.

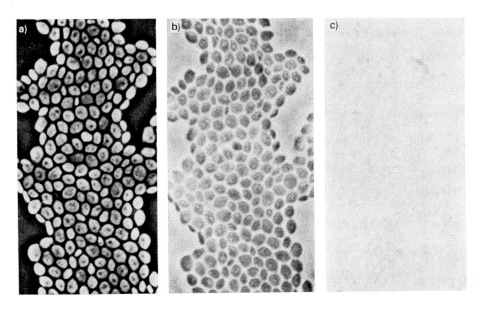

Fig. X. Monolayer of dry yeast cells obtained by evaporation from an aqueous suspension of yeast between a slide and cover slip. KFA highly sensitive negative phase contrast device with soot phase rings (a), typical positive phase contrast device with dielectric-metallic phase rings (b), and ordinary bright field microscope (c). Objectives $100 \times /1.25$ (oil immersion). Print magnification $900 \times$. Cross ref.: Subsection 5.5.2.

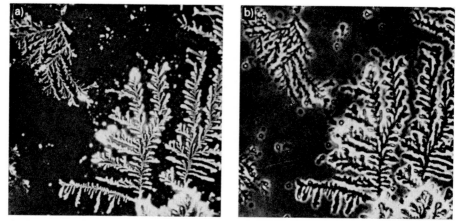

Fig. XI. A layer of microcrystals obtained from a solution layer sandwiched between a slide and cover slip. KFA highly sensitive negative phase contrast device (a), and typical positive phase contrast device (b). Objectives 40 × /0.65. Print magnification 400 × . Cross ref.: Subsection 5.5.2.

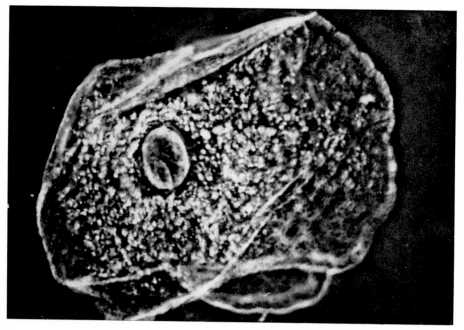

Fig. XII. Oral epithelial cell immersed in saliva. Highly sensitive negative phase contrast device (KFA). Objective 100 × /1.25 (oil immersion). Print magnification 1500 × . Cross ref.: Subsection 5.5.2.

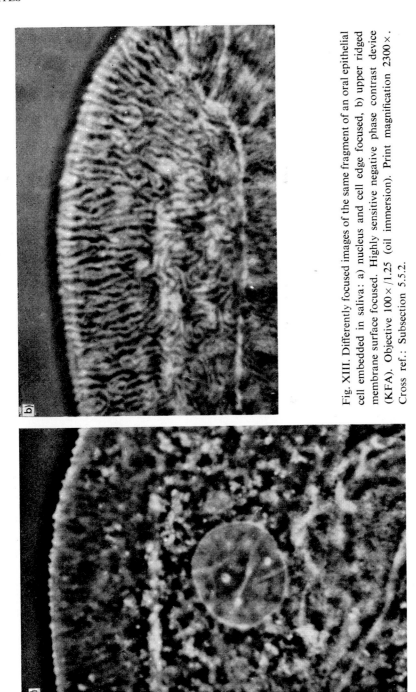

Fig. XIII. Differently focused images of the same fragment of an oral epithelial cell embedded in saliva: a) nucleus and cell edge focused, b) upper ridged membrane surface focused. Highly sensitive negative phase contrast device (KFA). Objective 100×/1.25 (oil immersion). Print magnification 2300×. Cross ref.: Subsection 5.5.2.

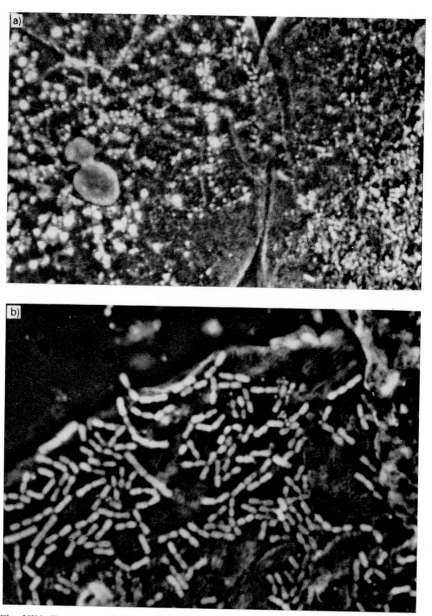

Fig. XIV. Fragments of two oral epithelial cells, one of which has a characteristic bilobate nucleus (a), whereas the other (b) has a fine membrane structure covered by a colony of bacteria. Highly sensitive negative phase contrast device (KFA). Objective 100×/1.25 (oil immersion). Print magnification 1700× (a) and 2600× (b). Cross ref.: Subsection 5.5.2.

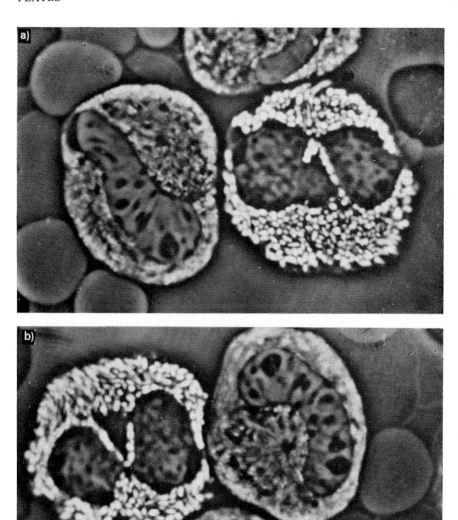

Fig. XV. Neutrophil and eosinophil granulocytes among red corpuscles "ghosts" in a fresh preparation of living human blood. Highly sensitive negative phase contrast device (KFA). Objective 100× /1.25 (oil immersion). Print magnification 2600×. Cross ref.: Subsection 5.5.2.

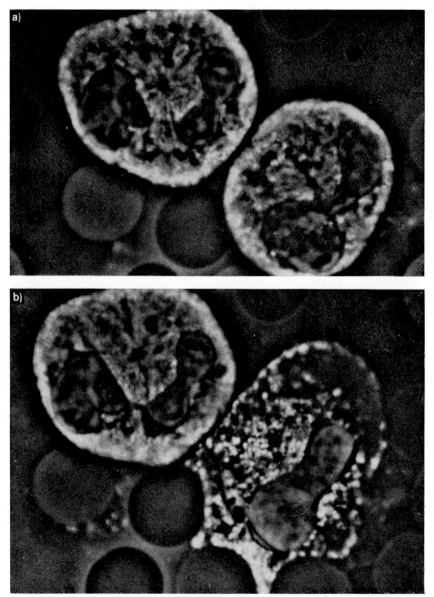

Fig. XVI. Two neutrophilic granulocytes photographed after a few minutes interval. The bottom photomicrograph shows that one granulocyte burst, Brownian movement ceased and cytoplasmic granules appeared sharp. Highly sensitive negative phase contrast device (KFA). Objective 100×/1.25 (oil immersion). Print magnification 2600×. Cross ref.: Subsection 5.5.2.

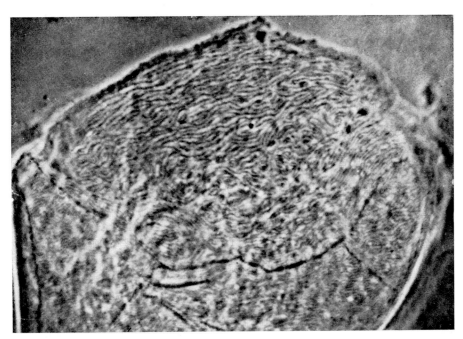

Fig. XVII. Ridged membrane surface of an oral epithetial cell immersed in saliva. Highly sensitive positive phase contrast device (KFS). Objective $100 \times /1.25$ (oil immersion). Print magnification $1800 \times$. Cross ref.: Subsection 5.5.3.

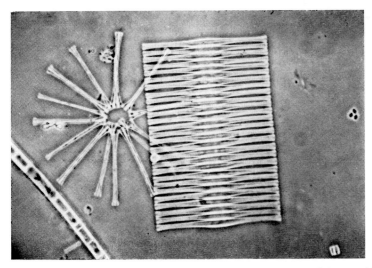

Fig. XVIII. Fresh-water diatoms mounted in Canada balsam. Highly sensitive positive phase contrast device (KFS). Objective $20 \times /0.40$. Print magnification $400 \times$. Cross ref.: Subsection 5.5.3.

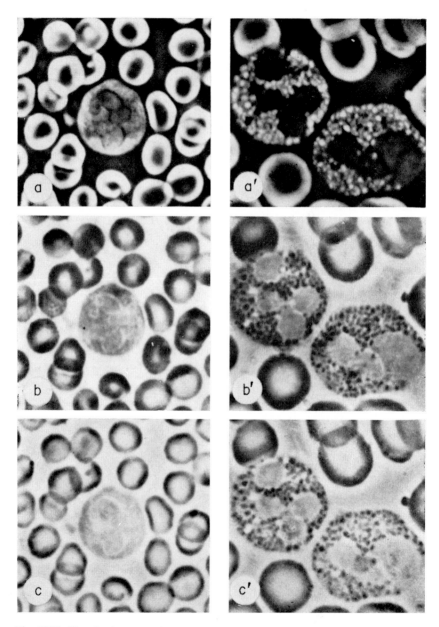

Fig. XIX. Unstained smear of human blood embedded in immersion oil and covered with a cover slip. Alternating phase contrast device (KFZ) set at negative (a, a′) and positive (b, b′) image contrast, and typical positive phase contrast device (c, c′). Objectives: 40× /0.65 (a, b, c), and oil immersion 100× /1.25 (a′, b′, c′). Print magnifications: 1400× (a, b, c) and 2000× (a′, b′, c′). Cross ref.: Subsection 5.6.2.

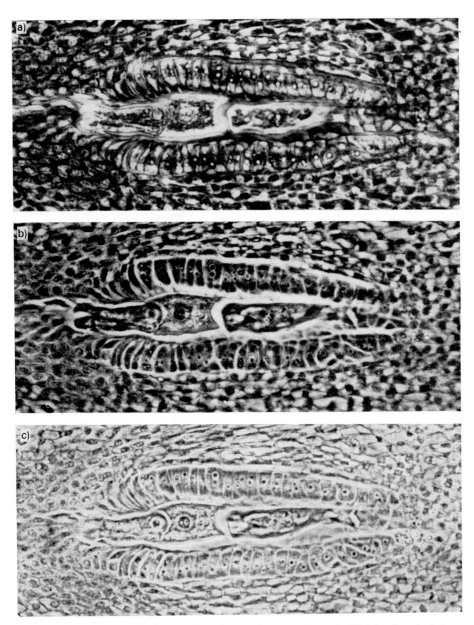

Fig. XX. Unstained longitudinal section of a sunflower grain, embedded in Canada balsam between a slide and cover slip. Alternating phase contrast device (KFZ) set at negative (a) and positive (b) image contrast, and typical positive phase contrast device (c). Objective $10 \times /0.24$. Print magnification $160 \times$. Cross ref.: Subsection 5.6.2.

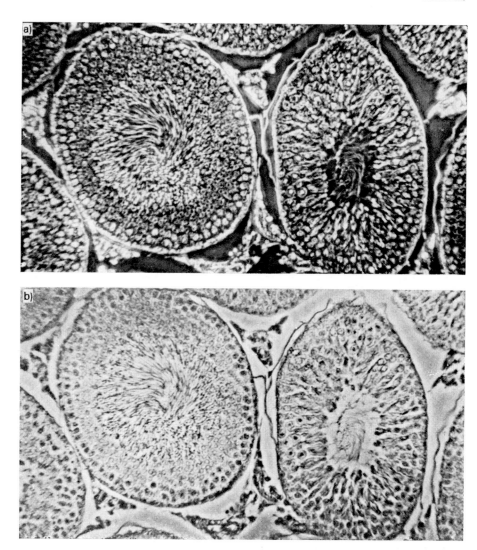

Fig. XXI. Unstained section of a testicle labule of rat, embedded in Canada balsam between a slide and cover slip. Alternating phase contrast device (KFZ) set at negative (a) and positive (b) image contrast. Objective 20× /0.40. Print magnification 290×. Cross ref.: Subsection 5.6.2.

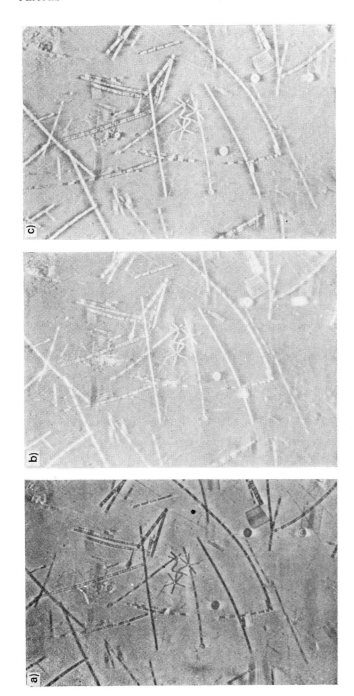

Fig. XXII. Fresh-water diatoms mounted in Canada balsam between a slide and cover slip (the same specimen as in Fig. XVIII). Variable phase contrast device with a single polarizing phase ring (KFV) set at optimum positive (a), and lower (b) and higher (c) negative image contrast. Objective $10 \times /0.25$. Print magnification $150 \times$. Cross ref.: Subsection 5.7.4.

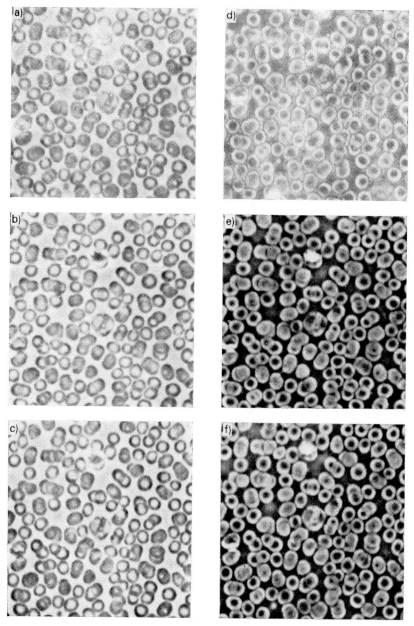

Fig. XXIII. Unstained smear of human blood embedded in immersion oil and covered with a cover slip. KFV variable phase contrast device set at different levels of positive (a, b, c) and negative (d, e, f) image contrast. Objective $40\times/0.65$. Print magnification $550\times$. Cross ref.: Subsection 5.7.4.

Fig. XXIV. Metallic thin film of 45% transmittance with a number of parallel clear slits (this is an amplitude specimen). KFV variable phase contrast device is set at pure amplitude contrast, i.e., phase shift ψ of the direct relative to diffracted light is zero and only the direct light intensity is varied. In this case, the reduction was to 35% (a) and to 15% (b). Objective $2.5 \times /0.06$. Print magnification $50 \times$. Cross ref.: Subsection 5.7.4 and Section 6.1.

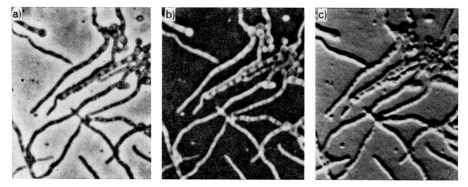

Fig. XXV. Spores of Penicillium chrysogenum (Agar culture) seen under the Amplival-Interphako microscope set at positive (a) and negative (b) phase-interference image contrast, and differential interference contrast (c). Objective $12.5 \times /0.25$. Print magnification $240 \times$. Cross ref.: Subsection 5.8.1. Photomicrographs reproduced by permission of VEB Carl Zeiss Jena.

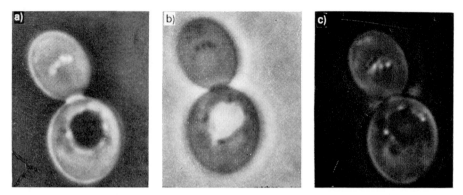

Fig. XXVI. Living yeast cells immersed in sugar solution seen under the Amplival-Interphako microscope set at optimum dark contrast for vacuoles (a), granules (b), and background (c). Objective 100× /1.25 (oil immersion). Print magnification 4000×. Cross ref.: Subsection 5.8.1. Photomicrographs reproduced by permission of VEB Carl Zeiss Jena.

Fig. XXVII. Stereopairs of different phase and amplitude specimens taken with the stereoscopic phase contrast microscope described in Section 5.9. From top to bottom: a) mechanically ground and chemically etched glass surface, next covered with an immersion liquid and cover slip (positive phase objective 40× /0.65, print magnification PM = 650×); b) salt microcrystal embedded in paraffin oil between a slide and cover slip (negative phase objective 20× /0.40, PM = 250×); c) another microcrystal of salt of the same specimen as above (negative phase objective 40× /0.65, PM = 450×; d) chromosomes of onion cells stained by Feulgen method and embedded in Canada balsam (amplitude objective 100× /1.25 with oil immersion, PM = 1200×); e) another fragment of the same specimen as above (amplitude objective 40× /0.65, PM = 450×); f) oral epithelial cell embedded in a concentrated water solution of sugar (negative phase objective 100× /1.25 with oil immersion, PM = 900×).

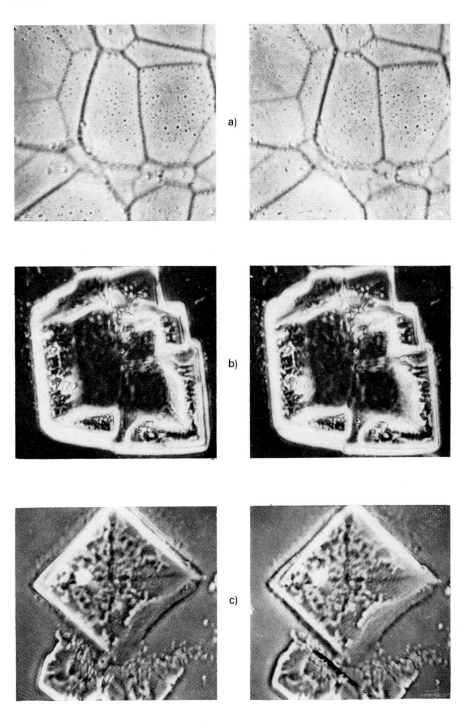

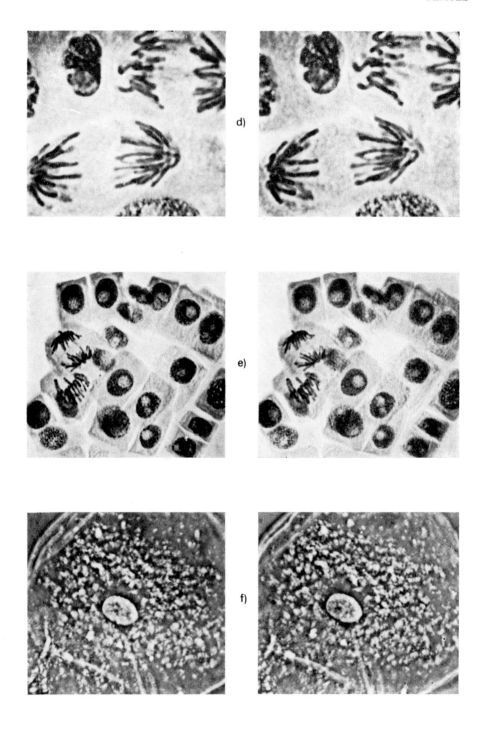

d)

e)

f)

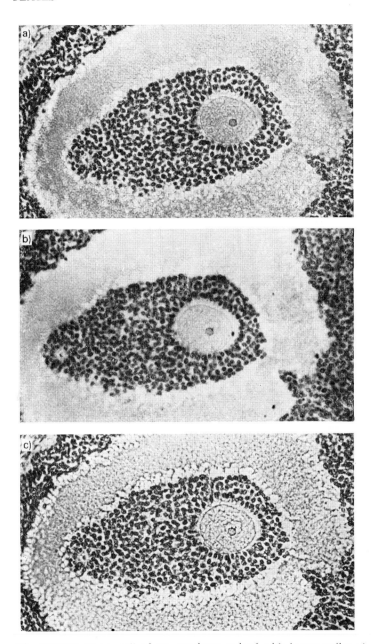

Fig. XXVIII. Ovary cell of rat, specimen stained with hematoxylin: a) amplitude contrast, b) bright-field, c) phase contrast (positive). Magnifying power/numerical aperture of objectives: $20 \times /0.40$. Print magnification $300 \times$. Cross ref.: Subsection 6.1.2.

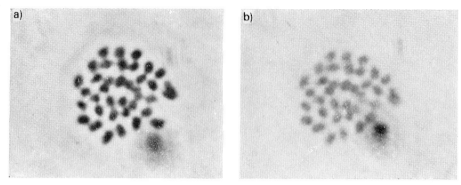

Fig. XXIX. Mitosis of sugar beetroot cell stained by Feulgen method: a) amplitude contrast, b) bright-field. Magnifying power/numerical aperture of objectives 100× /1.25 (oil immersion). Print magnification 3200×. Cross ref.: Subsection 6.1.2.

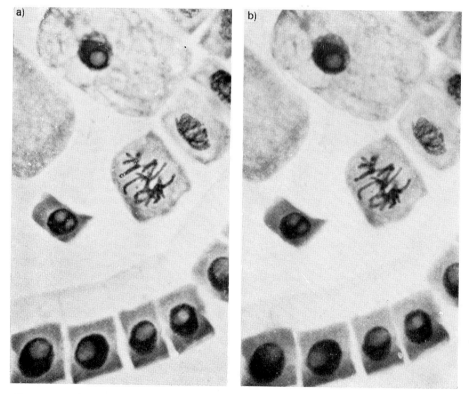

Fig. XXX. Onion cells stained by Feulgen method: a) amplitude contrast, b) bright-field. Magnifying power/numerical aperture of objectives: 100× /1.25 (oil immersion). Print magnification 700×. Cross ref.: Subsection 6.1.2.

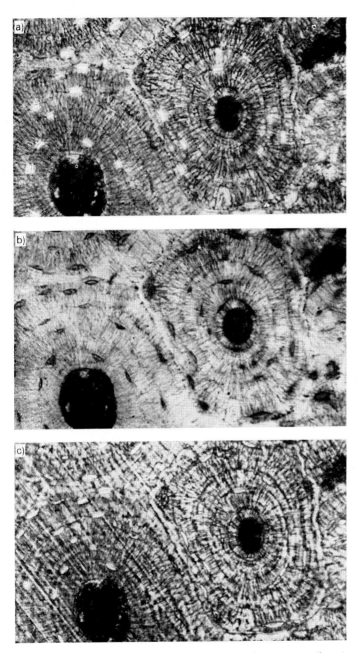

Fig. XXXI. Bone tissue, cross section stained with hematoxylin: a) amplitude contrast, b) bright-field, c) phase contrast (positive). Magnifying power/numerical aperture of objectives: $20 \times / 0.40$. Print magnification $280 \times$. Cross ref.: Subsection 6.1.2.

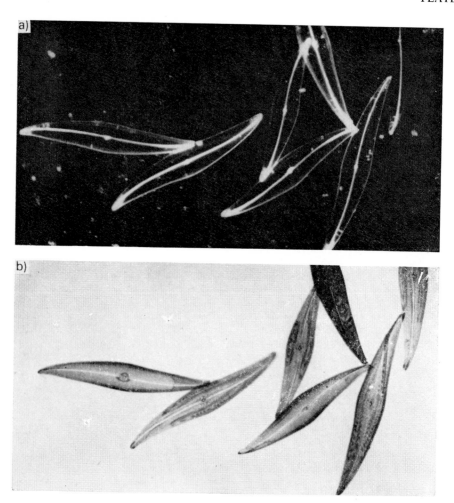

Fig. XXXII. Dark-field images (a, c, e) and, for comparison, bright-field images (b, d, f) of diatoms: Pleurosigma angulatum (a, b), Navicula lyra (c, d) and Pinnularia apulenta (e, f). Photomicrograph a corresponds to Fig. 6.9a, c to Fig. 6.9b, and e to Fig. 6.9c. Objectives: $10 \times /0.25$ (a—d) and $20 \times /0.40$ (e, f), print magnification ca $200 \times$. Cross ref.: Subsection 6.2.2.

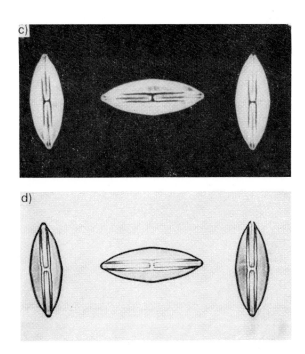

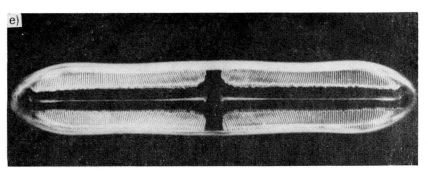

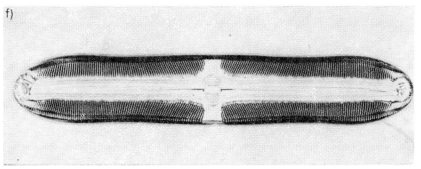

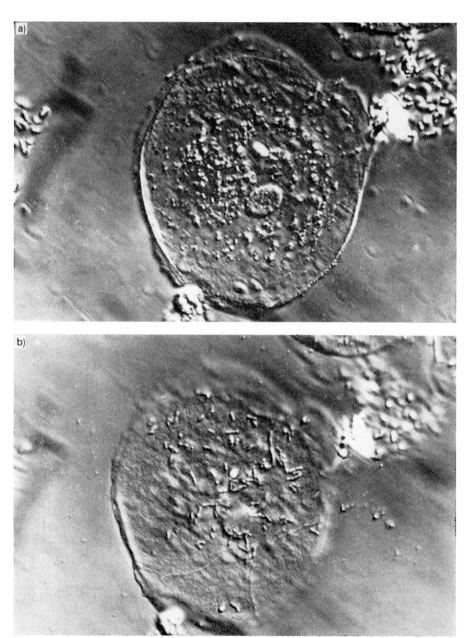

Fig. XXXIII. Human epithelial (cheek) cell seen under modulation contrast with 40× objective, demonstrating optical sectioning: a) nucleus and cytoplasmic granules focused, b) cell surface, cell membrane folds and adhering bacteria focused. Cross ref.: Subsection 6.4.3 (by courtesy of R. Hoffman and L. Gross).

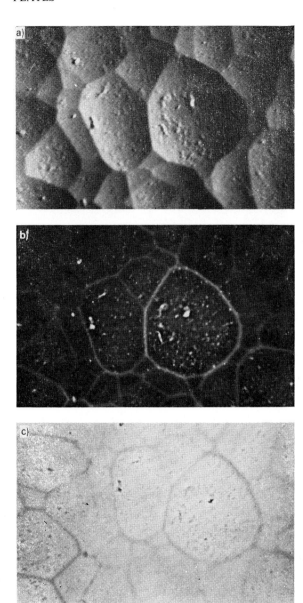

Fig. XXXIV. DIC image (a) and phase contrast images, negative (b) and positive (c) of a glass plate surface etched with hydrofluoric acid, then covered by a layer of immersion liquid (glycerin) and cover slip. Magnifying power/numerical aperture of objectives: 20× /0.40. Print magnification 320×. Cross ref.: Subsection 7.1.4. These photomicrographs and those of the following Figs. XXXV—XXXVII taken with microscopes manufactured by PZO, Warsaw.

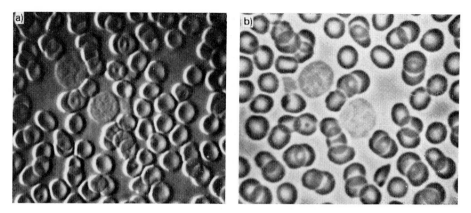

Fig. XXXV. Human erythrocytes (non-stained smear) embedded in immersion oil: a) DIC image, b) phase contrast image. Magnifying power/numerical aperture of objectives: $40 \times /0.60$. Print magnification $800 \times$. Cross ref.: Subsection 7.1.4.

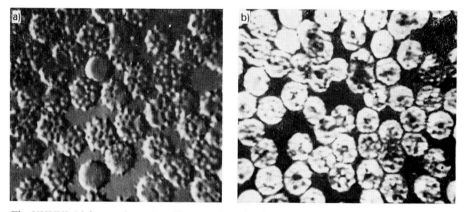

Fig. XXXVI. Living erythrocytes of human blood in slightly hypertonic medium: a) DIC image, b) negative phase contrast. Magnifying power/numerical aperture of objectives: $100 \times /1.25$ (oil immersion). Print magnification $950 \times$. Cross ref.: Subsection 7.1.4.

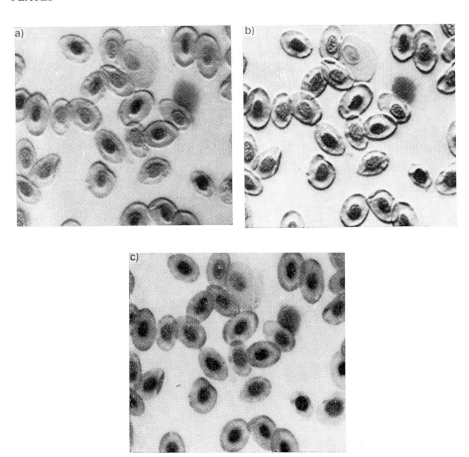

Fig. XXXVII. Smear of frog erythrocytes, stained with hematoxylin and embedded in paraffin oil: a) DIC image with crossed polars, b) amplitude DIC image, c) brigh-field image. Magnifying power/numerical aperture: 100× /1.25 (oil immersion). Print magnification 900×. Cross ref.: Subsection 7.2.2.

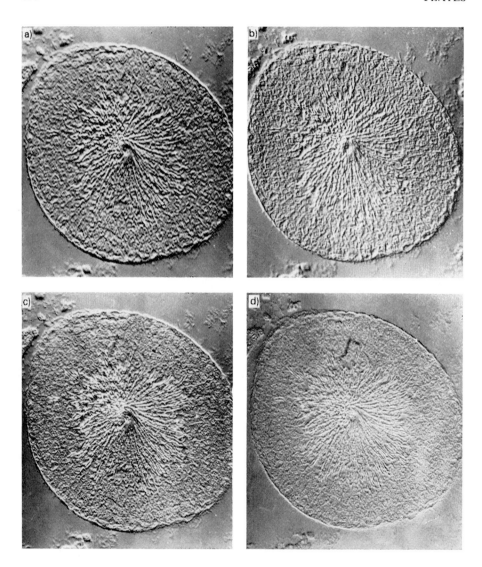

Fig. XXXVIII. VADIC images of a testicle labule of rat. Wavefront shear for the successive images is equal to about 2.3 μm (a), 1.7 μm (b), 1.0 μm (c), and 0.4 μm (d). Objective magnifying power/numerical aperture: 20 × /0.40. Print magnification 220 × . Cross ref.: Subsection 7.4.1.

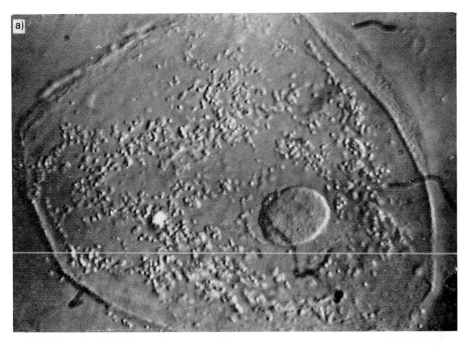

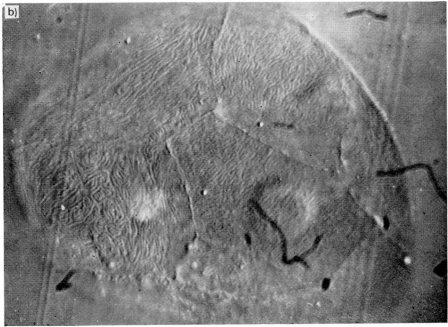

Fig. XXXIX. VADIC images of an epithelial cell embedded in water medium and focused at a) nucleus, b) upper surface, and c) lower surface. Wavefront shear is 0.2 μm. Objective magnifying power/numerical aperture: 100 × /1.25 (oil immersion). Print magnification 1750 ×. Cross ref.: Subsection 7.4.1.

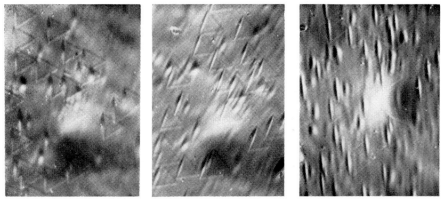

Fig. XL. VADIC images of triangle stacking faults in an epitaxial Si film. The same fragment is shown with different directions of wavefront shear. The amount of shear is equal to about 1 μm. Magnifying power/numerical aperture of the objective: 20× /0.40. Print magnification 500×. Cross ref.: Subsection 7.4.3.

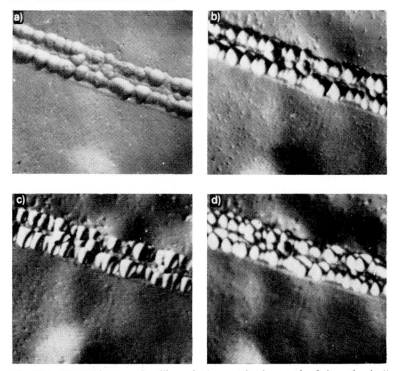

Fig. XLI. VADIC images of a silicon single-crystal substrate (wafer) mechanically polished and smoothly etched. Wavefront shear of the successive images is equal to 0.4 μm (a), 1 μm (b), 1.7 μm (c), and 2.3 μm (d). Magnifying power/numerical aperture of the objective: 20× /0.40. Print magnification 450×. Cross ref.: Subsection 7.4.3.

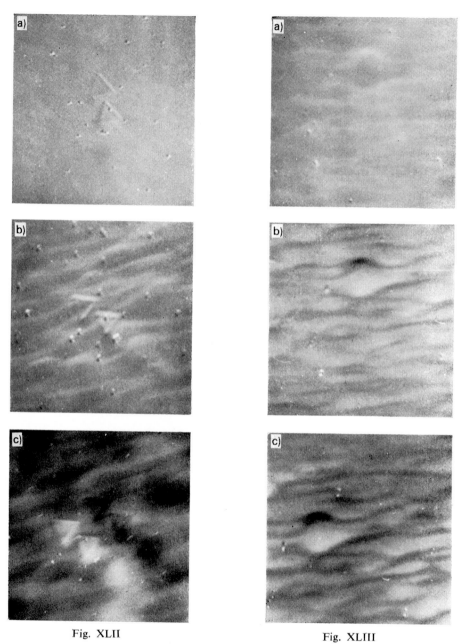

Fig. XLII

Fig. XLIII

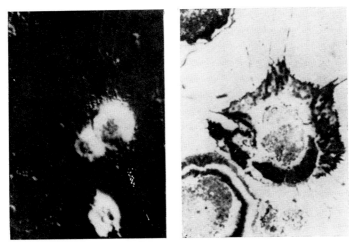

Fig. XLIV. DIC (left) and reflection contrast (right) images of a macrophage. Courtesy of Dr. W. J. Patzelt, E. Leitz Wetzlar. Cross ref.: Subsection 8.2.1.

Fig. XLII. Another region of the same specimen as in Fig. XL. VADIC images with wavefront shear equal to 1 μm (a) and 2.3 μm (b), and largely sheared uniform-field interference image ($s = 23$ μm). Magnifying power/numerical aperture of the objective: 20× /0.40. Print magnification 500×. Cross ref.: Subsection 7.4.3.

Fig. XLIII. VADIC images of another region of the same specimen as in Figs. XL and XLII. Wavefront shear of the successive images is equal to 1 μm (a), 2.3 μm (b), and 3.5 μm (c). Magnifying power/numerical aperture of the objective: 20× /0.40. Print magnification 500×. Cross ref.: Subsection 7.4.3.

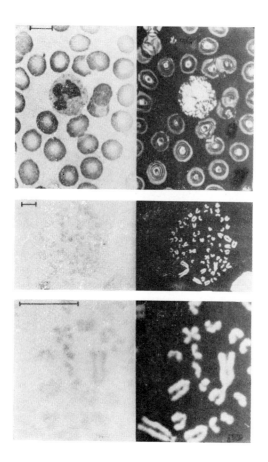

Fig. XLV. Blood smear and chromosome preparations shown in bright-field (left) and reflection contrast (right). Top photomicrographs show human blood smear stained after May-Grünwald and Giemsa; a neutrophil granulocyte with (green) reflecting granules, erythrocytes interference rings, and platelets are clearly visible in the RC image. Middle photomicrographs show Feulgen stained chromosomes of an aneuploid cell line of Microtus agrestis. Highly magnified fragments of the same preparation are shown at the bottom. The bars indicate 10 μm. Courtesy of Prof. Dr. F. Pera. Cross ref.: Section 8.3.

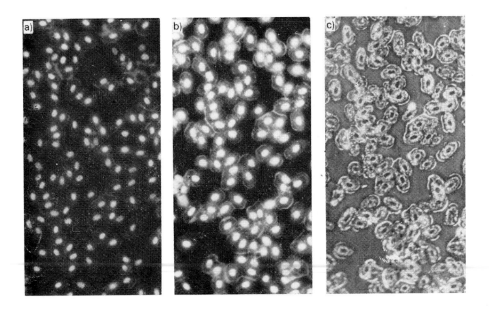

Fig. XLVI. Smear of frog blood stained with acridine orange seen under a fluorescence microscope (PZO, Warsaw) with phase contrast by using annular diaphragms made of thin interference
films; a) fluorescence image only, b) simultaneous fluorescence and negative phase contrast
image, c) negative phase contrast only. Magnifying power/numerical aperture of the objective:
$20\times/0.4$. Print magnification $280\times$. Cross ref.: Section 9.4.

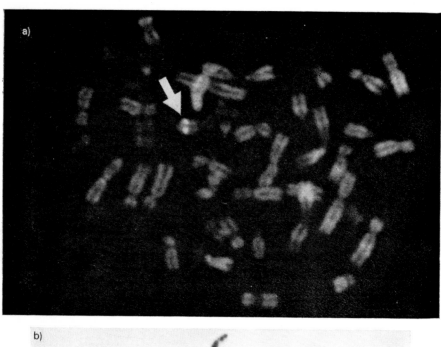

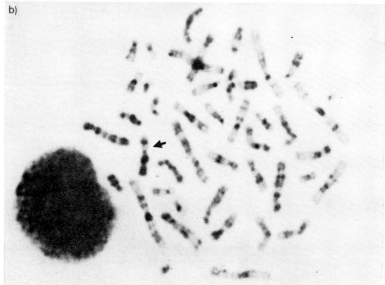

Fig. XLVII. Human (male) chromosomes from lymphocytes culture: a) fluorescent image
(preparation stained with quinacrine mustared), b) bright-field image. The white arrow indicates

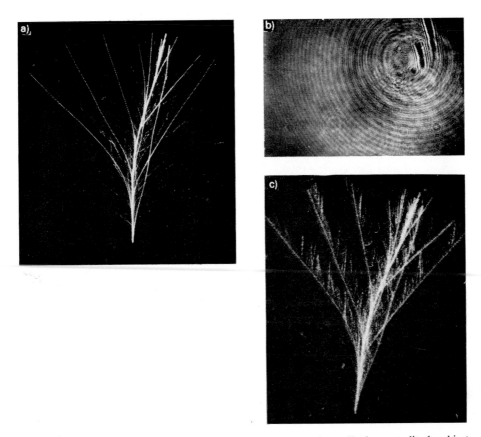

Fig. XLVIII. Illustrating lens-less holographic microscopy: a) small phase-amplitude object (a plumelet), b) its hologram of original size, and c) holographic image reconstructed from the hologram b. The object (a) and its holographic image (c) are shown magnified of several times. Courtesy of M. Szyjer (Central Optical Laboratory, Warsaw). Cross ref.: Subsection 11.3.2.

the Y chromosome whose fluorescence is relatively strong, while the black arrow shows the X sex chromosome. By courtesy of Dr. B. Kałużewski (Laboratory of Genetics, Medicine Academy in Łódź). Cross ref.: Subsection 9.8.1.

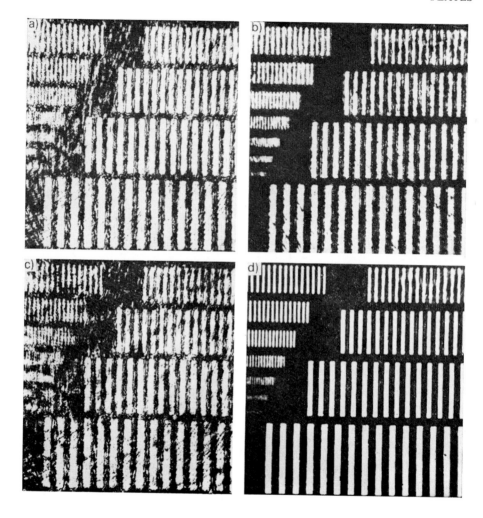

Fig. XLIX. Photomicrographs illustrating the performance of the holographic microscope shown in Fig. 11.22. Resolution target available from Ealing was used as a test object; a) holographic image without coherent noise suppression, b) coherent noise was suppressed during the process of hologram recording, c) coherent noise was suppressed during the reconstruction process, d) coherent noise was suppressed during both the recording and reconstruction processes. In this latter case, the resolution limit is equal to 1.6 μm (group 9 of test bars). Magnifying power/numerical aperture of the objectives: 20× /0.4. Print magnification 550×. Courtesy of Dr. R. Pawluczyk (Central Optical Laboratory, Warsaw). Cross ref.: Subsection 11.5.3.

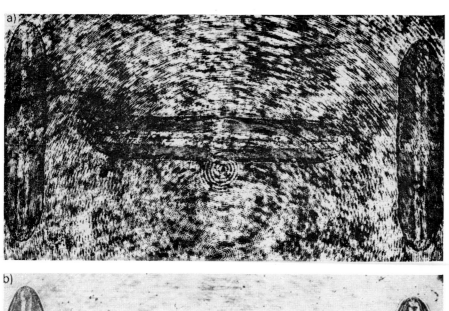

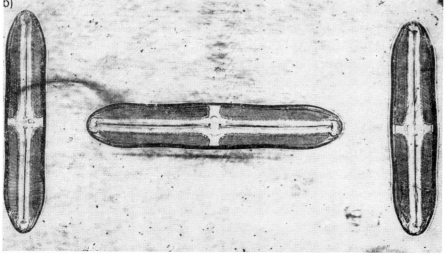

Fig. L. Photomicrographs of holographic images of diatom cells observed through the microscope shown in Fig. 11.22: a) image without coherent noise suppression, b) with coherent noise suppression during both the hologram record and the image reconstruction. The period of the fine structure of the cells is equal to about 2.5 μm. Magnifying power/numerical aperture of the objectives: 20 × /0.4. Print magnification 200×. Courtesy of Dr. R. Pawluczyk. Cross ref.: Subsection 11.5.3.

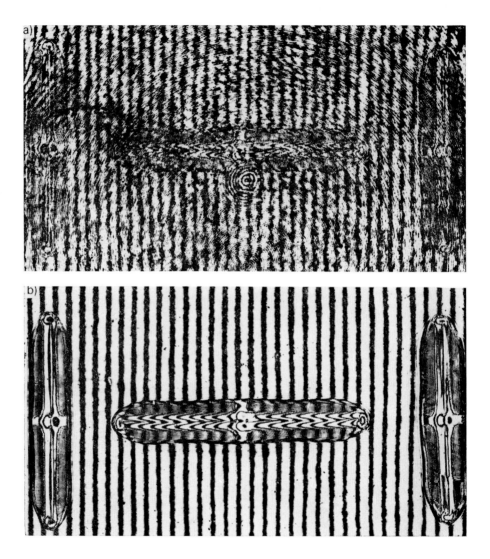

Fig. LI. Holographic interference images of the same diatom cells as in Fig. L. Photomicrographs were taken with the microscope shown in Fig. 11.22; a) holographic interference image without coherent noise suppression, b) with coherent noise suppression during both hologram recording and image reconstruction. Magnifying power/numerical aperture of the objectives: $20 \times /0.4$. Print magnification $200 \times$. Courtesy of Dr. R. Pawluczyk. Cross ref.: Subsection 11.6.4.

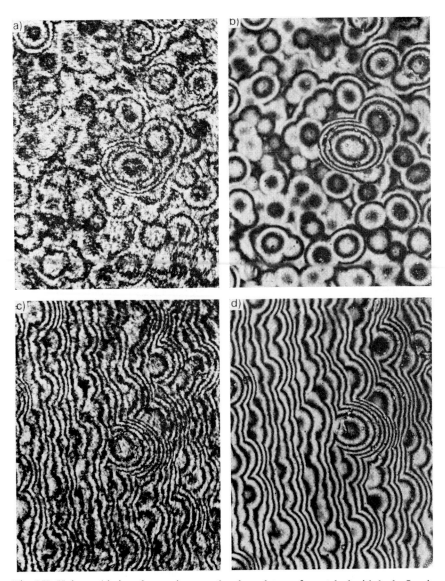

Fig. LII. Holographic interference images of a glass plate surface etched with hydrofluoric acid, then covered by a liquid (water) layer and cover slip: a) and b) uniform field interference images without and with coherent noise suppression, c) and d) fringe field interference images without and with coherent noise suppression. Photomicrographs were taken with the holographic microscope shown in Fig. 11.22. Magnifying power/numerical aperture of the objectives: $20 \times /0.4$. Print magnification $200 \times$. Courtesy of Dr. R. Pawluczyk. Cross ref.: Subsection 11.6.4.

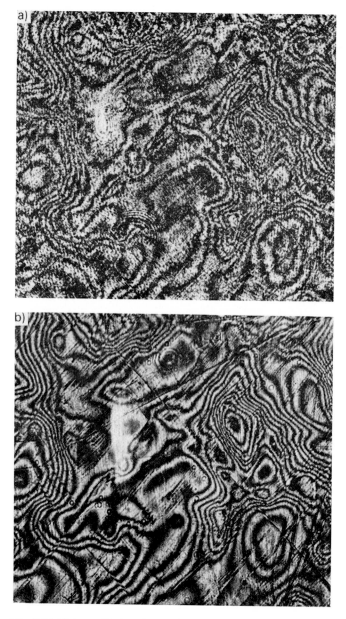

Fig. LIII. Holographic interference images of a rough light-reflecting surface (chromium plated metal): a) holographic microscope (Fig. 11.22) adjusted at uniform field interference without coherent noise suppression, b) like a but with coherent noise suppression. Magnifying power/ numerical aperture of the objectives: $20 \times$ /0.4. Print magnification $200 \times$. Courtesy of Dr. R. Pawluczyk. Cross ref.: Subsection 11.6.4.

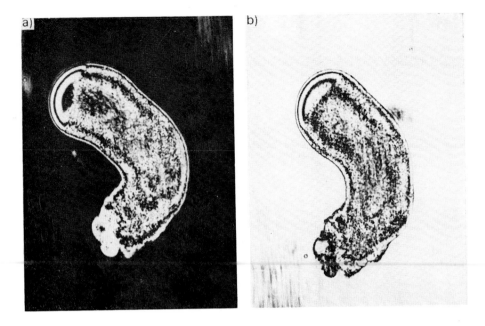

Fig. LIV. Holographic interference images of Amoeba proteus observed in real time by means of the holographic microscope shown in Fig. 11.22: a) uniform interference with dark field, b) that with bright field. Courtesy of Dr. M. Opas (Institute of Experimental Biology, Polish Academy of Sciences, Warsaw). Magnifying power/numerical aperture of objectives: 20 × /0.4. Print magnification 200 × . Cross ref.: Section 11.7.

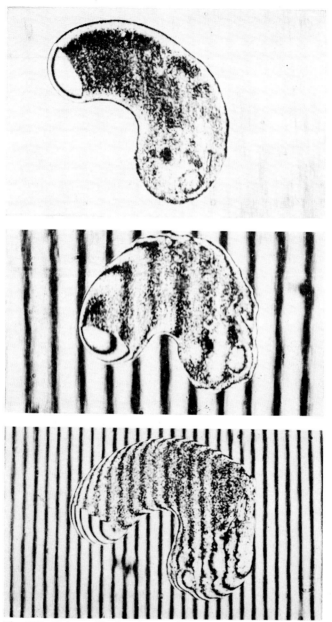

Fig. LV. Holographic interference images of a glycerinated amoeba observed in real time by means of the holographic microscope shown in Fig. 11.22: uniform field interference, and fringe field interference with different interfringe spacings. Magnifying power/numerical aperture of the objectives: $20 \times /0.4$. Print magnification $200 \times$. Courtesy of Dr. M. Opas. Cross ref.: Section 11.7.

Fig. LVI. Two frames from 16 mm motion picture holographic microinterferometry using the microscope shown in Fig. 11.22. Images of freely migrating plasmodia of Physarum polycephalum. Magnifying power/numerical aperture of the objectives 10× /0.25. Print magnification 150×. Courtesy of Dr. Z. Baranowski (Institute of Experimental Biology, Polish Academy of Sciences, Warsaw). Cross ref.: Subsection 11.8.1.

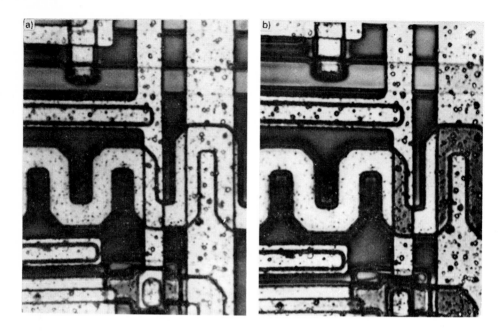

Fig. LVII. An integrated-circuit element: conventional (a) and confocal scanning (b) microscope image comparison. Image b shows the resolution improvement and depth discrimination property quite clearly. The width of the central meander pattern is 6.6 μm. Courtesy of Dr. T. Wilson (Department of Engineering Science, University of Oxford, England). Cross ref.: Subsection 12.2.1.

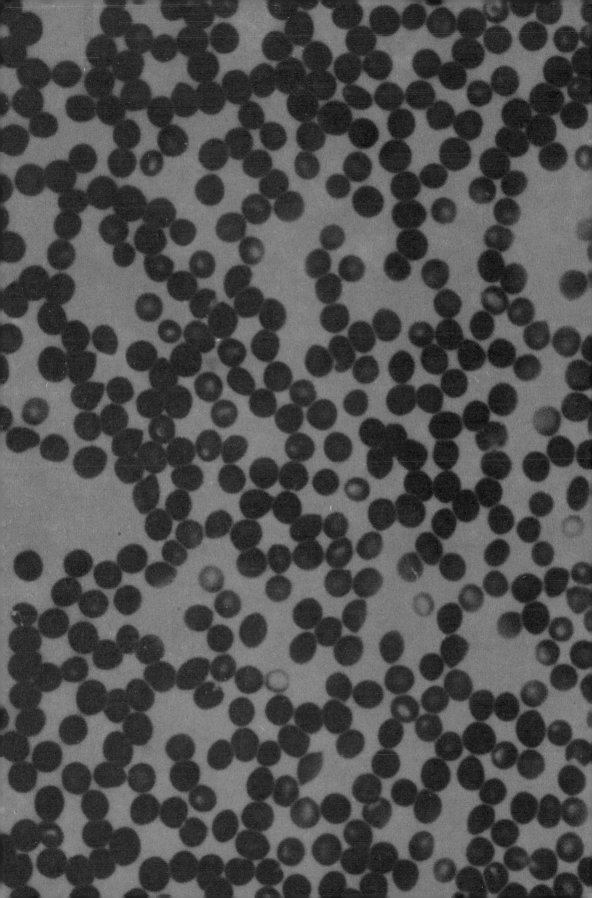